The Anatomy

of Judgment

The Anatomy

of Judgment

Philip J. Regal

University of Minnesota Press · Minneapolis

Published by the University of Minnesota Press
2037 University Avenue Southeast, Minneapolis, MN 55414.
Printed in the United States of America.
Cover design by Patricia M. Boman

Library of Congress Cataloging-in-Publication Data

Regal, Philip J.
  The anatomy of judgment / Philip J. Regal.
    p. cm.
  Includes bibliographical references.
  ISBN 0-8166-1823-2. — ISBN 0-8166-1824-0 (pbk.)
  1. Reasoning. 2. Judgment. 3. Methodology. 4. Science — Methodology.
  5. Cultural relativism. 6. Critical thinking.
    I. Title.
  BC177.R345 1990
  128'.3 — dc20   89-20374
                                                                  CIP

The University of Minnesota is an
equal-opportunity educator and employer.

CONTENTS

PREFACE

This is a book about the reasons we have problems making accurate judgments, even of ourselves, and about the paths that humankind has taken in our search for insight, knowledge, and sound values. It is about the spirit of the liberal arts. It is much about science, but the context it establishes is distinctly larger in scope than the philosophy, history, or practice of science. It is a book about the problems and possibilities of being human.

I began this book simply for college students in science, to discuss briefly but from a broad perspective how science and research fit into historical, anthropological, and philosophical contexts, its prospects and pitfalls. That was in 1979, when I was invited to be a visiting distinguished scholar and professor at the University of Western Australia and at the Australian National University. These opportunities gave me the time and freedom from daily concerns to outline this book and write the first chapters.

It soon became clear that there may be no easily accessible, comprehensive analyses of mind and judgment that begin to synthesize, even for the well-read citizen or scholar, if only in outline, important perspectives as diverse as neurobiology, cultural anthropology, sociology, and history. I came to appreciate that my own background is unusually broad and perhaps unique. It outfitted me first, to see that such a discussion is possible and worthwhile, and second, to write this outline/essay. So, when I could make time to return to this demanding project, I expanded the scope of the book.

Despite these perambulations through several objectives, and expansion for a more general and more sophisticated readership, I tried to retain the book's level of readability for students and its practical aims. I hope this still can serve as a sort of field guide to the intellectual trees and forest, and that it is engaging and at the same time gets into the elements of self-sufficiency and survival. It expects the reader to be an active, adventuresome person who is looking not simply for diversional reading but for ideas to guide his or her own intellectual adventures.

I am a biologist and get paid to spend my time thinking about how organisms function and survive. But in order to think more effectively, I have long pondered *ways* of thinking. Why does one intellectual approach get results, while another proves fruitless? Good logic or bad will not necessarily be the answer to such questions. Through my field study I have been privileged to work in remote places with non-Westernized peoples, including hunters and gatherers, and have come to respect and understand them much better than I would have from books alone. Just as I was disposed to think carefully about how different scientists see and think about the world, I was disposed to learn about how non-Westernized peoples in remote areas of Mexico, New Guinea, or Australia reasoned.

How, moreover, is it that field biologists who must work closely with such peoples will often develop considerable, if qualified, respect for their thinking and knowledge, when people with different contacts will usually miss seeing rich and balanced mentalities and instead see only ignorant or quaint savages who do not view the world "rationally" (in our colloquial sense of rational)? Was I seeing in the deserts of Mexico and the forests of New Guinea misperceptions akin to those that caused several of our great artists to die unappreciated and even ridiculed by their contemporaries? Generals to underestimate enemies? Thus I found myself not only reading and talking a lot on the subject, but looking closely at non-Westernized peoples' thinking processes and *comparing* them to those of scientists of different philosophies, and to those of government officials, businesspeople, missionaries, and tourists.

Years before conceiving this book, I did pre and postdoctoral studies in the interdisciplinary mental health training program of the Brain Research Institute at the University of California, Los Angeles. At the time I was busy studying and doing research in neurophysiology and animal behavior and, while I took courses in psychiatry and other human behavioral sciences, it was not at all apparent that those studies would one day converge with my field experience and interests in the history and philosophy of science and ideas to allow me to write this book.

This unusual background ultimately enabled me to think about the Western mind, and minds in general, from an unusual number of vantage points. I find myself thinking of any given interaction from the perspectives of neurophysiology, of multicultural comparisons, of a practicing scientist, of one involved in science policy formulation processes at the national level, and not least from the point of view of one with a dogged interest in the arts, history, and philosophy.

I write this book to share some of the pleasure that I have had in circling about the human mind and looking at it from different points of view, trying to single out the relevant questions to ask about it, and trying to grasp its nature.

But the book is not just entertainment. Thinking is a matter of survival and competition for each of us as individuals, and for the societies in which we live. It is a deadly serious business from which communities and nations may prosper or fall, wars may be won or lost, and individuals may become full-fledged actors in life or instead may become life's duped pawns. Good thinking skills *must be cultivated*, which represents an investment of time, energy, and thought. For some readers these points will be as plain as facts can be, and they will require no underscoring. But I have seen a growing, pervasive, and remarkable attitude that quickness of mind, or brightness, together with facts, are the only factors that determine good thinking.

We live in times when dozens of great and small, ever more sophisticated forces may act to manipulate our values, perceptions, and behavior. If we allow this we are not truly free—we are simply more oblivious or agreeable to our subordination than is someone who takes orders directly and unwillingly.

In a sense, every single one of us is already a proven failure at avoiding manipulation and thought control. We have been manipulated into speaking a particular language and into learning the manners and values of a certain society, and into feeling good and right about it, no matter which of a thousand societies it may be. This is adaptive behavior in that it gives us a place in a society and a chance to survive. What we should be concerned about are aspects of manipulation that are largely exploitation. But for now the point is that most of us have longer histories of being manipulated than of really thinking for ourselves. In the midst of these forces, how can we make a life that is truly best for ourselves? When we have succumbed to exploitation, how much of our gullibility is due to the limitations of our brains? To our lack of real desire to be free? To our belief system? To the counsel we keep? It is to the advantage of anyone to understand the dynamics of such things.

This discussion will first explore the personal need for and historical origins of the quest for good judgment in Western civilization. Then we will turn to modern neurobiological studies on the brain itself, to explain from that point of view why we cannot simply open our eyes and ears in order to see truth, and how we can so easily and convincingly deceive ourselves. Third we will explore how the nature of brain function presents difficulties for communicating with each other, and the nature and limitations of common sense.

After examining limits of our brains and common sense as instruments for seeing and communciating truth, we will consider the institutions we have invented to supplement the brain's functions and to transcend its flaws.

The arts, language, philosophy, science, theology, courts, and even shamanism and astrology deserve discussion from this point of view. And *from this point of view* I believe we need to ponder the question of how well we handle our modern intellectual institutions such as academic disciplines, universities, and science. Do we maintain a clear enough sense of purpose so they are not lost to other demands on them?

Above all, though, I hope the discussion is useful to the *individual* who has a will to understand and use the available institutions, no matter how fragmented and diverse their functions and organization can be, to claim the legacy of rational thought that generations of minds have devoted themselves to improving and to passing on to the future.

The Anatomy

of Judgment

CHAPTER 1
Critical Thought, Science, and Justice

Civilization advances by extending the number of important operations which we can perform without thinking of them.

Alfred North Whitehead

To have doubted one's own first principles is the mark of a civilized man.

Oliver Wendell Holmes, Jr.

Justice Is an Essential Motivating Human Concern

Justice has long been one of the most powerful of human ambitions. It may seem strange to argue for this at a time when the media daily present us with evidence of contentious litigation, compulsive competition, habitual and even cynical exploitation, and emotionally brutalizing relationships in general, whether familial or societal. Nevertheless, the great books of all civilizations reflect a deep concern for how one might be fair to oneself as well as to others. All cultures have developed systems of laws and customs that help define fair and decent conduct and resolve disputes within the society's daily life.

The fact of moral geography, that there are so many different systems of laws and conduct throughout human societies, illustrates how difficult it is to agree upon any perfect model of fair and decent conduct. With a proper road map, one can find within some border or beyond some sea standards of conduct to astonish the most active imagination. At the same time, this diversity illustrates that many workable models can be constructed, and that worldwide there has been the motivation to construct them. In this light, one might begin to understand how Daniel Webster could have claimed that "justice is the great interest of man on earth."

Western standards of public and personal justice have been profoundly influenced by science. This is quite wonderful in a way. To a large extent, science today serves power—business, agriculture, government, the health industry, the military. Yet the origins of modern science were firmly tied to Greek philosophy and religion in their search for the truth about the nature of things. The motivation was to decide upon just and fair standards for living.

One can still see the vestiges of this early union in our legal system. In this respect, the obvious contributions of forensic science are merely superficial. Fingerprints, blood tests, and wiretaps are not the issue. Science has contributed to the ways in which Western tradition insists that one *think* about physical facts in a court of law. There are rules of evidence believed to improve the quality of the decisions made; for example, to avoid the human tendency to assign guilt by association. Society has also learned from science that logic is not in itself sufficient to establish truth, nor is physical evidence. Judgments are ideally made by combining logic and physical evidence. Supporting the formal rules of the court are the standards for rational thought that our society strives for. In many societies it is and has been common sense that dreams can be taken seriously, but the modern individual restricts the use of dreams as judgmental evidence.

But my point is not to detail each contribution of science and philosophy to our legal system. It is first, that irrational thought and sloppy thinking can become a personal menace to any of us. If one conducts a thought experiment, and places oneself in the position of an injured party, one sees that it is easy to be seriously wronged or to wrong another or oneself when judgment is sloppy or based on incorrect ideas about how facts should be evaluated. Second, there has clearly long been a serious human stake in justice and in some search for truth, whatever social forces may inform the effort.

One might argue that a core concern for justice should be expected in a large-brained social animal with extended periods of parental care, family-bonding, and defensive and foraging alliances. If we are seriously misjudged by those we care for and have emotional bonds with, or if we discover that we have misjudged them, it can hurt deeply. Our emotional bonds are wounded. The pain, unless healed, can be as deep and as sharp as that from the death of a loved one or an important ally. If misjudgment leads even to damage of one's political position in or beyond the family, the damage is compounded.

It is today obvious that critical thought *can* help to make one a superior general, businessperson, banker, farmer, hunter, politician, or physician. This is so obvious in our age of science and technology that it seems quite curious that the most brilliant minds that probed beyond the status quo and engaged in the search for truth, when viewed cross-culturally and historically, have only relatively recently been motivated by such considerations. Practical, winning strategies and crafty intuition were apparently adequate to serve occupational concerns for much of human history.

Many of the great philosophers, though, sought to develop critical thinking skills to find truth (Truth) in order to know wisdom, the good life, the right way to live, the path away from vexation and suffering. Critical thought was a tool to better understand oneself and the world, to nurture oneself and one's family, to avoid pain and disappointment, to avoid being an unwitting fool, to explore the possibility of humanity. Such issues evidently motivated the epic searches for wisdom of whole intellectual and moral traditions. The varied paths taken in the epic searches for good judgment and wisdom will become clearer as this book's themes are developed. Science arose through such a search. What has been the result? This important question will be explored, especially in the last parts of this book.

Next, though, we will examine some aspects of individual and collective judgment. The stage will be set for understanding that even in modern times good judgment may be difficult because the power of self-deception is so awesome that it can trap not only individuals but whole organizations. It may help to think of an individual locked in a prison of mirrors that keep one from seeing beyond oneself, no matter how hard one tries to find an outside. In chapter 2 we will begin to explore the elegance with which the most brilliant of the ancient Greeks defied common sense and broke through the concealing mirrors, while laying the foundations for modern science, philosophy, critical thought, and much of the Western value system.

Without some understanding of those basic issues it is not simple to understand why improvements in human systems of judgment have been so vigorously sought throughout history and are important to every individual today. It is not simple to understand why human societies have generated such a rich variety of approaches to wisdom, nor why each of us has some stake in understanding the anatomy of our own judgment and that of others. It is not simple to be just to ourselves or to others.

In chapter 3 we will begin to dissect the anatomy of judgment. We will examine the biological features that chain us to a capacity for flawless self-deception. One should understand the brain, the physical machinery for this great task that we began to formalize some twenty five centuries ago. A consideration of this organ's properties, strengths, and limitations will help define some dimensions of the problem that humankind faces.

In chapters 4 through 7 we will continue the dissection of the anatomy of judgment, with examinations of social forces that can both aid and warp good judgment. These include social habits of perception based on common experiences, commonsense beliefs, philosophies, and language.

Avoidable Failures of Critical Thought in a Modern Community

Peter Reilly was in good spirits when an evening meeting of his youth group at the Methodist Church of North Canaan, Connecticut, ended. He drove straight home to the small cottage he shared with his mother. It was 1973 and Peter was eighteen years old. He was a good-natured kid, well liked and considered quite normal by the community, despite his mother's colorful habits and sometimes irritating sense of humor, which had made the family an object of tolerant gossip.

Entering the cottage, Peter found his mother largely unclothed and in a pool of blood. Her head had been nearly cut from her body, her legs were broken, and she had been sexually mutilated.

The police at once focused their investigation on Peter. They were suspicious of him because he remained dry eyed and well composed. Moreover, Peter had called an ambulance, initially thinking he heard his mother breathing, although from the severity of her neck wounds that seemed impossible to the police. This appeared to be a suspicious contradiction—had Peter first called the police and then finished cutting her throat? There was neither a drop of blood on him nor any physical evidence linking him to the crime. Indeed, there was an unexplained bloody footprint, fingerprints, and evidence of robbery that might have led the police to direct their investigation elsewhere.

During interrogation, the police suggested that Peter might have killed his mother during a frenzy and then blotted memory of the foul deed out of his mind. This idea unnerved him, as was evident both from his manner and from lie-detector responses. At the time, Peter regarded the police as his friends and, in fact, wanted to become a policeman someday, so when they insisted that his emotional response showed that he had definitely killed his mother and suppressed the memory, he came to believe them. Eventually, after sleepless hours of interrogation, lack of food, and psychological pressure, he began to "remember" parts of his supposed misdeed. An officer assured him during the interrogation, "Once you get this out, you're going to eat like you've never eaten before. . . . Well, once you get this out, Pete, you'll be able to sleep for a week." The worn teenager "remembered" enough gory details to write and sign a confession.

In effect, and probably quite unconsciously, the police put Peter through a brainwashing situation similar to those now so commonly used by political and religious cults and even a business or two. Exhaustion, hunger, and a controlled environment are combined with offered friendship and even love (see Conway and Siegelman 1979, and other references to mind manipulation in chapter 3).

Once booked and in jail Peter began to doubt his assisted memory of the crime and to suspect that he had been led to imagine murdering his mother. His doubts increased after talking with older prisoners who questioned him and determined that he was naive and credulous. They began to convince him that the police are only human and can make mistakes. They also convinced him that he needed a lawyer.

Peter had a permanent change of heart, but it was not appreciated by the police, the prosecutor, or the courts. Neither did the police then extend the investigation to other potential leads, nor did the state of Connecticut drop the case. Peter was tried and convicted, entirely on the basis of his signed and verbal confessions made when he had indeed believed that he could remember butchering his mother and even jumping on her legs to break them.

Families who knew Peter continued to believe, even after his conviction, that he was emotionally incapable of such an act. Moreover, they did not believe that he was physically capable of driving home from the church, slaughtering and mutilating his mother, and disposing of the murder weapon, all in some fifteen or twenty minutes and without getting a drop of blood on himself. Last, they were suspicious of the procedures and actions of the police and State attorney.

The neighbors' commonsense conclusion was that Peter was innocent. The authorities' commonsense conclusion was that Peter had killed his mother and conned all those good people into believing he was innocent.

Funds were raised and more supporters joined the cause. Author William Styron and playwright Arthur Miller helped to bring in talented attorneys and experts. A private investigator, James Conway, volunteered his efforts and found new evidence to prove that the time available to Peter was so little that he could not have committed the crime. Conway turned up a wealth of evidence pointing to another suspect, and showed that the state had withheld important evidence and had not acted on other facts that would have weakened the prosecution's case.

The state continued to the end staunchly maintaining Peter's guilt. But at a 1976 hearing the judge granted a new trial, noting that "a grave injustice has been done"—and it had been done in his own court. A new trial was never held. The prosecutor died of a heart attack after the hearing and his successor found withheld exculpatory evidence in his predecessor's files. The state dropped its case.

The case had eerie evocations of the Orson Welles film *A Touch of Evil*, where the police detective was so sure of his intuitions about a suspect that he compromised his ethics and integrity and harmed innocent people in an all-out effort to build a case and obtain a conviction. No one can say if the comparison here is completely fair, since the Reilly prosecutor is dead and cannot defend his actions. However, a one-man grand jury—a judge—later found widespread evidence that authorities at various levels had focused on vindicating their own and their colleagues' decisions and actions, rather than open-mindedly seeking the truth. At issue here is the limited ability of the human beings in "the System," using common sense and intuition, to locate the truth. Sometimes common sense is sufficient, sometimes it is not. One group's common sense is another's nonsense.

We cannot know all the factors that led the police and the state to prosecute so doggedly Peter and ignore other leads that later proved to be significant: ambitious publicity-minded officials, bureaucratic ass covering, group-think, and the like. But Peter's case still serves as an illustration of the paths down which common sense alone can lead the naive.

Throughout much of history it was common sense that someone would only confess to a crime if he or she had actually committed it. Even the clergy saw no difficulty in accepting a confession extracted under torture during the holy Inquisition. Even late in our own century, confessions have been routinely beaten out of people and used in courts of law in the United States, not to mention some other societies. The incidence of political torture in the world has been rising during the time I have been working on this book.

Modern society devotes awesome human and financial resources to its systems of justice. Yet it is possible for the commonsense reasoning of the Middle Ages to survive with vigor even among literate people in a well-meaning and prosperous state during the space age. The prosecutor for the state of Connecticut asked the jury to set aside any demands for hard evidence and to use their common sense in evaluating circumstantial evidence. After all, it was put, the boy's mother was an arrogant practical joker, an intellectually pretentious welfare recipient, a lesbian who had also once lived with a black man, and a drunk—all in a small New England town. Her fatherless son had good *reason* to hate her, even though he *claimed* to love her and to respect her manner and life-style. He never shed a tear over her body. He did confess to the crime before a number of people. Further, they had no other suspects in hand.

As a general matter, the judge was accurate in his instructions to the jury to use "your own common sense ... if the law were to impose on you the duty of regarding circumstantial evidence as insufficient ... the execution of criminal laws would be practically paralyzed as would be, also, the protection of civil rights." The jury did use common sense and convicted an innocent boy of a hideous crime, adding injustice to his tragic loss.

The Reilly case is a good example because it became probably the most famous case in the history of Connecticut and is therefore extensively documented. I might have chosen many other well-documented cases as examples, but this particular one also raises questions that can be touched on in other chapters: Was there something odd about Peter or, under the right conditions could anyone have been led to imagine such grotesque fantasies? Is Peter an example of someone who was too open-minded to alternative explanations of events — too willing to mistrust his own eyes and to consider seemingly improbable possibilities, such as the scenario laid out by the police?

Further, this case will be a preface to the Oedipus theme in chapter 2: apparently, the state was so blinded by its "clear-seeing eyes" that it — the organizations of men involved — chose not to pursue other clues. As Donald Connery put it, "I think it likely . . . that the authorities brainwashed themselves into believing that they had caught the true killer. Willfully ignorant or woefully unprofessional, they closed their minds to every development that should have inspired them to reexamine their conclusions." Assuming that the police, grand jury, prosecutor, courts, and so on, did brainwash themselves, were *they* all odd people or could any of us have helped to convict an innocent youth without any substantial evidence? Connery has written a commendable book (1977), but we will try here to go deeper than "willfully ignorant or woefully unprofessional."

These authorities, like Oedipus, as we shall see, look absurd in hindsight, though their case was not nearly so improbable as the situation that Oedipus faced. We can use the true Reilly case and the epic *Oedipus* drama to ask what traits of mental discipline or its lack cause people to stumble down more subtle, but often equally destructive, paths? We can find avoidable misinterpretations and misjudgments made by meticulous scientists because of the same psychology that has driven kings, presidents, religions, police, and government agencies into shame.

The human mind is capable of views of reality and systems of explanation completely satisfactory and satisfying to one's common sense and emotions even though they are utterly wrong. The brain automatically smooths over the logical or perceptual rough spots. Given an apparently reasonable explanation, small and large inconsistencies pale before the vividness and apparent harmony of the accepted view. Then there is no incentive for a reasonable person to question, to look beyond. In many cases it would even be emotionally or socially threatening to tamper with personal or social perceptions. We all do this from time to time, and so do judges and scientists, even though warning signs have been built into their professions. A person who has not been formally or informally trained to think critically cannot always be expected to make competent judgments, but we expect professionals to know the warning signs and pitfalls in their own discipline. Do ignoring the signs and stumbling again into known traps constitute incompetence in these professions? Often this is clearly so, but even the most cautious practitioner may on occasion make avoidable mistakes.

Some habits of thought can help us get past such problems, while other habits of thought only serve to rationalize difficulties and smooth away the frustration they might cause. Mediocrity is not laxity or inability so much as it is a system of information processing.

When we falter in our ability to see truths, can we partition the blame into causes? How much of the problem is caused by the limitations of the device of thought—by the construction of brain tissue itself? How much is caused by those aspects of ourselves that we like to believe are subordinate or at least vulnerable to will or conscious control—personality, mental discipline, beliefs, cultural processes, and conventions?

It is clear that professionals who should be interested in truth—scientists, teachers, jurists, philosophers—might have considerable interest in such questions. But so too should any citizen who desires *intelligent control* over his or her own life—not simply the physical freedom to choose colors, flavors, politicians, television channels, or lovers, but the true control that comes from knowing the meaning of choice and the short-and long-term implications of choosing. Peter Reilly did not understand the meaning or implications of his choice to submit to lie-detector interrogation without a professional and independent adult present. The police apparently did not understand the meaning and implications when they chose to "question" the eighteen-year-old in the manner that they did. Both parties were free to make choices, but even this freedom was dangerous without more fundamental appreciation of one's own limitations and potentials.

Mass Hysteria: A Society's Failure of Critical Thought

Peter Reilly was not the first youth to be pressed by officials into believing unreal things. Arthur Miller's involvement in the Reilly defense recalls the true incidents on which his play *The Crucible* was based. An incident in nearby Massachusetts had spun even further out of control back in 1692. Under the potent influence of peer participation, several children in the town of Salem began to have convulsions at winter's end, and the doctor and clergy were faced with an incident of group hysteria, which they could not explain and which they eventually came to see as the arrival of the devil. The girls were apparently led by questioning into believing that witches were harassing them. They soon developed the ability to recognize witches and began to name them, initially in modest numbers. The town was grateful for the discovery of the first few witches and this served to fuel the situation until it grew out of control.

Twenty-two innocent people died, scores were tried, and hundreds were imprisoned. Of the twenty-two deaths, nineteen persons were hung, two died in prison, and one was crushed by his neighbors while they tried to extract a confession by piling stone after stone upon him. The community as a whole was involved. We do not call such things bureaucratic bungling; we call this mass hysteria.

It is impossible today to say if, at the very first, the girls actually thought they saw witches or lied outright; some would argue for the latter. Should we assume that through fear or playfulness alone they would stand by for so long and condemn their town to a rampage of torture and killing, jailings, economic disruption, and savage ruin? Perhaps they initially, under psychological pressure and the influence of suggestion, like Peter Reilly, really thought they saw things. Later, at least some of them began to doubt that their visions had been real (and we know that some did try to deny the original stories), but by then the adults had convinced themselves of the truth of the girls' suspicions, and the ball was rolling of its own momentum. Perhaps the allegations started out as a bit of hysteria, a bit of delusion, and a bit of a prank. Behavior often defies neat pigeonholes. We may never know for sure.

The more interesting point is the disastrously poor judgment of the adults. They were apparently ignorant of the unreliability of information extracted from hysterical youngsters under pressure, or subsequently from adults under intimidation and torture. They were quite insensitive to group dynamics and to the manner in which fear could get out of control and set friends and neighbors even to killing each other. Their intellectual assumptions left them rationally helpless: they had no good and reasonable intellectual tools to decide if someone was really guilty or innocent once accused, or if an accusation was untrue and had been made from fear, pain, or malice.

They were living in the New World; they had had years of experience in self-government; the Renaissance had come and gone; Galileo and Shakespeare were history; Harvard had already been founded. The Inquisition, religious persecution, and tyrannical oppression were all supposedly distant. History had many relevant lessons to offer, but the people of seventeenth-century Massachusetts had not learned them, much as Connecticut officials nearly three centuries later would ignore the lessons of history in the case of Peter Reilly.

Our reason and judgment are not *automatically* good simply because we no longer live in the proverbial Dark Ages. We have made some progress as a civilization, but that does not ensure individuals and organizations against making unnecessary and stupid mistakes. Good judgment and critical reasoning ability are things that each person and organization must acquire through individual efforts. Our civilization only provides resources to learn and by no means provides guarantees. Societal norms do not ordinarily embody critical thought. Likewise, reason has never been able to free itself entirely of social myths and bias. Critical thought is a negotiation between the positive logical and data-based resources available for rational thinking, and the negative covert social influences on their use. This is a book about that negotiation.

The entire story of the Salem witch trials is too long to relate in detail, but it is of particular interest not only because it is one of history's most famous witch-hunts. Since it is well documented (because of the numerous arrests, trials, and executions), the interested reader has the opportunity of further independent study and reflection.

The witch-hunt began to grind to a halt only when Samuel Willard, the president of Harvard and the pastor of the First Church of Boston, was accused. The magistrates concluded that the girls must simply have been wrong. With over 250 years of hindsight in this case, we are inclined to agree. Of course, some few continue to argue today that witchcraft was actually present in Salem, but they have presented no coherent evidence, and on the other hand the episode can be explained by natural phenomena.

Massachusetts was at that time undergoing political disorder and people were uncertain about their futures. The idealism of Puritanism was also being strained for various reasons (Bednarski 1970). Thus, the Salem witchcraft trials are an example of the rule of thumb that groups of people have difficulties in maintaining rational progress during times of social or economic confusion. We might recall in modern times the rise of strong, charismatic leaders in the West during the world economic decline of the 1930s and the intellectual disillusionment that followed the senseless First World War. Germany, for example, was ripe for the picking by a leadership offering a simply understood set of national beliefs and goals, and a simply identifiable set of enemies. Indeed, Germany—one of the most highly developed nations on earth in terms of sophistication in letters, the arts, and science—was harvested.

> There was nothing new, nothing at all peculiar to Salem Village in the outbreak. Similar examples of mass hysteria and on a far more enormous scale had occurred repeatedly in the Middle Ages, and always like this one in the wake of stress and social disorganization, after wars or after an epidemic of the Black Death. There had been the Children's Crusades, the Flagellantes, the St. Vitus Dance, and again and again there had been

outbreaks of witchcraft. Sweden had recently had one, and on such a scale as to make what was going on in Salem Village look trivial.

Nor has susceptibility to "demoniac possession" passed from the world. A rousing religious revival will bring out something like what Salem Village was experiencing; so will a lynching, a Hitler, so will a dead movie star [such as Rudolph Valentino, Jean Harlow, James Dean] or a live crooner. Some of the girls were no more seriously possessed than a pack of bobby-soxers on the loose. (Starkey 1949)

It is necessary to clarify two points here to prevent possible misunderstanding. First, Starkey is not saying that religion necessarily leads to mass hysteria any more than she is saying that movie stars necessarily cause mass hysteria. Second, and more importantly, we use terms such as mass hysteria only for ease of communication, and rarely is anything remotely like the stereotype of clinical hysteria involved. One tends to think of the undesirable as pathological and then take the (apparently) logical step of assuming that groups of people engaging in undesirable behavior are individually and physiologically sick. But this is to fall into a semantic trap.

The fact that people make decisions and behave in ways that in our day and age may seem odd and emotional can seduce us into conjuring up meaningless theatrical images of wild-eyed or hot-tempered and impulsively out-of-control individuals. Images of emotional crowd scenes may be dramatic and useful to painters, writers, and film and television news producers to convey vividly their ultimate message that something is very seriously out of control. But true incidents of so-called mass hysteria always involve many sober, patient, well-meaning individuals who are simply using what turns out to be bad judgment. The Nazis who were "hypnotized" by Hitler were in many cases ordinary people who had bought into a philosophy and political system that turned out to be seriously flawed in its assumptions and to be destructive to civilized values.

The authorities in the Salem witch-hunt soberly and patiently deliberated the ground rules under which witches would be detected, tried, and punished. Careful written records were patiently kept, as they had been during the holy Inquisition. The clergy and civil authorities who methodically conducted the bloody holy Inquisition (actually there were three inquisitions) over hundreds of years were among the best educated and most rational people of their times. They were not illiterate "rabble." To their credit, relatively quickly the people of Massachusetts soberly and patiently appreciated that they had made serious mistakes. We blind ourselves to the lessons that history can teach us when we use terms such as "mass hysteria," "mass hypnotism," or even "craze," and let it go at that.

My point is not that historical or recent episodes of mass hysteria may not involve emotion or manipulation. The point is that the emotional components may be of a subtle sort and that the people involved may be, for the most part, not at all clinically hysterical, but often as sober, calm, and apparently well balanced as those we interact with every single day. They may differ primarily in their preoccupations and in the harm that they eventually do to themselves, their family and friends, or to society in general. Sound judgment is not simply a matter of staying cool and calm like some film folk-hero.

Many people have expressed bewilderment that the Nazi mass murderers, who were, after all, known to be brainwashed and in a state of mass hypnosis, could nevertheless laugh and love and cry; they could play happily with their children and enjoy fine music just as normal relaxed people do. This is only surprising or bewildering if we have fallen into the tragic trap of taking terms such as mass hysteria literally and have imagined irrelevant clinical stereotypes or machinelike characters from war films.

Poor judgment can, though, easily be exacerbated by emotional tension, as many detailed studies show and as anyone who has survived an ill-advised love-affair, financial difficulty, episodes of self-doubt, and so on, can testify. In this sense children and even adults can be manipulated if the powers that be, or circumstances, keep a large nucleus of them in a state of psychological stress or uncertainty. The threat of war, social or economic turmoil, or of an enemy within are classic devices that have been used to suppress clear judgment and provide emotional focal points and outlets for a population's anxieties. The threats may be real or contrived, but that is not the point here: they make people vulnerable to poor reasoning and to manipulation of their perception and judgment.

In the Salem mass hysteria there was a spectrum of diverse individuals participating—not uniform ranks of robots—with the sober, reflective, though tragically misguided and perhaps frightened, individuals who contributed to the debacle far outnumbering the young girls, several of whom may well have been hysterical and had visions.

The purpose of these first sections has not been to breed pessimism with the possibilities for progress in our understanding of ourselves, and for progress in defining and achieving social justice. Somehow, we have made much more progress than a cynic centuries back might have ever dreamed possible. Fragile, improbable, and meager as that progress may be, it has been accomplished.

Rather, I hope to have introduced some topics of proper and serious concern to any would-be freethinking citizen. I also hope to have made the first step in explaining how science should or does touch our intimate lives in perhaps even more fundamental ways than through technical applications such as curing diseases or accelerating communications. Witch-hunts and Reilly trials may never be eradicated, but the intellectual resources available improve, and we can at least ask more of ourselves and hope to learn more quickly to spot failures. Science has been inextricably bound to the ways in which most of us think and to the reliability with which each of us can wisely act. Its modes of thought have come to touch us all in one way or another and to affect not only the ways in which we invest our savings and pick mates and leaders, but our sense of fairness and justice. Logic had to be invented. Statistical thinking had to be invented. Analysis had to be invented. Experimental hypothesis testing had to be invented. The comparative method had to be invented. And, skepticism of the obvious and of deeply entrenched popular beliefs had to become established through a series of public scientific demonstrations and discoveries.

Yet, we will see that science has distinct shortcomings, and I refer not merely to the now-common criticisms of runaway technology, an indifference to feelings and compassion, or even to its use by power groups to control masses of individuals, as portrayed, for example, in George Orwell's *1984* and Aldous Huxley's *Brave New World*. These are important issues, but their overlap with the present discussion is small and even less than it might seem at first.

We must negotiate our relationship to science as a "way of knowing." Science is like an awkward, shallow-keeled boat in which we have set ourselves adrift on an indifferent and restless sea, and only with skill may we be able to catch the shifting, gusty winds of imagination and tack our way into calm, deep waters.

My discussions are not aimed at social reform. The social, economic, demographic, and political tides of social decline or change are much too powerful for me to hope to impact much. But the material covered in this book should be of some interest to individuals who wish to do more than drift with the tide, who would instead increase and sharpen their perspective on the rich heritage of intellectual tools for improving on one's powers of judgment. These issues are of interest certainly to the young scientist or scholar, but also to any person who aspires to think as clearly as is now possible, or at least to understand why clear thinking is so difficult.

CHAPTER 2

The Eyes of Oedipus, the Cave of Plato

Alas, the chronicle of humankind
adds up to nothing.
For who has won more than illusions—
happy visions only, dreams that shatter?
You are my example, Oedipus—
my warning to call no mortal blessed.
You towered over all in wisdom, glory, joy,
but now who is more wed to pain and agony?
You once brought peace into my heart;
you now bring darkness to my eyes.

Sophocles, *Oedipus Rex*

Humanistic Concern: The Common Thread in the Arts and Sciences of Ancient Greece

Great tragedy and comedy often have much in common. Both explore our frail grasp of reality. The deceived lover whose reality is an illusion may be a tragic or a comic figure depending on the artist's treatment of the subject. Likewise, the misled idealist or faithful follower may evoke compassion or ridicule. Great artists have been able to make us come to grips with the ambiguity of a situation: is this character good or bad? . . . wise or foolish? . . . was her action heroic or impetuous? . . . should she be judged with compassion or harshly?

As they debated how to be better people, the ancient Greek thinkers were deeply involved with questions of reality and of how to find it. We shall see throughout this book that this issue has also been a central concern for other, non-Western, cultures and their sages throughout history. Indeed, illusion and reality are central *human* concerns. Distinguishing between them remains central to the ability to make sound judgments and lasting progress toward wisdom. There is a distinct component of emotional, and frequently material, *self-interest* to truth and justice.[1] Various cultures have taken their own paths; in this book I shall explore these. But we begin from the point of view of enormously influential Western civilization and modern science and with an interest in tracing their roots to the rich soil of humanistic concern from which they grew.

The Western humanities and scientific ideals, as well as religion, from ancient Greece through the present, have been used to disguise and abet national, racial, economic, social, gender, and intellectual subjugations, and for such reasons they are now severely criticised by scholars. This will be discussed in due course. Yet there are strong universal human issues that have been vigorously explored in the West and by science once. A critical examination of these issues must consider their progress as well as their misuse.

It was in a climate of intense humanistic concern over reality and how to find it that the philosophers Socrates, Plato, and Aristotle, perhaps for us the most influential thinkers from those days, debated, studied, and wrote in the fifth and fourth centuries B.C. Socrates searched for reality and truth by questioning one's psychological qualifications to capture reality. If we would know the world around us, he argued, we must first understand ourselves. He tried to bring his students to understand how biases, preconceptions, muddled thoughts, and unexamined hypotheses can cause us to live by illusions and lose the trail in the search for truth and wisdom.

Socrates, "the gadfly," made many enemies for this. His simple life and self-righteousness were a living reproach to the community, as Plato presents him (but see Stone 1988). He was made to drink poison. Even today, gadflies can earn intense hostility.

Socrates' student, Plato, emphasized pure reason as a means for finding reality and he saw the "True" reality as being not in the material world in which we all live, but in an abstract world of forms that can be seen only through reason.

Plato's student, Aristotle, though he was the founder of formal nonmathematical logic, was apparently less exclusively cerebral ("rational") than Plato and put his own efforts into the task of actually observing nature as an important way of checking on the validity of one's ideas ("empiricism"). The first great naturalist, Aristotle aimed his research toward answering questions about the nature of things, and from this, answering questions about right and wrong and the meaning of life.

Aristotle's student, Alexander the Great, was as irrepressible as his teacher. The young military genius had a penchant for cutting Gordian knots as well as for patient analysis and dreams about a better world. While only in his twenties he conquered the known world to rebuild it his own way.

Whatever the shortcomings, and they were indeed major, what an interesting tradition of intellectual wonderment and idealism. Why? According to recent common wisdom, it is scientific to set aside analysis of historical causes and speculate rather that philosophy and science simply unfolded from some biological property of humans—simply from naked curiosity about life. But curiosity can be focused in many ways. We might comfortably spend our lives merely peeking in on our neighbors' lives, slipping into their bedrooms, exploring new markets, sampling new foods, or sitting through to the conclusions of films and books. It is much harder to understand why, within some cultures but not others, there developed traditions of investigating Babylonian astronomy and the parasites of sea urchins. Typically people are not intensely curious about the general nature of reality when they think societal norms already contain the major answers. The culture and subculture have a great influence on what people are or are not curious about in any serious way. Thus, for insights into the origins of philosophy and science we should look to cultural traditions and contexts, not merely into the physiology of curiosity and exploratory behavior.

Alexander's philosophical teachers understood that the question of reality has a lot to do with how we should lead our lives: what can we agree is right and wrong, or at the very least, how can we keep from making stupid mistakes? Life or death, happiness or guilt, sex, freedom or enslavement are involved. First we must know what really is, or at least which apparant realities are false. This was recognized to be a bread-and-butter, everyday issue then, a question artists worried about and dealt with. It is not certain why there was such very deep concern over reality and how to find it in those interesting days of creativity. As we shall see, especially in chapter 8, there have been actually worldwide efforts puzzling over reality. Be that as it may, issues of reality, the difficulties in perceiving it, and its moral implications were *formalized* powerfully in classical times in ways that formed much of the basis for Western philosophy, theology, science, and the arts, including drama.

The Tragic Situation of Oedipus

Sophocles wrote probably his greatest play, *Oedipus Rex*, about the king of Thebes. Oedipus was a wise, brave man, an irrepressible seeker of truth. He had good judgment and was able to see beyond the obvious. But those assets could not save him from a horrible fate—he unknowingly killed his father and married his mother.

The play has many dimensions but surely the moral dilemma is this: If avoidable ignorance or short-sightedness leads us to do great harm, can we shrug off the responsibility? What traits of character can lead to false reasoning and perception that can turn intended good into acts of evil? How can we claim to be, or aspire to be, people of action and to pursue ideals unless we can grasp the truth?

Those who do not know the play but who know of the "Oedipus complex" may mistakenly assume that King Oedipus of Thebes was a creature of improper lusts. However, in truth Sophocles leaves no room for a Freudian interpretation of the king's killing his father and pairing with his own mother.[2] Oedipus was raised by foster parents, the king and queen of Corinth, and quite explicitly did not know of his adoption or that Laius and Jocasta were his true biological parents. Even though Oedipus was a newcomer to Thebes, he was made king when he arrived because he displayed an extraordinary talent to see through riddles—to see nonobvious associations. Yet he ruled for the next fifteen years without seeing through his own preconceptions to the true connections between events that had preceded his arrival in Thebes. He had slain some men in a roadside quarrel at about the same time and in the same place where Laius, the king of Thebes before him, was found killed along with his aides. The truth, of course, was that Oedipus himself had unwittingly killed the king and that, moreover, Laius was in fact his natural father. So by marrying Jocasta, Oedipus had come to father children by his own mother. There was a considerable stigma attached to incest and patricide in that culture. When he finally had to face the truth, he became so distraught that he put his own eyes out with a brooch.

Oedipus was quite innocent in all this, by our present Judeo-Christian moral standards that are so concerned with beliefs and intentions. His *motives* would be more or less acceptable. He had seen not a king and his aides on that road, and certainly not his own father; he had seen only a band of ordinary, rude men. He had reason to be confident about what he had seen and so he never questioned his own interpretation of the event until he was later forced to.

Even if Oedipus was morally innocent by our standards, was he for practical purposes blameless? He could have done better. There had been warnings. Some will say he could not possibly have known the truth, but for the Athenian audience this would certainly not have seemed the case. Prophecy had told him that he would kill his father and marry his own mother. He did *not* ignore this prophecy. To escape the predicted misdeeds he uprooted his life and fled Corinth and the parents that he believed to be his own. And it was during this very escape, with the prophecy presumably burning in his mind, that he killed his true father. So he was not one to take prophecy lightly. Yet he did not take the prophecy seriously enough to question his view of reality and go on to explore what might at first have seemed improbable interpretations of events. Oedipus even ignored the esteemed blind prophet Teiresias, who revealed that Oedipus himself was the murderer of Laius. Oedipus became indignant and would not seriously consider this, leading Teiresias to exclaim sarcastically:

> You ridicule my blindness;
> Yet with fine eyes what can you see
> of the corruption in your life?
> Unknowingly you are your loved ones' enemy,
> you know not who you are, with whom you sleep.
> Who are your parents?
> Know then this:
> your clear-seeing eyes will darken.
> And where shall not your cries be heard
> as you in horror come to see
> yourself and what your children are.

Here, Sophocles touches on an important point: Whatever we see with our eyes, our personal realities can blind us to the truths that are before us.

There have been many interpretations of *Oedipus Rex*, from Aristotle to our own times. As children in school, among so many other myths, we learned that Oedipus's character flaw was in trying to escape fate or prophecy and defy the gods. But a close reading does not support this notion (Kaufmann 1968). Moreover, then *Oedipus Rex* would be only a routine morality play and not a brilliantly constructed tragedy. Sophocles' contemporaries (like Aristotle), who certainly did not look to the popular gods for truth in prophecy, would not then have held up the play as the finest example of tragedy.

How can one possibly heed prophecy when there are apparent contradictions? Can Oedipus, or modern people for that matter, be blamed for not accepting any prediction or prophecy when its reliability is uncertain? Prophets, wise persons, secular experts, and authorities often contradict one another. As Sophocles himself put it:

> Zeus and Apollo know earth's secrets.
> But has a prophet secret knowledge above mine?
> There is no certain test.

So, the Chorus thus implies, we simply have *no choice* but to use our own best judgment.

Did Oedipus err in ignoring the accusations and warnings of a prophet when they required him to question his own experiences, his own parentage, indeed, his own identity? The right course, with our hindsight and non-involvement we might think, would have been for the king and loving father simply to have accepted the "obvious truth" that the blind prophet presented. But then Oedipus would have had to sit and ponder:

Hmmm, perhaps this old blind prophet has something. After all, my wife Jocasta *could* be my mother. It would only be necessary to assume that my own parents were not really my own parents and that I was never the prince of Corinth at all! Well, let's assume that for a moment. Now, suppose Jocasta had a child—me—and for some reason (let's not try to guess why just now) I was adopted by the king and queen of Corinth, and for some reason all of this was kept secret. My goodness, this is getting complicated. But as Creon said only today, "Seek and ye shall find. Unsought goes undetected," and Creon is a brainy fellow. Let's think then. Perhaps those rogues that I killed were in fact Jocasta's former husband, King Laius, and his men. Then it would fit that I had killed my own father: How did the prophet put it so well? "Son, and husband, to the woman who bore him: father-killer, and father-supplanter." But the men that I killed were such rude and ragged fellows to my eyes, they were hardly a beloved king and his aides! Perhaps they were in disguise for safety's sake. But then there is the story, that I would have to explain, by the one man who escaped. He told the people here in Thebes that it was robbers who did the deed and that there were many of them. So it could *not* have been me! I was alone and certainly did not rob those rascals. Wait, ah! I need only assume next that the fellow *lied* and then it will all fit Teiresias's insulting accusations. But why should the escaped man have lied? Hmmm. Perhaps he was trying to save face. It would have looked terrible if one man had killed all but him and he had cowardly run away to save his own precious skin. But then everything would make sense! What a clever fellow I am. I must find someone to try out my fine new ideas on.

But then if Oedipus had come to this point in his deliberations, he might well not have gone on, thinking:

Wait, if I suggest this to anyone, they will think that I've gone nuts! What if I have? Perhaps it is a lot simpler to assume that Teiresias is confused. After all, the old fellow talks in riddles.

Oedipus was neither cursed, damned, stupid, nor afflicted, and he did not have a complex. He was an ordinary man, though one with an extraordinary passion for the truth, caught up in an intricate and inherently deceptive situation in which good intentions and even an exceptional ability to see through riddles were inadequate to avoid an escalation of misdeeds. What gives the story great power is that it could have happened to any of us. Virtue is no protection against the curse of illusion, and each of us carries the potential for that curse in our brain structure. We are all capable of such things; a wise king is no exception.

Oedipus was *the State*, but also what a very *personal* matter for any of us to misperceive. No matter how innocent one might be of the circumstances leading to the misjudgment, it would be difficult for most of us to stand back at an emotional distance from the revelation, call it a simple accident, and try to excuse ourselves from any guilt or responsibility. In retrospect there were so many clues, so many warnings; yet their meanings escaped Oedipus, although he was relentlessly seeking both the killer of Laius and the truth. Then to discover that love is sin, that one's dear and beautiful children are abominations, that he, the protector of the city and the seeker of truth and justice, was himself the killer and the cause of his city's ills—this would present an agony for nearly anyone. What matter that one's intentions were good? "If pride be honored and its violence condoned, what good the sacred and harmonious dance?" So even if today we might not condemn Oedipus, who can blame him for condemning himself?

Then Oedipus put out his own eyes, an event more significant than mere sensationalist gore. There is the implication, as blind Teiresias had suggested, that Oedipus's eyes had deceived him. Moreover, he could not bear to look upon his deeds once forced to face them; he felt that there was no beauty left to look upon in the world.

There is a certain pathetic quality about such reasoning, as though it were the eyes and not some aspects of personality and reason that were at fault, as though one could blot out false consciousness by smashing the senses. One recalls the tradition of killing the bearer of bad news, confusing the messenger with the message. But as soon as Oedipus blinds himself, it is clear that vision itself was not to blame for the tragedy, but rather reason and character were involved—how Oedipus had thought about what he had seen.

Again, should Oedipus have known better? Was not Homer, the great culture hero of the Greeks, believed to be blind? Did he not see vividly and clearly nevertheless? Does this example of Homer not prove that the mind and not the senses is the responsible element in the human equation? Or, perhaps given Homer's example, it seemed logical for Oedipus to blind himself. Perhaps Oedipus thought that, like Homer, he would see more clearly without sight to mislead him.

The Unreliability of the Senses

The emphasis upon the deceitful senses in *Oedipus Rex* recalls Socrates' arguments in Plato's *Republic*. Since Socrates never published, we know him only from reports of his students such as Plato. Plato/Socrates argued that the leaders of the model state should be trained in gymnastics, music, and practical matters, but they should be both soldiers and philosophers and so must first and above all be trained in reason and calculation. Since the senses are imperfect and a given sense will give contradictory information, leaders must learn not to depend on them: we cannot rely on that which seems obvious, on apparent reality, on common sense, we must seek with our minds beyond what our senses tell us and search with Reason for the Truth.

Of course, science today confirms that the senses are imperfect. There are many energies, sounds, colors, tastes, and textures that we cannot hear, see, taste, or feel without special instruments. Socrates lacked this information, but he nevertheless could reason that there were serious difficulties with the senses. For example, no one sense is in itself perfect. Vision will obviously not tell us if an object is soft or hard, touch will not tell us if an object is blue or yellow. Moreover, a sense will contradict itself, as when a finger feels soft compared to a rock but hard compared to a feather. Is the finger hard or soft? This is a question for Reason, not for the sense itself. The same problems exist with large and small, heavy and light, and so on. If we press two adjacent fingers on our hand tightly together, he continued, the eye sees both two objects and one object at the same time. The sense of vision cannot tell us if there is one object or two, and Reason is required.

There will be, Plato assured us, resistance to this essential process of questioning the senses and the obvious—resistance to the elevation of Reason. Socrates introduced the now-famous analogy of people raised from infancy chained in a cave, who could see only the shadows of themselves and others, and who came to believe that the shadows on the walls were in fact the realities of life. The philosopher might break loose, come out of the cave, and be at first blinded and confused by seeing, unexpectedly, real objects in the light of day, by seeing the objects that had been casting the shadows, and even new objects. But he would become used to this world of light. However, if the others in the cave had given names to the shadows,

> and if there were honors and praises among them and prizes for the ones who saw the passing things most sharply and remembered best which of them used to come before and which after and which together, and from these was best able to prophesy accordingly what was going to come . . . . (*Republic* VII)

Then the philosopher who knew the Truth would become alienated from his former associates and their values as they talked of their shadow world with conviction, and indeed he would appear foolish in their eyes if he were to challenge their world.

> Wouldn't they all laugh at him and say that he had spoiled his eyesight by going up there . . . ? And would they not kill anyone who tried to release them and take them up, if they could somehow lay hands on him and kill him? (*Republic* VII)

(Plato probably primes us here to comprehend better why Socrates was eventually sentenced by a jury of 501 citizens to die.)

Not all of the ancient thinkers had quite the same emphasis. Some stressed the unreliability of the senses and others stressed the fraility of the mind and of Reason. But there was general recognition that there are serious problems with taking apparent reality at face value. Indeed, Plato attempted to eliminate the material world from consideration as the fundamental reality and he convincingly argued that Reality was a world of "Ideal Forms" in a move, for better or for worse, with profound implications for all subsequent philosophy, science, theology, and commonsense values in the Western world. Much of all subsequent intellectual activity involved attempts to resolve Plato's idealistic denial of the importance of material reality with Aristotle's empirical position that much can be learned from a study of life.

Lucretius, writing for a Latin audience in the first century B.C., insisted that the senses cannot be at fault in errors of judgment since reason is ultimately based on input from the senses. And:

> not only would all reason give way, life itself would at once fall to the
> ground, unless you choose to trust the senses and . . . that host of words
> then be sure is quite unmeaning, which has been drawn out in array against
> the senses. (*On the Nature of Things* IV)

Lucretius was an Epicurean philosopher, belonging to a group that stressed so-called materialism and a philosophical path intended to free one from superstition and the fear of death, and that resulted in peace of mind.

> Vain is the word of a philosopher which does not heal any suffering of man.
> For just as there is no profit in medicine if it does not expel the diseases of
> the body, so there is no profit in philosophy either, if it does not expel the
> suffering of the mind. (Epicurus, *Fragments on Philosophy*)

The good life was to be found in the moderate, indeed disciplined, living of it and not in abstract contemplation, or in attention to groundless fears and devotions. As we shall see, somewhat similar materialist movements also developed early in India and China. These were relatively unsuccessful reactions to state religions that were growing in power and were convincing subjects to ignore conditions of misery and injustice and focus instead on ideal truths, afterlifes, or salvation through religious rituals (e.g. chapters 8, 11; Chattopadhyaya 1986).

Epicureans were reacting to the dualisms of mind versus body and of material illusions versus the eternal reality of the abstract forms, that Plato and others had set up. In their view it was not life-as-it-appears and the deceitful senses that stand in the way of the good life, but empty imaginings and troubles of the mind. Yet, Lucretius was well aware that experience can in any event be misleading.

> The ship in which we are sailing, moves on while seeming to stand still; that one which remains at its moorings, is believed to be passing by. The hills and fields seem to be dropping astern, past which we are driving our ship and flying under sail. The stars all seem to be at rest . . . and yet are all in constant motion, since they rise and then go back to their far-off places of setting. . . . In like manner sun and moon seem to stay in one place, bodies which simple fact proves are carried on. And though between mountains rising up afar off from amid the waters there opens out for fleets a free passage of wide extent, yet a single island seems to be formed out of them united into one. When children have stopped turning round themselves, the halls appear to them to whirl about and the pillars to course round to such a degree, that they can scarce believe that the whole roof is not threatening to tumble down on them. . . . Again, although a portico runs in parallel lines from one end to the other and stands supported by equal columns along its whole extent, yet when from the top of it it is seen in its entire length, it gradually forms the contracted top of a narrowing cone, until uniting roof with floor and all the right side with the left it has brought them together into the vanishing point of a cone. . . . Then to people unaquainted with the sea, ships in harbour seem to be all askew and with poop-fittings broken to be pressing up against the water. For whatever part of the oars is raised above the salt water, is straight, and the rudders in their upper half are straight: the parts which are sunk below the water-level, appear to be broken and bent round and to slope up and turn back towards the surface. . . . And when the winds carry the thinly scattered clouds across heaven in the night-time, then do the glittering signs appear . . . to be traveling on high in a direction quite different to their real course. Then if our hand chance to be placed beneath one eye and press it below, . . . all things which we look at appear to become double . . . ; the light of lamps brilliant with flames to be double, double too the furniture through the whole house, double men's faces and men's bodies. . . . Many are the other marvels of this sort we see, which all seek to shake as it were the credit of the senses: quite in vain, since the greatest part of these cases cheats us on account of the mental suppositions which we add of ourselves, taking those things as seen which have not been seen by the senses. For nothing is harder than to separate manifest facts from doubtful which straightway the mind adds on of itself. (*On the Nature of Things* IV)

Science, Knowledge, and Wisdom, Then and Now

The unreliability of the senses, or of evident reality, was a fundamental issue for many of the ancient Greek intellectuals, whether they were in modern terms artists, mathematicians, or philosophers. It was an issue having very much to do with right and wrong, justice, government, and other social institutions. Western philosophy and theology were in turn fundamentally influenced by Greek thinking, and particularly by the writings of Plato and Aristotle. Much later, Western science grew directly out of philosophy when a specialized, empirical approach to questions of reality proved its worth.

This history is why most scientists today earn a doctor of philosophy (Ph.D.) degree. This is why Isaac Newton would give one of his great books a title such as *Mathematical Principles of Natural Philosophy*. ("The whole burden of philosophy seems to consist in this . . . to investigate the forces of nature, and then from these forces to demonstrate the other phenomena.")

With science, a new way of thinking evolved that has influenced all Westerners to one degree or another. Today one can find books that treat separately Greek art, science, mathematics, philosophy of government, and so forth. But these were far from being such separate and disrelated subjects for the Greeks, as they are for us today. The search for Truth and Reality, through what we today call science, philosophy, the arts, and logic, was for them deeply involved with issues such as justice, progressive social goals, and the meaning and proper conduct of life. Can we talk of good and bad, of right and wrong, of proper social conduct, if we cannot rely on the senses or on what seems to be obvious? Can we rely on reason? How do our character and habits of thinking and seeing affect our reason? How can we avoid the tragedy of Oedipus?

When we do turn back to the roots of science, philosophy, mathematics, law, and what I will for lack of a better term call serious art, we too often make the mistake of representing the older outlook as being as compartmentalized as our own. Of course, what scholars today are doing by compartmentalizing Greek thought is trying to trace our own highly specialized contemporary approaches to knowledge back to their origins. And, modern scholars may be doing their best in coping with a plethora of problems in language translation and concentrating on producing a technically coherent translation or analysis of some text for students in a narrow field of study. Not least, of course, the Greek philosophers themselves, in trying to be logical and precise, were beginning to construct analytical categories, just as here we have seen Socrates arguing strongly for a separate intellectual consideration of the senses and of reason. The point is that the search for truth continues today, but in a number of highly specialized disciplines that have little awareness of their origins or of their relationships to each other or to central human questions.

Much of what I want to do in this chapter is to introduce a sense of the vitality, unity, and harmony of the Hellenic approach to knowledge and wisdom. I do this, not merely as an academic exercise, but in the belief that it remains today a vitally relevant human priority. Modern science has made enormous progress in learning about the nature of humankind and the universe. The effort to find truths beyond what seems obvious is also a primary focus in the serious arts, legal and political theory, and other disciplines. Yet science and the humanities are often seen today as antithetical and indeed as competitors, and they seem to feed one another only unconsciously and inefficiently, if not grudgingly. In part this results because the recent social functions of science have become complex and confusing, even to its practitioners, as science has become increasingly a source of greater power to the medical establishment, industry, commerce, government, and military, and ever less a part of the liberal arts in its traditions, memory, administration, and priorities.

We tend these days to learn as children that science emerged if not from simple curiosity, then from humankind's desire to control nature.[3] But this simple commonsense view is a half-truth at best. The deepest and most nourishing roots of ancient and even early-modern science may well be in the search for justice and wisdom and not for material power.[4]

The issue at this point is not at all, though, whether the modern trend to cultivate science mostly for power is right or wrong, but to understand the original nature of science, its vital contributions to our thought processes, and its continued relevance for our hopes and efforts to improve our means of judgment, and to our ability to live together peacefully.

Moreover, we all face the increasing specialization of knowledge, pressures for economic relevance, increasing management of science, and the like. What are the social costs? How will the judgment process of the individual scientist or interested citizen be likely affected? The matter is not nearly so simple as the often-discussed competition for funds and resources between pure and applied science.

The confusion of functions broadly affects the training of most young scientists and other scholars, research strategies, hiring and tenure standards at universities, teaching orientation, philosophies of granting agencies, and dialogue within the scientific community. It thus affects the agenda and thought processes in the academic community, and in turn influences what is passed on as scientific thinking to the generation in apprenticeship, both inside and outside of science.

In universities, science was once taught as part of the liberal arts (the studies appropriate for freemen, or enfranchised persons, in contrast to a vocational education). Today there are few who would think to teach it this way, or who could, and *liberal arts* is usually as anachronistic and vestigal a term as *doctor of philosophy*.

There is a terrific burden on the shoulders of any individual who would like to steer an independent course. It is also difficult for the young scholar to steer one's own career intelligently if one approaches it not merely with ideals, but with illusions, and without realistic orientation. Practitioners in each field of study seem usually to begin their careers without much awareness of how or why their field of study came to exist as an activity worthy of human energy, with little historical or philosophical context in which to place their discipline. They may understand too little of the dynamics of the institutional structures that shape consciousness and values in their discipline. They are thus ill equipped for their intended trek through unexplored territory, and even the best may end their careers wondering where they might have gone had they thought when they were younger about boots, a compass, a map, and a hat for the sun.

The citizen can best decide how to benefit from science (as *critical thought*), this potent creation of Western culture, if one understands the nature of the enterprise of science and the nature of the attitudinal and institutional changes within it. We all need science and scholarship today as much as ever for the contributions that they can make to help us avoid misjudgments, and the young scientist too needs to understand the strengths, limitations, and avoidable obstacles in the chosen profession.

Misjudgment and disillusionment have certainly not been transcended in our time. But many mistakes are avoidable and are not simply spontaneous happenings like earthquakes or radioactive decay. There is *much* we can do better. Too often, as children we are told that life is beyond control. But that is another half-truth at best.

Each of us will have his or her most convincing examples of avoidable misjudgment: a disastrous love affair, the cataclysm of war, disillusionment with a political or religious leader or movement, a financial misjudgment. We do not have to turn far beyond ourselves to big events and wonder, for example, how Hitler could have been so stupid as to ignore the lesson of Napoleon, and get bogged down in a winter war in Russia.

If scientific thinking is one form of modern prophecy, as some like to view it, endless disastrous examples of the ignoring of prophecy abound: the young man who could not believe that he would become a statistic and rushed into battle, the speeder, the drinker, the smoker who sees him or herself as too healthy to become a cancer statistic, the young couple who could see all the problems and were not going to let having a baby affect their lives as it had those of their friends.

Today the full dilemma of Oedipus remains, since when the degree of reliability of any prophecy is less than certain, one must weigh it and be willing to ignore it. We might say that with us, as with Oedipus, the optimal balance between pride and humility is still difficult to find and remains disharmonious; the distinctions between reliable and false knowledge, between courage and rashness, are unclear. We have difficulty in drawing the dividing lines and in balancing out our perceptual and intellectual strengths and limitations. The potential may be present to act decisively and wisely, yet we may fail to properly identify the necessary facts. Realities can become illusions as contexts change, and we may not know when that happens. So we rush along through life too quickly and too certain of the reliability of what can be shifting points of reference.

OEDIPUS: Must you always talk in riddles?

TEIRESIAS: Were you not unexcelled at solving riddles?

OEDIPUS: You mock the gift that is my greatness?

TEIRESIAS: Your great fortune was your ruin.

Sophocles, *Oedipus Rex*

In the following pages, I will expand on the idea that science is one way by which we have made enormous progress in overcoming our limited senses. Is the message to be that by constantly improving our catalog of truths we can eventually discard fantasy and build better lives for ourselves? No. If only the task were so simple and straightforward. Scientists too often promote such ideas, but they are unrealistic.

If nothing else, while it could be unquestionably valuable to catalogue reality, paradoxically, the process of so doing would require fantasy and imagination. Oedipus might have been able to grasp reality if he had been willing to engage in some seemingly outrageous fantasies. There are too many values of fantasy for us to abandon it. These and related problems are deep and complex and require some discussion. We dare not yield to the common desire of our day to see the matter in dualistic terms, to see science and fantasy, or reality and fantasy, as antagonistic forces.

NOTES

1. Of course, we have intrinsically *conflicting* emotional self-interests. Denial, myth, and injustice can also be in one's self-interest or in the self-interest of a social group or manipulation party, as will be clear through the course of this book. This intrinsic conflict makes for a difficult landscape to negotiate—the point of much of my discussion.

2. For completeness, it should be pointed out that, in his reference to Oedipus, Freud only dealt effectively with the power of patricide and incest as dramatic devices. Is there something in us all that makes us cringe at the difficulties that this ancient fellow fell into? Freud was dealing primarily with the nature of the completed crimes but not the motives that led to them, the circumstances of their concealment or discovery, or with Oedipus's punishment by self-inflicted blindness. Whereas Freud claimed to have dealt with more than the common fascination with the myth, he has been properly criticized (Kaufmann 1968.).

3. Even Aristotle *seems* to support this. He begins his *Metaphysics* suggesting, "All men by nature desire to know." But he soon reveals that deeper motives had emerged at least by his lifetime: "The science which knows to what end each thing must be done is the most authoritative of the sciences, and more authoritative than any ancillary science; and this end is the

good of that thing, and in general the supreme good in the whole of nature." Certainly, at least as early as the Pythagoreans, a Greek search for knowledge was tied to the desire to know how to live wisely.

4. What would the Greeks have thought of probing the secrets of nature in order to acquire power? To them nature was gods, and they would have to consider whether or not the gods would be angered at having mortals looking closely into their private affairs. Aristotle noted, early in his *Metaphysics*, that philosophy is honorable for it seeks only to know and has no utilitarian end. Proof of this, he claimed, is that people only begin to seek such knowledge after the comforts and necessities of life have been secured. The Divine Power cannot be jealous, because the quest is honorable. Philosophy seeks only to know the nature of all things and their nature is divine. Therefore, he argues, it must be "Divine Science."

The Greeks did not turn much of their philosophy/science to practical ends. As children we were told that this was because they simply could not see practical applications, as they were not practical people. It seems rather that there would have been serious religious objections for some of them, at least, to a philosopher's being in the business of turning discoveries into profits. This would be prying into the gods' secrets to gain money or power, using the gods for petty mortal ends. In any event, one gathers that Aristotle would see our society's present insistence that science should serve utilitarian and economic ends as being dishonorable. Would he even see a danger of angering the Divine Power, the Prime Mover? Would Greeks see the present dangers of nuclear proliferation, pollution, economic destabilization, social problems, and such as warnings that we must return science to a "pure" and "honorable" search for Truth? This is impossible to say, and I present it only as an interesting question to play with, to remind ourselves how different Aristotle's point of view was from ours.

CHAPTER 3

The Illusion Organ

I've got the whole place wired for sound, a sort of very elaborate intercom or walkie-talkie system, so I can dictate to my secretary ... any time of the day and night. ... Has it ever struck you ... that life is all memory, except for the one present moment that goes by you so quickly you hardly catch it going? It's really all memory ... except for each passing moment. What I just now said to you is a memory now—recollection. ... I'm up now. When I was at the table is a memory now. ... When I turned at the other end of the terrace is a memory now. ... Practically everything is a memory to me, now, so I'm writing my memoirs.

Mrs. Goforth, in Tennessee Williams's
The Milk Train Doesn't Stop Here Anymore

The Brain Must Construct a Subjective Reality

We are normally not conscious of our internal organs. We cannot determine the path of blood flow through our heart, or the complex chemistry of our livers by simply thinking about them. We do not even have the sensory nerve endings in these organs to give us any precise information. The brain is another such organ, remote from our consciousness. Its mechanism is not something that we can see simply by mental concentration or know by common sense. It has taken intense scientific research to obtain only an outline of how the brain works, and many of the details of its mechanism remain obscure.

There have been surprises in the study of brain function. For many of us the biggest surprise might be that the brain constructs its own reality. Normally, in the waking brain, this reality seems to be anchored to the physical world in some way such that what appears to be a red book to me will appear to be a red book and not a goldfish to you. Yet numerous experiments and other sorts of evidence reveal that the brain constructs its own model of reality and does not simply mirror the physical sensations that bombard the senses. It *cannot* be strictly and completely objective.

Let us first ask why the brain should construct a reality when common sense would seem to suggest that a simple unadulterated passing on of pristine information into the brain, and the literal representation of this information, would be a far more straightforward or easy way to function. (We can ignore for the moment the question of who or what would be inside the brain to view such information or representation, with *what* inner eyes would see it, and how *those* eyes would function. Potential infinite regress, from a physiological point of view.)

A simple mirroring of reality would *not be desirable* for many reasons. Our brains actually do us a service by reshaping the information that comes from the nerve endings. Think carefully about the sorts of physical information that actually reach our sense organs and about the construction of the sense organs themselves. We have "blind spots" in our retinas that the brain fills in. We blink, but we are not plunged into blackness each time. The vague forms of night are enhanced, somewhat as modern computers enhance poor-quality photographic images, so that from shadows we see trees and other dim objects (or imagined monsters if we are scared). We can see familiar objects by a glance at their outline, without prolonged examination. We can hear discrete sounds in a jumble of vibrations. We see the world right-side up, even though the image is inverted on our retina. We move our eyes, and yet the stationary objects that we see do not appear to move.

I prefer it this way. I would rather live with the realistic illusion of a stable room than actually to experience the room move about as the image in fact shifts about on my retina. In other words, the sense organs themselves would give a very distorted picture of the physical world if their physiological input were literally represented in consciousness. So the nervous system modifies the truthful though misrepresentative information to produce a representative illusion of reality that is much closer to the world of physical objects than is the actual information available to and from the senses.

This may seem paradoxical at first, but the effect of the transformation is a little bit like techniques in recent Western art that we have come to take for granted and in which today we no longer see a paradox. An accurate portrayal on a canvas of two persons of their actual equal size standing at different distances from us would look odd and unrealistic because we would normally see the farther person as smaller. So the Western artist, especially since the Renaissance, uses perspective and paints the more distant individual as smaller, when in fact they are the same size. Inaccurate information about relative size, when presented on a flat surface, creates an illusion that more convincingly represents reality than would an attempt to be "honest" about true size by painting the two individuals as identically sized images on the canvas. More about such things later, but first let us consider the implications of time itself.

Imagine yourself to be on a busy street. The rush of cars and people is something that you can be aware of only because of your sense organs: you can see and hear the cars and trucks roar by, you feel people as they occasionally brush you, the smells are unavoidable (fortunately, the taste buds are spared sharing in the experience).

A car does not slow to stop at the intersection and a pedestrian who sees this hurries out of its path. What does the pedestrian actually see? The image of the automobile stimulates the retina of the eye, and nervous messages travel to the brain, granted, but is the nervous system told of a speeding car? The car's position is changing each instant. At any given instant there can be no actual physical perception of the car in a previous position since that event of the previous instant no longer physically exists. If the car had hit the man, at just the instant that his body was shattered he would no longer exist as a whole, living human. Physicists assure that Einstein and relativity can liberate us from this constraint only in the literature of science fiction. The past *no longer exists* and we cannot exist in it or experience it. This is a simple point with profound implications.

Our only "physical," or sensory, contact with the real world can possibly be at the exact instant that a sense organ is being stimulated. Our conscious experience must consist almost entirely of memories of real experiences. Our subjective consciousness of continuous experience is an illusion in terms of actual input, but an illusion that better matches physical change than do the incredibly brief impulses alone.

Rapid bursts of nervous impulses from the sense organs enter the brain like so many volleys of coded machine-gun fire. At a given moment the input is like a snapshot of the world outside. Somehow the brain lays each new snapshot in proper sequence with the memories of the others immediately before it. It is difficult to resist thinking of motion picture film or television, where instants of reality are captured in sequence. Multibillion-dollar industries are based on this illusion of motion and many of us spend large portions of our life watching films or television — they are now a fact of life. We do not call them "stage illusions" anymore, as people once did, but that is what they are. When we see the static pictures rapidly projected in sequence onto a screen, our minds fuse the separate images and without conscious effort produce the illusion of witnessing continuous happenings. Indeed, the motion picture illusions are possible because they conform to a process that the brain is conducting all the waking hours. The brain blends the brief frames of a motion picture and eliminates the discontinuities; at the moment that any given frame is received by the brain, we are conscious not only of that frame but of previous frames, their sequence, and the action implied therein.

In life we may be conscious of a complex, changing world as our brain creates the illusion, but we cannot sense or transduce change directly; we can only sense the instant. Our minds can only infer change from differences between the direct sensation of the instantaneous present and the memory of a second ago, a minute, an hour, a day or more past.

Perhaps we can also infer change from sensing the equivalents of motion-caused blurs in photographs. And nerve cells can register that they have just been changing. But, discussion along such technical lines would lead beyond the simple point at hand: a consideration of time itself reveals that the mind has to construct its own internal reality and does not simply capture and witness events. Without memory and mechanisms for relating memories to one another and to the instant of ongoing perception, conscious experience as we know it would be impossible. In a way it is like the problem of trying to see a whole room through a tiny crack where the angle of vision is very small. By sweeping our head about we can "see" most of the room, but only in our minds, since at any instant we can only physically see a small bit of the room.

The Temptation of the Generic Philosophical Argument

By saying that the mind constructs its own internal realities, I do not mean that these are necessarily wrong, are fantasies, or are very different from objective physical realities—particularly in the elementary domain of physical objects and their movements. This point requires repeated emphasis here, since there have been so many doctrines that would distort the facts and with which the reader will have had formal or informal contact. It would be tempting to try to stuff this discussion into old pigeonholes. There is no returning to solipsism though, or to the naive interpretations of Bishop Berkeley, for example.

Experience can be vivid when we are dreaming, yet we usually are not aware that we are then in an imagined fantasy. It has long seemed reasonable to wonder if perhaps all of life is a dream. Indeed, I know of no way to prove to the otherwise convinced reader that I am not merely a product of his or her imagination—or they products of my imagination. Countless people have pondered such questions through the ages, from ordinary people to philosophers, religionists, nihilists, swindlers, and psychopaths passing as any of the former.

I refer to the *generic philosophical argument*, since it is a powerful if cheap ingredient in countless discussions. If the standards are severe enough one can never *prove* anything, since one cannot even prove existence to *complete* philosophical satisfaction. People will probably always deal with this question, for at least a few minutes in their lives.

Are we all, the reader and myself included, the dream of a butterfly, as the ancient Chinese philosopher asked? Can we be absolutely certain that this too often cruel, chaotic, and unfair world is not simply the sadistic or irrational fantasy of some obscure (insect or noninsect) mentality? Perhaps our life is all simply a no-win game, a sinister joke, whose punch line is that the virtuous will be punished and the evil ones rewarded. Dare we then opt for following some notion of virtue? How can one be absolutely certain?[1] Or is all of the unfairness and chaos really the brainchild of a nice, friendly (insect or noninsect) mentality who is simply testing our faith for one reason or another?

I no longer worry about such things. The range of such possibilities for anyone with even moderate imagination is simply endless, and trying to place more value on one possibility than another by reason or faith cannot be justified. It is too easy to invent such scenarios in profusion. The time I spent pondering such questions as a child got me nowhere at all, and then slowly I discovered that just about no adults really take this supposed philosophical dilemma seriously, at least when it comes down to their pocketbooks. Its advocates may heatedly proclaim sincerity, but see what happens if you borrow $1,000 from them and try to pay it back in imaginary currency.

Surely many billions of people have pondered the question of reality, yet none has been able to prove that life is a complete illusion. Those who followed the hypothesis that the world really exists, and who sought and found its rules, learned that we can feel our way along with our imperfect senses if we are careful. This empirical approach has conquered diseases such as polio and smallpox, and has shown us the hidden side of the moon. Of course, the misuse of science has also invented astonishingly powerful weapons, contributed to the population explosion, pushed our civilization to balance upon a few shrinking resources, and plunged us into mass communications that affect us in ways we scarcely comprehend. Nevertheless, the point is that acceptance of the physical world has led to a growing ability to predict and to manipulate it, however unwisely we have used that power. I am not advocating that the goal of science should be manipulation. But this power offers evidence that the reality of physical experience is the most relevant hypothesis.

Evolution Produces Adequacy, Not Perfection

Obviously, part of the time each of our normal healthy minds does construct internal realities that *are* great fantasies, as for example, in dreams, and we can be seriously misled by doctrines, propaganda, social pressures, self-deception, and skillful manipulators (Conway and Siegelman 1979, Marks 1980, Martin 1986, Mackay 1932, Tabori 1959, Watson 1980, Randi 1975, 1980, Houdini 1924, Huxley 1958, Key 1973, Jahoda 1969). But when we are awake there is usually a good correlation between the way we "see," or sense, physical objects, and the subjective physical world of others, and to a large extent even the way that instruments measure many aspects of physical objects. For the most part, the brain is a practical organ, a steady workhorse of a sort. It works for our survival and does not mislead us by mere whimsy. Indeed, from an evolutionary point of view, this is what one would expect.

In the millions of years of finding food, mates, and shelter, during primate evolution and before, it was apparently important to develop a "mental world" that simulated the physical world. It is difficult to see how a highly misleading mental reality could do anything but harm an active animal, and thereby lead to its extinction. Fantasy, in its time and place, may have great advantages for humans, but this point deserves its own discussion. During most of our waking hours our survival is served best by a mental reality that closely corresponds to the physical world.

However, I once took more comfort in this position than I do now. And that is why I use qualified phrases such as "for the most part." There is overwhelming evidence from nature that *natural selection results in the evolution of adequately designed creatures, but not in perfectly designed creatures*. So we might have expected that we would have a mental reality that adequately reflects the physical exterior under most circumstances but that has distinct operating limits, that can be "overloaded," and that is not necessarily adaptive in every situation. Indeed, this seems actually to be the case, as we shall see.

We should not have a simple faith that millions of years of evolution have honed the brain to perfection and that our mental realities will consequently always reflect external realities. But we can certainly minimize blind reliance on this organ if we understand its inherent strengths and limits and our cultural means for augmenting, and corrupting, its services. The negotiation of this augmentation is, of course, a major *explicit* theme of this book, and I will argue that it is also an *implicit* theme that recurs in Western philosophy, science, art, and law. But let us return now to some examples of sensory modification.

Levels of Information Modification by the Brain

The first level of the modification of environmental information is obviously the *selectivity of the sense organs*. The eyes cannot hear, as Socrates noted, and the ears cannot taste. Moreover there are sounds, colors, and tastes that are outside of our ability to experience. There are light levels that are sufficient for some animals to see by but that we are unsensitive to—and so on. Next, there is what is known as *peripheral organization* or filtering of information. This means that some processing of information takes place in the sense organ itself before it travels to the brain, so the brain sees a censored or abridged version of the original. Nobel Prize winner H. K. Hartline demonstrated *lateral inhibition* in the eye of invertebrates, which turned out to be a common feature of sensory systems. Basically, a highly stimulated cell inhibits the activity of its less intensely stimulated neighbors. This competition creates the perception of sharp boundaries and distinct sounds and touches within what are really less distinct spectra of energy input to the senses.

Next, various sorts of feedback dampen energy input to keep the sensory cells from being overloaded, and such *regulation of input* gives a distorted idea of actual energy levels in the environment. For example, the eye adapts to darkness both chemically and mechanically. In bright light our pupils close down and we cannot see exactly how much brighter it has become. It is difficult to guess what a light meter will read. Similarly, we are protected from loud sounds by tiny muscles in the middle ear that adjust the input.

Then, at some deeper level in the brain, two or more sources of information combine to modify our perception of a stimulus. Consider *central regulation* of perception in vision. If one gently slides the eyelid with a finger enough so that one eyeball moves, one sees the visual field move. If only one eye is open, then the scene and its objects appear to move. If both eyes are open, then two sets of objects may be seen to separate where there was one. This happens because the image of objects on the retina moves when the eye moves, a different set of cells is stimulated, and we see this as a change in position. Of course we can reason that the objects are not in reality moving, because we know that our finger is moving the eye. Why does the object not appear to move with normal eye movements? In normal eye movements, an unconscious compensation is made for the brain's subconscious instructions to the eye muscles.

If one next turns one's head, keeping the eyes always looking straight out from the sockets, the image of objects in a room again moves across the retina to new positions on its surface. But our consciousness does not see the objects change position relative to ourselves as it did when we pressed the eyeball, even though the eye sees an equivalent change in the position of objects on its retinal surface. In both cases similar information may go from the retina to the brain, so how can we see the world so differently in the two cases?

The short answer is that in normal vision the brain takes feedback information from muscles (proprioception) about the actual position of the eyes and head relative to the body, takes also information about what it has asked the eyes and head to do, and adds this information to that received from the retina, to construct a mental reality that is close to what is actually happening: one is conscious that the body is moving and not the world.

In experiments by von Holst, the eye muscles were paralyzed by a temporary chemical nerve block. Then the subject "willed" his eyes to move, but of course they could not. Yet since the brain "expected" them to move, the effect was that predicted by the scientists: The visual scene was reported to shift and move. It moved in the direction opposite to that in which the eyes were willed to move. In other words, the brain automatically shifted its internal reality to compensate for an *expected* movement of the eyes (that *could not* occur), producing the illusion of a moving external world.

Without such proprioceptive or other supplementary information, it may be difficult to "see," or sense, things "correctly." If we are seated completely enclosed in a large cylinder on which stripes or scenes are painted and it is rotated *around* us, we may falsely see the background scene as unmoving while we, the smaller object, seem to rotate (Bullock et al. 1977).

By now we have carried examples of sensory modification deep within the brain, where the distinctions between "seeing" or "sensing" and "being aware of" in many cases become difficult to agree on. One person may argue that he perceives the sun to pass overhead. Another may argue that she has fully grasped the concept of being situated on the surface of a huge spinning ball and perceives herself to be moving beneath the sun. Yet all that either can really sense is the position of the sun and the memory of where it was earlier—most of their perception is internal and is highly conditioned by their beliefs and reason.

It might be appropriate next to delve at length into research on mechanisms that operate at this more central level and that will highly modify perception, such as selective attention, denial, rationalization, motivated inattention, cognitive dissonance, and cognitive consonance. But I wish only to mention them for now.

To the extent that perception involves consciousness, it is bound to be a complex phenomenon involving feedback between higher and lower levels of information transfer and processing. Fearful reactions to snakes, insects, and various real and imagined threats may be so rapid as to be thought of as

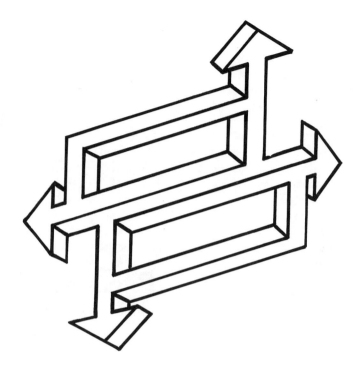

Figure 1. The image is confusing because the brain automatically tries to impose order on the lines on the paper, and it runs through one unworkable interpretation after another.

reflexes. One may brake the car to miss a cat before there is time to think about it, and yet the "reflex" is clearly rooted in some higher opinion and thought about the value of cats, and might be modified, beginning with a change in opinion. I find the data and experiences to be most in agreement with "active" theories of perception as reviewed and stated most eloquently by R. L. Gregory (1970, 1974, 1978; Gregory and Gombrich 1973. See also Maturana and Varela 1987, Watzlawick 1984). The brain takes a very active part in shaping its input and interpretations of events. The operation is as though the brain has, through experience, formed sets of hypotheses about the external world and constructs subjective physical models of reality based on these. Then the brain scans for sensory input that quickly verifies, rejects, and modifies these hypotheses and models in an ongoing process.[2]

To some extent reality is always being tampered with by our nervous systems, not only when the magician entertains us, when artists such as Salva-

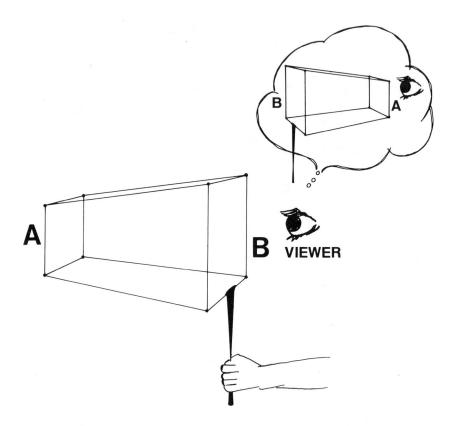

Figure 2. One can study the automatic effects of imagination on reality construction using a truncated pyramid built of stiff wire. View with one eye shut. Invert mentally. Twist about with handle. Joints will appear to bend in unreal ways. (Pyramid may be about 6 inches deep, A to B.)

dor Dali, M. C. Escher, or René Magritte present us with ambiguous creations, or when the psychologist brings out ink blots to probe the predispositions of our minds. The construction of internal reality is a continual process in the human brain, except perhaps for those dreamless periods during sleep. Figure 1 is confusing, because the brain is actively trying to impose order on the lines on the paper, and it runs through one unworkable interpretation after another.

Optical illusions are of considerable interest to psychologists, because their study offers insight into the ways in which the brain organizes information. While many examples could be given, a simple three-dimensional "Necker cube" (actually, a truncated pyramid) is particularly interesting and it is well worth the time to put one together with wire and solder or adhesive tape and play with it (figure 2). I made mine about ten inches in the longest dimension, but size is not critical. Hold the device by the handle and view the smaller square A through the larger square B (with one eye shut). Then *mentally* invert the figure so that A appears closest to you and B farthest: imagine that the structure is inverted. One can hold this interpretation of the structure with only a little mental discipline.

Now twist the structure about by the handle. It will appear to adjust in what at first seems to be unpredictable ways. Interestingly, there seem to be fluid joints where the handle joins the cube or at the corners of the cube—a plasticity appears in what is in reality a rigid structure. The brain modifies the image to make it all visually consistent with the premise that we have insisted on—that the smaller square is the nearer square.

The Subjective Illusion Is Regarded as Perfectly Normal

The optical image of the world on our retina is inverted by our eye lenses, yet we see the world as right-side up. If we wear special glasses with inverting prisms, then the world does appear upside down. However, after a few days of wearing the inverting lenses we may see the world as "normal." The brain, to some extent, internally corrects the gross distortion so that our view of reality partially matches that seen by our other senses and past experience.

A person with such glasses may *claim* that the world became normal, but when questioned closely about the details it becomes apparent that the reconstructed world is actually quite odd by ordinary standards, and the experience has always been difficult even to describe. Objects may be upright or shift to inverted, depending on circumstances. There may be contradictions of the rightness and leftness of separate objects. Parts of the new reality may seem to be unreal, but only at times.

There is not complete correction, and this is quite interesting. In fact, one *accepts and becomes comfortable with a very bizarre new reality*. Perhaps this should be no surprise. We do something like this when asleep, in accepting dream fantasies or in adapting to bizarre experiences in our waking lives. This is a very revealing phenomenon and we shall see implications of it repeatedly.

A main point first is, again, that the brain does seem "to try hard" to correct for the glasses and to make a normal world out of upside-down input. We should not think of these constructions and corrections as "mere" corrections. To say "merely" can provide semantic security if there may be emotional or philosophical difficulty with the findings. The facts remain, though, and they and their implications must be dealt with. *Corrections* is a small word that can conceal profound goings-on. Skyscrapers are corrections for a lack of land on Manhattan Island. Modern democratic forms of government are corrections for tyranny. Animality is a correction for not having the capacity for photosynthesis. A correction may take place by minor mechanisms or it may take place by conceptually profound and unique reorganizations. We should not try to dismiss the facts or make the phenomenon minor by saying that they are merely corrections.

Similarly, we are tempted to dismiss false perceptions semantically, as unusual errors—illusions—some extraordinary set of mysterious or special circumstances where our supposedly normally truthful senses are "tricked." There is an enormous and fascinating literature on illusions, and their formal study goes back to the midnineteenth century (Coren and Girgus 1978). Illusions are merely situations that denude bits of the *routine* business of the thought organ. Yet most of us think of them as being quite rare and separate from ordinary experience. It is as though by placing false perceptions off in a special category we need not question our working reality—the foundation for stability in the lives of many of us. Possibly this is why we no longer classify motion pictures as stage illusions, or refer to the use of perspective in painting as illusion. Perhaps they have become so much a part of our ordinary lives that it becomes difficult, or even threatening, to blur too freely the line between the categories, reality and illusion.

I do not now want to make too much of our chauvinism about our particular views of reality or the possible reasons for this, except to note that the chauvinism is there and does influence us. The *Honi phenomenon* is a simple and interesting example to ponder in this connection. Psychologists can study the perception of visual distortion in a number of ways and one way is to ask a subject to look through a lens that distorts the visual field somewhat as a fun-house mirror does. A graded series of lenses ranging from no distortion to unmistakable distortion provides a scale for measuring an individual's threshold for seeing distortion. Of course, most of us will not see low levels of distortion because eye and brain compensate for them, somewhat as a computer can enhance and sharpen inferior photographs. This is why you cannot tell much about the quality of binocular or camera lenses just by looking through them—your eye and brain make the images better than they really are; a true test involves actually measuring the distortions of standard test patterns transmitted through a lens.

The Honi phenomenon can be studied quantitatively by measuring thresholds for visual distortion for individuals viewing objects through graded series of distortion lenses (Wittreich 1959). In such studies the distortion threshold was remarkably high when viewing pictures of loved ones compared to neutral objects or persons. In other words, we are particularly inclined to compensate for what should be the perceived defects in those to whom we are attached, and we really do not "see" some of the imperfections. Is this a defense against some threat that the imperfections would be to the relationship, or merely that the internal reality is idealized to an extent and does not allow for some introduced imperfections? This is not clear, but it is clear that the Honi phenomenon is not caused simply by familiarity, since it also pertains to amputees and authority figures and perhaps to other emotional situations as well.

Imagine the Destruction of Imagination

What would be the reality projected to our consciousness if our senses were unmodified? What would the world look like to us if we could take a pill that would dissolve away the internal, reconstructed, reality so that we might "see" only what our unadulterated sensory nerve endings can tell us?

Let us first take only half of a pill. Objects would blur as lateral inhibition ceased to sharpen their edges. Black holes would appear in the world as the blind spots ceased to be filled in. Thin, black veins would be seen through the world, due to nerves and blood vessels that cross the retina. If one were in a room and looked to the left, the scene of chairs, walls, windows, pictures, tables would all together fly to the right. To know that the world is not really moving, it would be necessary to reason that such shifts are associated with eye or neck movements. As the effects of the pill took hold, the room would invert and we would see it upside down. Trying to catch a playful kitten would become a tedious intellectual effort in a seemingly moving inverted room of objects with blurred edges and dark veins and holes in the visual field. To make things worse, it would take some time actually to see the details of fixed objects and of the kitten, since their characteristics could not be imagined from their outline. Instead we would see many blobs unless everything slowed down and we could scrutinize details.

All of this would be nightmare enough, but implicit in the example is that memory continues to function and produce an illusion of the experience of existence through time. If we last strip away this illusion by taking the second half of the pill, then reason becomes impossible, there is no "before," no action, the kitten does not run, but at the instant of reality it is a frozen form (or a blur). It would be nearly impossible to catch the kitten because, as by looking at a single photograph, the kitten would have no clear history and so there would be no way to visualize its trajectory and grab it, not where it is, but where it would be in the time that it would take the arm to reach out.

But, of course, there also could be no "aware" observer to become frustrated, since without memory no consciousness of ego or purpose could be possible.

Obviously, our life as we know it would be impossible this way. If we could not even catch a kitten efficiently, then imagine our naked and vulnerable ancestors trying to fight off a lion. Surely we would have gone extinct millions of years ago. Many of the sorts of illusions that we live with and require must be very very old; indeed there is evidence throughout the higher animals of most of the compensatory mechanisms that are so much us.

Our senses and brain have distinct limits and quirks and are apparently not constructed in the image and likeness of an omniscient god. Philosophers, artists, judges, politicians, reporters, ministers, scientists, and citizens find themselves not with the brains that they might like, with brains that would make their jobs ideally easy for them, or even with the brains that they had assumed they had. Instead they find themselves with nervous systems based on neurological designs that once helped fish to catch squirming worms and reptiles to leap on fast prey with a minimum of confusion.

We should not shrivel at this thought. The fish and reptiles were enormously successful, or they and we would not be here today, and they owe us no apologies. Indeed, we owe them some gratitude. The illusion organ is a marvelous device by any standard. It admirably simulates, albeit selectively, a rich slice of our external objective world in a few handfuls of soft, wet tissue and corrects for a number of considerable distortions due to the physical properties of the sense organs and even the energy fields themselves.

Moreover, evolution's devious route to efficiency may prove to be serendipity. Somewhere in the course of human evolution the illusion organ became expert at making models with enough substance and continuity that they could be twisted about without shattering. Some form of mental play was possible—one part of the internal model with another—creative rather than simply reflexive imagination. Our illusion organ can generate unanchored internal models of reality that we can play with, turn over and fold inside out, twist apart and put back together in new ways. We do indeed imagine many things that are not or can never be, but we can also imagine things that are, but go unseen, and things that can be, but would otherwise go undreamed of and unhoped for. Evolution has handed us a double-edged sword with which we often surely harm ourselves. Yet in the long run our world of illusion can, if well managed, also prove to be a far more valuable gift than would truthful senses alone.

Constructions of Ego

Who or what is the self-conscious "I" that is aware of the model subjective world, and that does the twisting about of the model world's structure in imagination? This is a profound question to which I have no firm scientific answer, which is not to say no probable natural answer. Experience in my own culture suggests a convenient way of thinking about the problem. The "I" may be thought of as a part of the internal model—an actor upon a miniature stage—and the stage and action are seen through the actor's eyes.

Persons and objects on the stage are anchored to reality through sensory input, the brain constantly comparing its models to the behavior of objects and persons in the objective world. The model of ourselves is likewise anchored to reality through a sensory perception of our own bodies, behavior, and subjectivity.

We are each the main actor, "I"—when we are awake and functioning in the physical world, that is. In dreams, unanchored to physical reality through sensory input, we may be other people on the stage, or stand apart from our ego image. Sometimes in a given dream we switch identities back and forth (Grunebaum and Caillois 1966), but the rules of my culture's script do not normally allow this when we are awake.

In some other cultures the rules are different, and commonly one may even be an animal or object in one's dreams. Members of the community, and particularly the shaman, may become animals or other persons or multiple persons and/or spirits when awake. There may be multiple souls or egos in a single individual. (This is similar to those in our culture who believe that the devil has made them do something against their own tendencies, and who negotiate their behavior among multiple forces believed to be within themselves.) These rules for self-conscious being, different from those in a scientifically informed culture,[3] may have important social functions in healing, hunting strategies, maintaining morale, group memory, and cohesion (chapter 8).

Sometimes Even Physical Reality Can Be Uncertain

I have referred to perceptions that are anchored to the external objective world in the case of physical objects, and it is very useful to keep in mind that even here the length of the anchor line varies and that we may float relatively freely from time to time, so to speak. Dreams are perhaps the commonly experienced state in which the anchor cable comes loose. During sensory deprivation, with external sensory input cut off, a normal awake person may soon begin to hallucinate and experience physical events that have no basis in the actual situation (Solomon 1961). The common factors in sleep and sensory deprivation seem to be the diminishment of external input to the brain, and this again suggests that the illusion organ is an active device that must "track" external reality to reflect it, and is not one that is merely entertained or molded by external events. Degrees of unanchoring of the illusion organ also can occur in hypnotism, during physical or psychological stress, in disorienting environments, during vision quests or religious ecstasy, in some forms of mental illness, and in some drug-induced states (LSD, mescaline, etc.) (Harner 1973, 1980; Halifax 1979; Ross 1974; Taylor 1979).

Events at Fatima, Portugal, in 1917 are interesting to consider in this connection (Pelletier 1951). In brief, some 30,000 people reportedly saw the sun spin about in the sky, then plunge to the earth, and then return.

Of course, despite the weight of testimony, the sun did not physically plunge to earth. The results would have been catastrophic even in a Ptolemaic universe. The rest of the world did not see such an event and many present, including newspaper photographers looking for a good story, did not see it. The Roman Catholic Church has, indeed, never officially accepted this aspect of Fatima as an official miracle, despite pressure from the faithful. Fatima is of interest to us here because it emphasizes again that while seeing may well be believing, believing can indeed also be seeing. Obviously this underscores the difficulty in using only the senses or intuition to perceive reality accurately. Moreover, it emphasizes that large numbers of observers can be mistaken.

Fatima, however, was an exceptional happening and was unusual by several standards. It occurred in a deeply religious country during a disquieting moment of political instability. World communications were improving and concern with the communist revolution in Russia featured prominently in the events in Portugal. World War I was upsetting to much of the world, Portugal included. Anticipation of a miracle had built up over months and the hopeful, old, and infirm had traveled for miles; apparently the lack of accommodations resulted in exhaustion and hunger for many. Through intent or not, many of the factors that have been traditionally associated with vision quests were present at Fatima. So it cannot be used as evidence that normally beliefs can shape our physical reality. But more modest examples exist.

In one very simple study by Ronald Stansfield, a number of science students were asked to watch water dripping and to describe the shape of drops *exactly as they saw them* — no interpretation. It requires a high-speed camera to determine the shape, yet many of the students claimed to be able to see the shape (Dixon 1978).

Having taught anatomy, I find this example quite in agreement with my own experience. In dissections, it was not uncommon for students to sketch what they should have seen rather than what the specimen actually showed. While sometimes lazy or insecure students would make up their reports and drawings, in most cases I am reasonably certain the students were having difficulty distinguishing between what they actually could objectively see and what they did subjectively see on the basis of expectation. In biology one must usually learn "to see" through a tedious process of trial, error, and instruction. Learning to distinguish closely related species, for example, can be difficult and exasperating. But one develops a "search image" for subtle shapes, colors, and patterns that the neophyte invariably overlooks. Such examples, I suspect, are usually a case of seeing not too much, but too little — of seeing structure as noise, order as disorder, much as themes in jazz may go undetected by the unaccustomed ear.

Here, competition in science is good, for if one merely imagines things, others will pounce on him or her and ridicule the error. Indeed there is much emphasis in systematics on finding objective criteria that are not dependent on experience and expertise. Often the taxonomist first decides that species are distinct on the basis of an experienced eye and of subtle features that are meaningless when communicated to people who do not share his or her experienced perception. Then after hours or weeks of searching and study, he or she will find one or more objective features that can serve as a guide for nearly *anyone* and that history shows confirm the original judgment. When one cannot find such objective features, the specimens may be left on a museum shelf labeled "undescribed" for years, known only among experts by word of mouth.

Difficulty, Not Relativity Implied

It is obviously in the domains of the *classification of*, and *values placed on* physical objects, the perception and interpretations of *events*, of *ideas*, that subjective constructions can least be automatically verified accurately and are of most concern here. It is with these that the issue of self-deception emerges as a tragic issue. These are issues that will be developed throughout this book. But I wish to end this chapter by clarifying that the above discussion of nervous-system operations only underscores the *difficulty* of objectivity and only superficially gives comfort to pedestrian relativism.

Anthropologists and biologists are both familiar with the fact that the learning of words can be equivalent to learning to perceive given classes of objects, events, or patterns at various levels, from the sensory to the cognitive (chapter 7). Words and language shape the way we look at the world, for better or for worse. It may be true that, as Conrad put it, "words, as is well known, are the great foes of reality," and yet also true that "the only living language is the language in which we think and have our being" (Machado). For if language interacts vigorously with perception, then it must destroy possible perceptions as it creates others. In viewing certain

Figure 3. One interpretation tends to obscure another. Is a monster eating the fish, or are two birds sitting on it?

ambiguous paintings one cannot easily appreciate alternative views simultaneously, and the mind's eye fixates on one, or at best oscillates between one image and another—one view seems to exclude the other, pending some deliberate mental effort (figure 3). We may define ourselves and the world with the templates of language, while at the same time these veil much reality. One point of view normally excludes another.

Culture modifies our perceptions, as figure 4 illustrates. While most Westerners will see a window above the head of the woman in the center, many Africans, accustomed to objects being carried on the head, will see a fourgallon tin there. Try to develop an African perspective. It may help to cover the T-shaped structure that might suggest the corner of a room.

Perception's rules may vary from one culture or another, from individual to individual, or from circumstance to circumstance. This appears somewhat arbitrary or "relative," and this may seem to give obvious support to

Figure 4. Cultural experience modifies perceptions. Most Westerners will see a window in the lines drawn above the woman's head. People from cultures where things are carried on the head may see a four-gallon tin. (Courtesy of UNESCO Press, Paris)

pedestrian relativists. It can also give support to absolutists, who argue, "If no one can convince me that there is one physically demonstrable and correct way of looking at things, then *my* way must be right because I feel so strongly about it." But simply because each of us sees things differently, it does not follow on the one hand that all strongly held beliefs are equal and can be given the same weight, or on the other hand that there is only one Truth. And, it does not follow that we are justified in adhering to a point of view simply because we feel strongly about it. To the contrary, the history of science and careful scholarship has shown time and time again that many strongly held ideas and perceptions must be rejected on the basis of careful observation, weighing of facts, experiment, and reason. Among those views that survive critical scrutiny, many can be built on and have considerable, precise, predictive power.

There may in theory be "other," nonscientific, valid languages for describing reality, but Aristotelian biology, for example, is not one of them. And it is simply not valid to return the Earth to the center of the universe. Likewise, the strongly held perspectives of alcoholics and other compulsive persons in denial are simply dangerous to them and to the mental health of their loved ones. If science is anything, it is an accumulation of hard evidence that not all opinions and perceptions are equal—many are demonstrably wrong.[4]

Human Nature

What is the Nature of *Homo sapiens*? Good or bad? Vicious or kind? Selfish or altruistic? Game players, gene spreaders, worshippers, followers, power-mongers, warriors, economic puppets? A mirage? We shall return to this question. One thing is clear by the construction of our brains: we are dreamers as much as observers and thinkers. Thus we can be visionaries. But whatever our *intentions* at any given moment, we are capable of living in delusion to profound degrees. We are *stumblers* because (metaphorically) we will see light where there is dark, and hills where there is ocean.

NOTES

1. Throughout the text I may say that such-and-such is a fact or true in all but the most technical philosophical sense. This difficulty in being *absolutely* certain of *any* experience, of *any* proof of anything, of being absolutely certain that we are not the dream of a butterfly or that we only imagine that we exist, is what I have in mind. If one insists on mathematical, logical proof of reality, objectivity, and so on, then all reality itself and all forms of proof must remain *technically* hypothetical. Science is *technically*, philosophically, irrational because it is not a system for finding the ideal Truth or Certainty that can be verified by pure Reason alone.

The Greek philosophers recognized that we cannot completely rely on our senses for truth, and they developed formal logic. But history has shown time and time again that despite the beauty and attractiveness of *pure* Reason, it will not necessarily bring us any closer to Truth than will intuition or authoritarianism. The Bengali poet Sir Rabindranath Tagore noted, "A mind all logic is like a knife all blade. It makes the hand bleed that uses it"—a point to ponder in the present context.

There are ancient, quite standard, cookbook arguments that we can know nothing for certain and so we must depend on faith. Or that we can know nothing for certain and so all is chaos and we should grab what we can while we can. Or that we can know nothing for certain and so one person's opinion is as good as another's. And so on.

These crazy lines of thought become convenient rationalizations to justify all sorts of convenient behaviors, and it is all too easy and too tempting to get sucked into them. They can be the stuff that cults are made of, too. When times get bad and people begin to question and reconstruct belief systems, we should expect to hear evermore questioning of reality—some of this is deeply instructive but some of it quite cheap, including arguments along these lines.

An academic philosopher has professional reasons to endlessly question reality, much as a mathematician explores an odd theorum. For the rest of us, science has shown the value of knowledge based on critical thought and disciplined experience: there are simple ways to understand and agree on such once-mysterious things as what lightning really is and why our hearts need to beat to keep us alive.

2. In *The Man Who Mistook His Wife for a Hat*, Oliver Sacks writes of Dr. P., a dazzling, alert musician and teacher in a music school. Dr. P. was by no means crazy. *He* was quite happy with himself, but *others* told him that he should see a doctor. "Magoo-like, when in the street he might pat the heads of water-hydrants and parking meters, taking these to be the heads of children; he would amiably address carved knobs on the furniture and be astonished when they did not reply. At first these odd mistakes were laughed off as jokes, not least by Dr. P. himself." But even as his condition advanced, Dr. P. continued to see nothing odd about himself. "Well, doesn't everyone make mistakes?"

Dr. P. would focus only on small details of complex things and then imagine the whole from those. Thus, in looking for his hat he grabbed his wife's head and tried to put it on himself. "His wife looked as if she was used to such things."

Dr. P. had a serious problem, to be sure, perhaps some form of "cortical blindness." But underneath its bizarre appearance, the condition is only an exaggeration of what we all do and it is not that hard to understand. For present purposes we can think of functions of the brain that construct internal reality, and of other functions that try to verify and correct these internal constructions, by comparing their details to details in the physical world. Most of this verification is done in fractions of a second, and we are usually unconscious of it as our eyes dart about and our attention scans here and there. As I write, is that a sheet or a lamp at the periphery of my vision? A slight reflexive movement of the eye establishes in a split second that the white folds are those of a lamp shade. The vague and improbable image of a sheet evaporates. Is that a cat or a slipper against my toe under the table? I may automatically poke it with my foot to make sure that the softness is not warm and alive and this may take a second or two. I hear an odd sound and find myself listening for a minute that reveals that it is the wind and not people talking outside, as it first seemed.

Over the years, much of Dr. P.'s internal world came to be organized by sound and smell cues—and he was better at rapidly verifying the correspondence of his internal reality to external reality in terms of these than in terms of visual information. Apparently there was damage to his brain functions having to do with visual verification. Or possibly it was attitude that was amiss, rather than organic damage.

Consider how especially vulnerable we all are to making mistakes when we are beyond the domain of simple physical objects and are in the more abstract domains of interpretation of events and classification and arrangement of physical objects and of ideas. It is much harder to verify components of internal reality that involve more abstract elements than simple physical objects. If our first impression of someone is that they are a jerk, it may take much more than a minute or two to correct this view, since jerkiness is not a physical quality. In the end we may even misjudge them permanently. In more abstract domains any of us may operate like Dr. P., using details to imagine what is going on in the world, and not infrequently we are as wrong as he and must correct ourselves when we can. Though since most people are visually competent, his mistakes look more odd to most of us.

It is fascinating, and makes a very important point again, that Dr. P. became so comfortable with his internal realities and bizarre habits of perception. "Well, so I thought my foot was my shoe, so what? Everyone makes mistakes now and then, ha, ha. OK, maybe I do need glasses. Well, if it pleases *you* I will see a neurologist." Sacks is a neurologist, not a cognitive psychologist, and he did not point out that we all do such things to *some* extent. Dr. P.'s pathology only shows us how strange these habits look when one can see them exaggerated and at a distance.

As I write, it occurs to me that, to a writer who tries to communicate serious new ideas, such quirks of the human brain present considerable obstacles. One learns from experience that small details—words, and phrases, style, examples—can trigger associations in some readers' minds that may cause them to imagine that they already understand much of what the author is trying to say. If one is not writing in a tradition that is clearly within some pigeonhole, this is especially a problem. The writer also faces the additional challenge of the reconstructions that will occur during memory, after the reader has put the book down and goes on each day to try to accommodate the new thoughts within an established system of socially reinforced beliefs and categories. Will the new thoughts evaporate, or be reclassified as readily as Dr. P.'s wife became classified and perceived as his hat once he had his hat in his mind, or then once again reclassified as he found that he could not lift her onto his own head?

3. Even in a scientifically informed culture our waking being may be *subconsciously* more complex than the single-actor model. For example, in self deception part of us must know some of the truth in order to keep it from the actor (references in Martin 1986).

4. It should be mentioned early that there are completely conflicting uses of the terms "relative" and "relativism." To the anthropologist a custom or belief can only be well understood and valued *relative* to other aspects of the culture or world view. It could be professionally crippling not to be able to see the relativity within belief systems and to see the inherent logic and functional validity of different cultures' perspectives on life. In this sense "relativism" has a very positive meaning to me.

In this sense it would be semantically correct to say, for example, that scholarly medieval theories of gravity, diseases, and poverty and modern theories are equally relative. Their theories had their places in maintaining their system, ours have places in maintaining our system. It is quite relative whether women feel pressured to wear high-heeled shoes, or to bind their daughters' feet and deform the bone structure.

A few anthropologists would wander beyond this to argue that all customs and beliefs not only make sense relative to their own culture's inner logic, but are equally right and true. This extreme view converges on what I will here call "pedestrian relativism," a popular cousin of nihilism, in which it is argued that anyone's truth is as good and right as anyone else's. Anything goes.

It cannot be so that customs are arbitrary, and while pedestrian relativism may appear consistent with the first sense of the word, it actually negates it. Conduct a thought experiment and mix beliefs at random in an imaginary culture. Cultures collapse when anything like this happens. Moreover, how deeply can any advocate really believe that "anything goes"? Parents, even those who are relativists, eventually act to influence their child's beliefs and so, while they may respect the child, it is demonstrable that they do not really respect all beliefs.

I suspect that most pedestrian relativism comes when people are a) comfortable with conforming to any set of ideas and customs by which they think they can survive and prosper socially; b) adapted to a life of changing commercial trends and styles; or c) persuaded that the very idea of scientific truth implies Platonic Truth and a single cultural norm (it does not), but that relativism is the politically egalitarian alternative.

The last, though, can be a self-defeating, romantic hope. If every culture and subculture's theory of poverty is presupposed to be equally true, for example, how could society ever develop a reasoned and united effort to minimize poverty? The same with theories of illness and so on. "Divide and conquer" is an old principle. If people cannot establish the basis for dialogue and agree upon common interests, they will be manipulated by larger interest forces. Without the possibility of appeal to reason, the hegemony of power politics would be complete and there would only be opposing interests, deceptive cunning, and might. The powerful would automatically prevail. There could be no Gandhi, Anthony, King, or Voltaire showing, with reason, some paths between a common sense of human decency and a better world. There could be no long-lasting egalitarianism.

On the other hand, there is the difficulty that a rational consensus can also become co-opted by ideological, economic, or class-interest forces. (Any motivating principle will become co-opted if someone or some class with cunning can find a profit or advantage in it.) History tells only too well how the mission to spread Western rational values was heavily co-opted by the forces of colonialism and social and economic exploitation.

Anthropologists have attempted to cut through some of the confusion over "relativism" with the concepts *emic* and *etic*. The etic point of view would be that of the anthropologist, admittedly burdened with an unknown amount of cultural baggage, and using the language and basic assumptions of science. The emic point of view would be that of the culture under study with its own categories, rules, and basic assumptions. Ideally *one does not subvert the emic and treat the etic as absolute*. One participates in a dynamic intellectual interplay. One elaborates the structure of the emic and uses it as one might a triptych of mirrors to gain better views of the organization of the etic so that *it* can in turn be better refined and improved.

Anthropologists have written extensively on this difficult issue. See Geertz (1984), Gellner (1988), Harris (1979), Oliver (1981).

CHAPTER 4

Inner Realities

Can you think of anything worse one can do to anybody than take away their worship? . . . It's the core of his life. What else has he got? Think about him. He can hardly read. He knows no physics or engineering to make the world real for him. No paintings to show him how others have enjoyed it. No music except jingles. No history except tales from a desperate mother. . . . He lives one hour every three weeks—howling in a mist. And after the service kneels to a slave who stands over him obviously and unthrowably his master. With my body I thee worship! . . . Many men have less vital with their wives.

Look . . . to go through life and call it yours—your life—you first have to get your own pain. Pain that's unique to you. . . . All right, he's sick. He's full of misery and fear. He was dangerous, and could be again, though I doubt it. But that boy has known a passion more ferocious than I have felt in any second of my life. And let me tell you something: I envy it.

. . . I go on about my wife. That smug woman by the fire. Have you thought of the fellow on the other side of it? The finicky, critical husband looking through his art books on mythical Greece. What has he ever known? Real worship! Without worship you shrink, it's as brutal as that . . . I shrank my own life. No one can do it for you. I settled for being pallid and provincial, out of my own eternal timidity. . . . "Oh, the primitive world," I say. "What instinctual truths were lost with it!" And while I sit there, baiting a poor unimaginative woman with the word, that freaky boy tries to conjure the reality! I sit looking at pages of centaurs trampling the soil of Argos—and outside my window he is trying

to become one, *in a Hampshire field!* ... *I watch that woman
knitting, night after night—a woman I haven't* kissed *in six
years—and he stands in the dark for an hour, sucking the sweat
off his God's hairy cheek!* ... *Then in the morning, I put away
my books on the cultural shelf, close up the kodachrome snaps
of Mount Olympus, touch my reproduction statue of Dionysus
for luck—and go off to hospital to treat* him *for insanity.*

Dr. Dysart in Peter Shaffer's Equus

Fashions in Passion

In several small towns in the mountains of New Mexico the people are
friendly enough, but they guard their privacy and are suspicious of inquisitive outsiders. They are concerned that their religious practices will be
viewed unsympathetically and perhaps eventually suppressed. These are not
the small Mormon-derived sects in the western United States who still quietly practice polygamy. These are the Brotherhood of Penitentes, or Brothers of Light. Their religious practices were brought to the New World by the
conquistadores and date back much further than the polygamous Mormons.
The brotherhood dates from the Third Order of St. Francis of Assisi, founded
in 1218. These isolated Spaniards keep alive vestiges of practices that were
once widespread throughout Europe but that are today almost gone from
the Christian world.

I refer to religious flagellation. This practice is best known to historians
from the Flagellant Craze that began around Perugia, Italy, in 1259, when
thousands of penitents marched through the streets, nearly naked, whipping
themselves and begging God for mercy. The colorful if unsanctioned movement spread quickly over much of Europe, included wealthy and influential
people, and as it grew it became of concern to the Church. The popularity of
its violent practices was eventually repressed by Rome.

Religious mortification of the flesh exists today and in various forms, as is
well known. During the capture of the United States embassy in Iran in
1980, the world witnessed Islamic fundamentalists beating their own bare
backs before the television cameras during holy days. The practice of walking on hot coals has been transported from India to Fiji, where it has even
evolved into a familiar tourist attraction. Religious castration still exists
among some sects in India. Various forms of circumcision survive in religious and secular contexts, despite the questionable medical value of this
custom.

It is not clear how religious flagellation has happened to survive in New
Mexico. These faithful were at the fringes of the Spanish Empire and of

Church control and their isolation was presumably a factor. But the whippings do in any event continue to exist in our midst, and what I wish to stress is how alien the practice now seems to most of us, even though it once made good sense to many Europeans. We may be able to follow the reasoning, that suffering leads to salvation, and we may respect the depth of their faith, but for most of us it is difficult to enter the intense mind-set of the flagellants, share their feelings, fears, joys, and to visualize and react to the world and to oneself as they do.

I recently read in the newspaper of a man whose house burned. He had been severely injured and his family had been killed. He said that the suffering only strengthened his belief in God. In the Old Testament Book of Job, the proposition is well laid out that God may test the faithful with seemingly unfair abuse. Satan dared God, who allowed him to tempt and torment Job. Job remained faithful and was eventually richly rewarded. The tenaciously faithful burned-out man may have been using Job as a role model. Justice is to be expected not necessarily soon in life, but eventually.

It is only a small step in one's thinking from accepting and enduring shame and suffering from above in order to reach salvation, to self-inflicting of suffering as a proof of one's devotion to God. I can follow the words of such reasoning, but I cannot readily enter into the world view and feelings of the flagellant. Similarly, I can grasp the logic to "There be eunuchs, which have made themselves eunuchs for the Kingdom of Heaven's sake" (Matthew 19:12), but this passionate mind-set too is difficult to recreate in my own head. It is not easy for most of us to develop *empathy* and that is why the Penitentes of New Mexico have learned that they have little to gain and much to risk by satisfying the curiosity of outsiders. For they are viewed then largely as objects of curiosity and not as rational human beings with deep feeling and beliefs, pride, conviction, humor, love, honor, and anger. Some psychic dimension of these faithful is circumscribed by a logic and a set of feelings and perceptions that were widely shared and appreciated only a few hundred years ago. Now it seems extreme and alien to all but a few. The living fact of the Penitentes, and our alienation from this aspect of their lives, underscores that the minds—the images, thinking, and passions—of people within a society change through time, just as language, manners, and clothing fashions do.

It is interesting to recall the graphic art of Dürer and Bosch in this discussion of individuality in internal realities. Bosch is particularly famous for his fantastic paintings and seemingly unrestrained imagination. His famous *Garden of Delights* is a panorama of the extraordinary to us today. Bizarre and mythical animals mingle with nude humans engaged in strange activities. Great bubbles enclose some of the people on the landscape, while animals dressed as people and half-human, half-animal creatures enter into the action. Musical instruments become implements of torture, and great knives and strange devices lay about on the landscape.

Hieronymus Bosch might seem to have been an artist who simply let loose his dreams and capacity for fantasy. He was sometimes mentioned as an artist of mere fantasy, or of satire. But this is not so and we cannot view his work as we do the fantastic creations of many of today's artists. Bosch was a respected religious artist of the fifteenth century; his work was commissioned by churches and was often, indeed, copied. He lived in an intellectually progressive city of the time, 's Hertogenbosch. He was a member of a Catholic society, the Brotherhood of Our Lady, and is recorded to have done several works for them. So there was a substantial religious demand and intellectual character to his art. This may seem odd, considering its erotic and apparent fantasy content. But after study, scholars have appreciated that most of Bosch's images are actually symbols from Christianity, alchemy, astrology, demonology, Gnosticism, and other beliefs that were powerfully meaningful in his time (references in Orienti and Solier 1976). His paintings were thoughtful commentary on complex issues in the magical/spiritual world of 's Hertogenbosch and other cities of the time. It was a world resulting from vigorous intellectual activities of the European tradition. Yet we can scarcely begin to grasp it today.

Since its earliest days the Church has had to deal with diverse interpretations of the Bible, and the Middle Ages were vigorous in this respect. But even beyond this, the people of 's Hertogenbosch lived during the years when mind-sets of the Middle Ages were having to deal with the intellectual explorations of the Italian Renaissance: Luther would soon lead the Reformation. The first Inquisition had declined in the north of Europe, yet witchcraft trials remained common, and the Spanish Inquisition was expanding. Burning at the stake, drawing and quartering, and breaking people's bones and leaving them displayed in public to die in anguish "took place in all the squares of Europe more or less every day" as Robert Held details in *Inquisition* (1985). One could be tortured or have one's life ruined over one's beliefs. People were deadly serious about salvation, and many were willing to pit their own best judgment against the authority of a clergy in Rome that they saw as incompetent, corrupt, or both. Beliefs were weighed seriously, strongly held, lived by, and died for. Nearly every material object was charged with supernatural significance.

It is difficult now even to imagine how to empathize with people's thoughts and feelings about life in the midst of such strong metaphysical inner worlds and profound intellectual and spiritual crosscurrents. We are tempted to pigeonhole Bosch's art as dream-fantasy or as some sort of great satirical joke because most of us have no other pigeonhole for it, and because it is essentially impossible for us to share the intellectual and emotional experience of his community. Indeed, scholars can recognize many of his symbols and see that they deal with a great constellation of serious belief systems and metaphysical issues of the day, but even they cannot really explain the passions of his paintings in detail. The present gulf between our world and his is profound. The beauty, intelligence, conscience, and yet the mystery and alien realities of his paintings give us a vehicle to ponder the separateness of mental worlds that can exist even within the life of a single great civilization.

The spiritual life of Europe was rich and diverse until only the last few hundred years. From ancient times the gods of the Celts, Goths, Greeks, Romans, and Jews gave people diverse eyes for looking at the world and for loving, living, and dying. Physical objects were charged with spiritual life.

The Church early sought to bring its own diverse sects and schools of thought and passion into line. But Rome was only partially successful. Witches and alchemists and the religions of the north and east added their views and symbols to the diversity of mind-sets that more or less openly expressed themselves in Europe. Then, gradually, this diversity of mind was pushed into the shadows. The rise of scientific thinking, printing and improved communications, and political centralizations all played a part in this advance of modern thinking to the center of a still-crowded stage.

How diverse can inner worlds be, what mechanisms allow this diversity, and what mechanisms can counter diversity and bring people's minds into conformity?

Social Synchrony and the Individual

Since our brains construct their own realities, as discussed in chapter 3, one might *expect* wide diversity in individual perceptions of life. Indeed, without strong counterforces to direct mental realities along similar paths, each person's reality would become free-roaming imagination. There are no obvious theoretical boundaries on dreams or imagination and we cannot place theoretical boundaries on attitudes, opinions, worldviews, perceptions, emotional triggers, tastes, passions, and such. But to a very large extent the conscious waking perceptions of individuals do in fact depart from such an unrestricted abstract baseline and are synchronized within any one of the thousands of diverse human cultures. They are directed along similar paths, and individuality is in practice much constrained.

Some of the synchronizing forces are easy to understand. We do not live our waking lives in a state of complete sensory deprivation. We all must survive and mature in the same basic world of physical materials and energies and we have more or less common sensory experiences with heaviness, dimension, color, sound, temperature, and so on.

Next, we are social creatures, and the individuals within a culture or subculture have similar experiences with particular sorts of objects, events, and even values by virtue of living in the same local environment, following our peers and leaders around, and imitating one another's selective activities.

Beyond this, in the learning of a culture's body and verbal language, one's attention is further drawn to particular parts of the environment. The curious and approval-seeking child learns to organize life's enormous complexity into classes of objects and events. Language, moreover, teaches the group's categories and hierarchies and puts boundaries on their categories. It communicates positive or negative emotional reactions to objects, events, and concepts. One function of language is to synchronize between individuals, their attention on, participation in, and experience of the material world, and this can bring about enhanced overlapping of individual realities. Belief and ideology are embodied in language and also produce an experiencing of life from fixed perspectives.

Further, ritual, art, myth, literature, and electronic media extend these functions of language and provide common experiences related to the mind-set and values of the group. Common participation in events with extraphysical meaning can again increase the degree of overlap between mental realities, as one can see by reflecting on one's own experiences or by carefully observing the experiences and reactions of others. It is one thing merely to photograph and record a ritual, like a tourist; it is another to anticipate, participate, recall, and interpret an event.

One result of this is that for individuals in some culturally isolated subgroups, individuality in perception might even seem inconceivable because their friends and relatives have such similar backgrounds to their own, leading to highly synchronized perceptions. Individuality may become a concept encompassing little more than surface mannerisms, talents, habits, and opinions. Indeed, many people restrict themselves to close relationships within subgroups because it is difficult and uncomfortable to communicate deeply across divergent mentalities. Most cultural isolation is thus self-imposed in my society and in others that I have known.

Art, including literature, functions in part as a means of producing synchrony between individuals. The paintings of Bosch and others helped an intelligent community of people to move beyond cold reason and to deal in bold and distinct images with the problem of how to come to grips with the complex spiritual powers that constituted their world. His paintings helped individuals to grasp and focus on ideas that were thought to be vital at the time.

It has been claimed that science is a system that can turn private truths into public truths. This view has much merit and is a main point to which I will direct the last part of this chapter. But it is important to appreciate that art and various usages of language are other solutions to the same problem of communication between sometimes profoundly separate mental worlds. Humankind has invented several ways of dealing with this problem. Several are needed because life and human issues are intrinsically complex. The driving force that ultimately selects for the preservation of the successful inventions may be an enhanced ability for individuals to interact and be empathetic, and the emotional and economic advantages that various levels of social stability offer to the individual and to the group.

Art can have many functions, including entertainment and decoration, helping us to look at ourselves and the world more insightfully. Some types serve as symbols of social status, or as investments, and it can be used to sell cigarettes, chewing gum, and ideology. But art can also be very important in helping us to synchronize our individual mental realities. The beauty of a heaven or the anguish of a hell, even the glory or the filth of war, romantic love, or high adventure may mean little without specific art themes or stories to focus on, visually or verbally generated imagery. Even such personal experiences as dreams can be verbally communicated in some degree of detail, and more can be communicated with paintings or film. Fiction allows us to relate common ethical dilemmas and their resolutions to the problems and reactions of a particular mythical or fictional character. Stories can become part of the internal realities of individuals in a group and serve as useful paradigms and models, helping the individual to grasp and discuss, or to unconsciously acquire, social norms.

Recall in this sense the story of Job: the just man who suffers. How should a good and faithful believer react to severe misfortune? How should one then think of one's God? Endure, remain faithful, and eventually one may be rewarded; this is the message. Of course, there are other very different interpretations and paths for one to follow that can be based on the Bible, and this is one reason why so very many contending sects have proliferated within the three great biblical religions. But to point out that religious stories do not guarantee absolutism is not to deny the power and usefulness of a great story or work of graphic art.

One could argue that the boundless potential of our brains for imagination even *requires* art and ritual (and, recently, science) for social stability in any large or complex culture—or a strong police control on the other hand. This is an important issue and one worth pondering. How much diversity of individuality in perception and action could various sorts of societies tolerate before the society must reconstitute or deteriorate? Even the highly diverse North American society is yet loosely held together by values and role models reiterated through mass media and the school system, and our reputed individuality is balanced by a widespread and strong, if paradoxical, conformity. Even in pluralistic societies people relate and live much of their lives within fairly homogeneous subcultures and do not try to relate to, or probe deeply into, the cultural diversity with which they could interact. Their world is not as diverse and confusing as it might be and their confusions and conflicts are only a small subset of those possible.

Theoretically unbounded internal realities do become much constrained in fact. In some teachings of Eastern philosophical disciplines such as Buddhism, it is claimed that wakefulness is a form of dreaming, is something less than complete consciousness. To the "truly enlightened," most of us seem to be acting in, and our lives are overwhelmed by, one dream story or another. We function together effectively to the extent that we are living out compatible dream stories. (Sometimes one has a relatively vivid feeling for this over a few days or weeks on returning to one's own culture after having functioned within a foreign one for some time.)

In a sense that one can grasp scientifically, the concept of the waking dream must be true. Much of our waking pattern of attention, emotional responses, perception, and habits comes from the subconscious just as dreams do. We are rarely alert to more than a sliver of the possible ways of examining and living in the world that we are potentially capable of. We carve out this sliver by rules of perceptual and intellectual habit learned for the most part unconsciously from experience, imitation, language, and training; and the rules become imbedded in our subconscious so deeply that they can be incredibly difficult to see, alter, or remove. Once imbedded, these powerful rules of habit in seeing, reacting, thinking, feeling, and behaving that carry us through life are like a creature with a mind of its own, as difficult to control consciously, or escape from, as any dream is. Our perception and activity while waking are thus limited and constrained to a small reality, and this process is strongly the product of the mysterious subconscious, the source of dream creatures.

I have laid out the general terms of a discussion, but the matter of individuality in perception remains difficult to define precisely. Few would disagree that there can be individuality to our feelings and opinions about material things. Few would disagree that it is difficult to have experiences with material things that do not include feelings and opinions. Few would disagree that experience of physical things is individualized in this sense. But if it were asserted that our individual perceptions involve differences in the characteristics of material objects that we usually believe are inherent and objective properties, most of us would not agree so readily. Most of us need to believe that our feelings and opinions are our own but that physical reality is objectively seen by us. Any hint of complication can be quite threatening. It may call into question the soundness of our self-control, our reason and action, and it could reduce our all-important respectability and social status among our peers. An object may inspire beauty or ugliness, rapture or fear, and we allow this sort of individuality if we have any sophistication at all. But we may at first be skeptical that physical perceptions vary in any large way. Well, we allow for colorblindness, dyslexia, madness, and visionaries. But common sense treats these as exceptions to the rule. Unfortunately the matter cannot be dismissed quite so easily.

The world does not consist of a few isolated simple objects like chairs, books, and dogs. Life is complex and varied. We do not focus on and define each object and event each moment as we live. Perhaps we *could not*. There is, then, much potential ambiguity. Boundaries between categories and events are not always inherently clear. Nor is it always clear what constitutes an event or a series of events. Real life in this complex, changing world takes us beyond the simple theoretical class of circumstances wherein we can all pause to study objects and agree that each dog has four legs, two eyes, brown hair, etc. The opportunities for beliefs and attitudes to interact with sensory perception of the physical world thus multiplies. Memory is malleable and open to suggestion and reconstruction, and much of experience is memory, as discussed in chapter 3. Modifications of memory may occur at any of the stages of aquisition, retention, or retrieval.

In walking along a busy street, someone fearful of dogs or attracted to them could focus on the dogs and pay only the slightest attention to the people, houses, and colorful flowers. There exist conditions for some of the scene to be constructed individually. The partially imagined portion of the the street will use prototypes from conventional experience. For example, the dog-fearer could imagine that he or she saw roses where there were actually gardenias, but one will not typically fill in with flowers that have no basis in the previous experience of waking reality. If a friend tells us she saw roses instead of gardenias, we pass it off as a mere mistake, almost without thinking about it. But if the friend tells us she saw flowers that were orange owl's heads with large eyes that kept blinking hungrily at her, then most of us in my culture will intuitively conclude that her mistake has exceeded acceptable limits and will begin to think of her as crazy. We may even begin to reconsider friendship. Misperceptions are so common that unless they include socially unusual images we scarcely think about them. As a result we tend to imagine that our perception and memory are much better than they really are.

Many of the initial misperceptions are unconsciously corrected in an instant by the hypothesis-testing brain. These are so much a part of the ongoing process of recognition that we would not normally think of misimpressions as errors, although they are. Many misperceptions are not corrected and do become part of our conscious or semiconscious perception. Let us further probe the sense in which a dog-fearer and a brave flower-lover really see the same street. What of social interactions with the initial perception? If we frighten someone about the dogs on Elm Street, will he tend to focus his attention mostly on the dogs and experience less of the other details of the neighborhood? Can one person's spoken recollection of the flowers—his suggestion—influence the memory/perception of others? The magician diverts our attention so that what we "see" taking place is strongly influenced by what he or she leads us to believe. Many scientific studies show that suggestion can influence perception by causing us to focus on one thing or pattern rather than another, by altering our recognition of the pattern that we are attending to, and by modifying memory.

Our pasts are gradually being modified or confirmed in an ongoing process that is significantly influenced by our friends and superiors. "Oh, you say those were gardenias not roses? I suppose I was mistaken. . . . Oh, yes, I remember now; they were gardenias." We all tolerate some sorts of mistakes and scarcely think about them when we or others make them, but there is a line in our culture that separates acceptable from unacceptable mistakes. It is acceptable in some societies to mistake a shadow for a spirit, but this is not acceptable for most literate Westerners. Our sense of social acceptability would come into play and we would automatically and unconsciously correct or alter the vision, possibly laugh off a fleeting consciousness of it if we are among trusted friends, or more likely keep the incident to ourselves out of fear of looking silly or being thought mentally unbalanced.

Witnesses to crimes may frequently misidentify suspects. In an experiment designed by Professor Robert Buckout, viewers of WNBC-TV in New York were given a good closeup look at a mugger in a simulated crime and then asked if they could recognize him among six in a lineup. Of the 2,145 viewers who called in, 1,843 of them were wrong. This and other experiments and actual cases are discussed in detail in *Eyewitness Testimony* (E. Loftus 1984). The book is a good introduction to the psychology of mistakes in perception and memory under complex, changing, or emotional conditions. Loftus also discusses the worrisome matter that any of us, including police and juries, rely heavily on and may be highly influenced by eyewitness accounts from persons with stereotypic sets of mannerisms that in our culture supposedly indicate reliability, even though history and studies show that this trust of personal style is unreliable.

When we see things that cannot be confirmed to exist, or when we see them incorrectly, does this faulty perception take place at the instant of perception, or is it a result of faulty memory a second later? This is hard to say. Certainly memory is malleable and open to suggestion; this can be carefully studied (e.g., Loftus 1979, 1984). But how can we study the instant of perception? *Sometimes* hypnosis may help a witness remember details of a crime, which suggests that initial perception is accurate and that memory is the weak link in the chain. But how often is this the case and how widely can we generalize from selected newsworthy examples? Difficult questions. It is even difficult to know how well our vocabulary for discussing such things corresponds to the actual organization of mind under realistic conditions: terms such as sensation, perception, cognition, memory. Specialists must attempt to define, classify, and study. But we should not become falsely secure that we deeply understand the organization of mind simply because there are some terms available for conversation.

If we expect to see things that do not really exist outside of our own minds, we can actually mentally see them under some conditions. Confusing or fast-moving conditions can, for example, open the door for enhanced misperception. There were days when people believed that they saw frogs fall from the skies during rains. Now we know they emerge from the ground to breed in puddles. Is it possible that people actually "saw" imagined frogs fall from the sky? In the course of many years of fieldwork, people have related to me, with firm conviction, many observations of plant and animal behavior that could not happen. When younger, I assumed they were having a joke at my expense. But over the years, this view did not hold up to examination. Now I am inclined to accept that if people believed that they saw it rain frogs, then some of them may actually have seen/imagined frogs fall during a downpour. Their minds did see the frogs.

The same may apply to hoop snakes or ghosts. A person does not have to be insane to see things that do not exist, and despite much interest in ghosts, we have no proof of them. Most ghosts have been seen in dim light when the brain must work particularly hard to create a mental world based on little reliable sensory input. This is why scientists, and magicians such as Houdini and Randi, who have studied the question and who have found much deliberate fraud, deception, and self-deception, regard it likely that the beliefs of sincere and sane individuals can help them to see ghosts, rather than that ghosts do exist in an objective sense.

How does one study this murky area of questions about individual differences in perception of the physical world? While there has been considerable progress, the subject is difficult. A strong cheese may be loved by one person and hated by another. Does each person actually sense the same flavors but react differently to them, or does each selectively sense different components in the chemical mixture? It is difficult to be sure. I find that people who do not like jazz, or some classical pieces, often do not hear the themes and their variations. When the themes and variations are pointed out, they will say "Ah!" as the patterns are perceived. Perhaps they were hearing all the sounds but not the *patterns* that make the music interesting to those who enjoy it.

If one pays careful attention to plants and birds, one may really begin to see and hear more when the eyes and ears are aided by intellect. Little spots and subtle shapes or sounds that one did not notice before will, once learned, become the cornerstone for focus and for seeing other details of a newly learned species. That is, once one learns a small clue that helps to distinguish different species of pines or sparrows, then one begins to notice other distinctive features of their anatomy, ecology, or behavior. (Many independent critical studies confirm that these species are indeed distinct and are not simply fabrications of the mind.) The point is that sometimes it takes us considerable effort to learn to see differences that eventually become quite obvious and that research confirms were really there in the first place and do distinguish real species. With scientific understanding, one's perception is genuinely enriched. This is the nature of knowledge of any sort; one could cite many everyday examples as well as those from science and art.

Even among scientists there is not some single way of experiencing the world, and most certainly our particular understandings of the world—our "beliefs" if one prefers—influence our perceptual experiences of it. I seem to actually "see" more diversity of life, more activity, more processes in a walk through a forest or desert than my beginning students do. And the biologist may see the world differently from the geologist. I know a small bit of geology, and the more I learn, the more complex and more delightful the landscape becomes. A mountain is something with a history and I see in its rocks and ravines not only the forms and textures that we all love, but the pages of its history. In reading these pages from the rocks, I see many subtle details that would have escaped my attention if I were uneducated. Similarly, a good physician will see details in our bodies and behavior that a physicist may not notice, but that are real and will help to reveal our state of health and the nature of disease.

Cross-Cultural Issues

Psychology, psychiatry, and biology are much the poorer for a lack of multicultural perspective on the mind. How often have we seen a book with a title such as *Sexual Nonsense of the Human Male*, only to find that it is pitifully culture bound, dealing perhaps with middle-class North American males, and does not begin to appreciate the true scope of its issue?

Our beliefs can influence inner realities in profound ways; it is instructive to contrast some cultures. One of the Seri Indians that I worked among in Mexico was famous for having gone to the moon. The Seri knew that it was possible to go to the moon, since the Yankees had done it, and now their tribesman had done it. Apparently he had gone in a dream. But this distinction between dream reality, or other subjective experiences, and objective reality was not important to them in the same way that it is to us. My culture is deeply rooted in rational ideals and scientific models, and it is second nature for most of us to want to know if the Indian objectively or subjectively went to the moon. The Seri do not make this the same matter of deep and immediate concern. I know that they can make the distinction, but I am not certain what form this normally takes. Anthropologists have never agreed on how to deal with questions of this nature.

> Is the act of what is seen and felt while dreaming an act of belief or a mystical experience? In our societies, the answer to this question will vary according to the circumstances of the dream and the people questioned; in the societies called primitive, generally the question will not even be posed. (Levy-Bruhl 1978)

In New Guinea there is an astonishing variety of cultures but commonly, at least, the line between what *we* would think of as objective and subjective is not a priority issue for the people, and it is difficult to see that they easily draw the same distinctions along these lines that we are trained to draw from childhood. Our objective and subjective categories are both real worlds for them, and for them these worlds interact to be sure, so the line between them is not compulsively drawn. But can they tell differences if they must? They can, in some sense.

While in Papua New Guinea, I discussed this issue with B. G. Burton-Bradley, head of the mental hospital in the country and author of *Stone Age Crisis: A Psychiatric Appraisal*. He embellished our discussion with the story of one of his patients. The man had dreamed that he was copulating with the wife of the Chinese owner of the general store that he worked in. Some of his friends came in and laughed at him. He was outraged and when he awoke he went and killed with an axe the friend who had laughed in the dream. His people realized that there was something wrong with his thinking, that one does not so compulsively kill in revenge for an insult in a dream. They brought him to the hospital rather than punishing him according to their strict laws. They were making distinctions between different sorts of their realities, but Burton-Bradley was not certain just how the lines were drawn and what they meant.

I have quoted (cautiously) from the *Notebooks on Primitive Mentality* of the controversial Levy-Bruhl. These notebooks are very interesting, if only in that they show this influential scientist after six major books and forty years of study, still not quite able to come to grips with this difficult subject. It *is* a difficult subject, and that is the main point one needs to grasp now.

Levy-Bruhl discussed his conceptual difficulty that the Orokaiva may regard a ghost and a corpse as being one and the same, though the ghost may be experienced miles away from the corpse.

He went on to discuss a report by a man named Grubb of an Indian who asked him for an indemnity for stealing pumpkins from the Indian's garden. Grubb insisted that he *had not* taken the pumpkins and that he *could not* have done it because he was 150 miles away at the time. But the Indian persisted because he had seen Grubb do it in a dream, and it did not matter that Grubb was *also* 150 miles away. The Indian was being *logical* enough; he had seen Grubb do it, he simply did not accept it as a physical impossibility that Grubb could be in two places at the same time. He could have been a very intelligent person, but he may have appeared stupid or crazy to us because his beliefs about physical possibilities did not cause him to reject certain of his own experiences as we would. Indeed, in some societies the dream world is regarded as being more significant than waking reality.

I should mention next that some people may not make sharp distinctions between what they have seen for themselves and what a trusted friend or member of the tribe has told them. The collective experience can have deep implications for the mental world of the individual. For example, I was talking with a bright, articulate, and outwardly Westernized fellow in Papua New Guinea and the subject turned to ghosts. He had seen ghosts with his own eyes. "With your very own eyes?" I asked skeptically. "Absolutely," he assured me. Then he began to tell me in great detail what he had seen. But I interrupted him and pressed for details on the observation conditions themselves. Where was he standing, how old was he at the time, how good was the lighting, was he fully awake? Without the slightest embarrassment he went on to describe how it was his grandmother who had seen this ghost. He trusted her and her eyes were as good as his own—they were his own eyes. His own visualization of the ghost was as vivid as other experiences that he called his own. I really do not think he intended to mislead me. Rather, it was as though he did not firmly conceive that one's mind and the minds of others can play profound tricks, and he did not have the habit to be watchful and skeptical, so he would not normally make the effort to sort out which things he had literally seen with his own eyes and which things he had visualized on the basis of trusted information from a relative. His grandmother's experiences were part of his world in some intimate way.

Indeed, one would not make such distinctions unless there were reasons to know that such distinctions can and should be important. I try to make such distinctions reflexly, usually unconsciously, following models and lessons from my upbringing and schooling. They are second nature to me now. In my society, our history has taught most of us the usefulness of at least attempting to make the difficult distinctions between objective and subjective, hearsay and firsthand experience. Our courts, though, must still today remind witnesses and jurors to distinguish hearsay from personal observation. Certainly not all Westerners always keep track of such distinctions. In such respects my friend in New Guinea, though Western educated, seemed to have retained a partially "tribal" mentality. Indeed, the mixtures of indigenous belief systems, fundamentalist Christian theologies, and scientific thought modes among individuals in Papua New Guinea was common, relatively conspicuous, and fascinating; my experiences there helped me to be aware of how diverse and synthetic thought modes are in my own society.

One should not expect, and workers do not agree, that there is a characteristic "primitive mentality" or a single "modern mentality." These terms are cautiously used working generalizations to announce a mystery, not to explain or circumscribe it. The distinction between primitive and modern mentality is artifical and often exaggerated or misunderstood.

The individual's subjective world has been notoriously difficult to penetrate, and this subject remains as a major challenge to science. Anthropologists as a group have regrettably all but given up dealing with the subjective on a cross-cultural basis (though they discourse at length on the methodological problems) and this is one reason why too little humanity and individuality shines through the ethnographic literature. Indeed, it is in the context of this profound difficulty that some have proposed an answer to the mystery of how famous author (*The Teachings of Don Juan, Tales of Power*) Carlos Castaneda's examining committee at UCLA could have approved his shamanistic accounts as a Ph.D. dissertation. A strong case can be made that they did not personally accept that Castaneda had literally turned into a bird, flown about the countryside, and such during the course of his research. But perhaps they hoped to shake the anthropological community. Were they, by endorsing the dissertation, agreeing that the elaborate and interesting subjective worlds in alien cultures can be illustrated and communicated in some rich and meaningful ways, though perhaps not through conventional means? Was the doctoral committee protesting, in their certification of the physically impossible, the timidity in confronting the human mind of the science that claims to study humans?

Honest Communication Can Be a Deceiving Mirror

It has been hoped that world peace can be achieved if one day we all speak a common language. The cautionary response, though, is that it is much too easy to deceive ourselves into thinking that we do really deeply understand one another simply because we talk and interact. Surely there is an important point here. The ability to converse is not the same as the ability to enter another's mental world. One learns through long and sometimes very painful experience to return to the alert child's awe at the mysteries of another's mind.

I emphasize again that the nature of the human brain, as a variably synchronized illusion organ, is a force to produce mental individuality. In theory we could all have mental lives profoundly different from each other—as different as a party of lunatics might differ among themselves. It is remarkable that common waking experience and language do interact to partly synchronize us and thereby allow viable social groupings with complex routine interpersonal dynamics and stability. Yet, how fragile our interactions can be, and when we see in another familiar words and reactions we can fail to see past this to the depths where major differences may abound. "No one would talk much in society, if he only knew how often he misunderstands others" (Goethe). Even husbands, wives, and children in a family may sometimes be scarcely able to fathom each other's views and feelings. "We seek pitifully to convey to others the treasures of our hearts, but they have not the power to accept them, and so we go lonely, side by side but not together, unable to know our fellows and unknown by them" (Somerset Maugham).

It is very difficult to know the mind of another. When we can share some basic emotional responses or bonds, some communication, we may too easily assume that we know the other's mind. Take a simple case first. Some people are sure that they can communicate with their pets and that the pets perceive situations and understand them essentially as the pet owners themselves do. Others would argue that the mental worlds of the pets and the owners could be quite different. We can program a computer to carry on a convincing level of conversation with us, but it is tedious to argue that the mental world of the computer is like our own.

Do porpoises, chimpanzees, and perhaps some other animals with large brains conceptualize the world much as we do? They probably do not have elaborate philosophical opinions as educated adult humans do. But do they have humane feelings and some sort of common sense, beliefs, plans, hopes and awareness (see Griffin 1981)?

When I was a graduate student at UCLA, I had a job assisting a scientist who studied porpoises, and the research group tried to be scrupulous in not attributing any higher intelligence to these creatures before hard scientific proof became available. Each person, though, had had one or more experiences that tended to force the issue and create at least the private suspicion that these creatures have some sort of advanced minds.

My own first experience was simple but memorable. Our first captive porpoise was netted one morning and by early afternoon we were unloading it into a circular tank on campus, releasing the straps that held it in a transport sling. The porpoise was "anxious" (in my terms) and intuitively, as

I might comfort a frightened child, I began to stroke and fondle its flippers and head region, even though I had no reason to believe that my actions would be effective in reassuring or comforting it.

Once released, the porpoise swam quickly around the tank only a few times and, as though finding no escape and having the sense to realize this, it slowed down, raised its head out of the water and looked us over. Then it swam directly to me and positioned itself so that I could again fondle it. I did.

Porpoises are distinctly built to race forward, so it took a dramatic effort for it to struggle to remain next to me in the tank. It would repeatedly lose its balance and had to swim off, but then it would return immediately to the awkward position in which I could continue to stroke it. The situation was unnerving. As I stared into the relentlessly probing eye of the bullet-shaped hunk of muscle and as we struggled to maintain contact under those physically difficult conditions, the clear probability seized me that there was a mind behind that eye, a mind fundamentally aware like my own, a mind that could read my own concerns and that could trust and hope, and put trust ahead of fear.

I have handled many other species of adult wild animals and never before or since had another such experience. Tamed animals will, of course, learn to behave to please us and will form social bonds with us. But the porpoise's reaction seemed quite different in character, if only in that it was so spontaneous and carefully directed to the individual who had shown particular concern for its "feelings." Wild animals will typically first exhaust themselves trying to escape from a cage and then only slowly learn to trust a keeper who feeds and cares for them. The porpoise seemed able to draw a quick conclusion. Everyone had been very careful and gentle in handling the animal, yet it seemed beyond doubt to have recognized my deliberate attempt to reassure it as something more, that it had sensed my empathy for its vulnerability and responded to my intentions and commitment to care for its emotional comfort. The porpoise was in a totally alien situation, few hours from the wild. It is unlikely in those days that anything with hands had ever touched it in nature, and yet it seemed able to interpret my efforts correctly.

This is not the kind of moving experience that one forgets. And yet, on sober reflection, how could I know what to make of it? The presumed communication, no matter if real, should not give me assurance that porpoises have minds that are like my own. At present this is unknown as a fact. The porpoise could have a mind indeed far advanced over mine, or quite different, or very much the same. Just so with humans. Some of us feel that when we can communicate with another we must have similar minds. We feel a deep kinship when we communicate.

A couple in love may feel that they understand each other's minds extremely well when conversation and emotional exchange is intense. But how often, when the relationship dissolves, perhaps after years of marriage, do we hear them say in exasperation that they came to realize they were living with a stranger? We commonly have strong feelings on such issues on the basis of very slim and often questionable evidence. But we cannot yet literally know the mind of another human. It astonishes me that the majority of us would kill whales, porpoises, and chimpanzees without any pangs that we may be destroying intellects as interesting and potentially sophisticated as our own, and yet at the same time take it for granted that a president or a friend or a fetus have a subjective reality quite like our own, simply because they are human or perhaps can communicate to us. Perhaps so, but perhaps not. How can so many of us jump to conclusions on such difficult questions? The question with regard to higher animals is open. And the question with regard to differences between humans is also difficult.

Philip K. Dick, the highly imaginative explorer of inner space, used an ingenious and entertaining idea to stimulate our imaginations to probe this issue. In his science fiction novel, *Eye in the Sky*, a group of visitors fall through a proton beam when an observation platform breaks at a bevatron facility. One of them, scientist Jack Hamilton, gradually discovers that the laws of science do not apply anymore and that new rules govern the universe. A cigarette machine is empty but produces cigarettes miraculously. His automobile repair manual is filled with prayers, instead of mechanical instructions, for fixing the car. "Negroes" are bug-eyed, walk with a shuffle, and say "sho-nuf." He learns that the sun is small and moves around the earth. Jack's liberal friends have become ugly and unclean.

It is soon clear that the rules by which the entire universe operates have indeed changed, and while there is a learnable pattern and order to life, it is difficult for Jack to adjust, since causation is now so different from what it had been for him. After several adventures Jack finds Arthur Silvester, a bigoted old man and one of the people who had fallen through the proton beam with him. The proton beam had so energized Silvester's brain that the stereotypes by which the old crackpot had perceived the universe to function had become those that actually organized events in the world in which Jack and the others now had to live. Jack was forced to live in another person's perceptual world, subjective physical reality, and he and others found it a tyranny.

Silvester is killed and the hope is that life will return to normal. But Jack soon learns that his wife has become sexless. Large blocks of space in the newspaper are blank. People are distressed to realize that they must now exist in the mental world of another of the energized visitors who had fallen through the proton beam.

Edith Pritchet has a mind that represses unpleasant things, including sex. People and things that are offensive to her simply disappear when she sees them, and this denial creates frightening chaos for Jack and his friends as things and people that they value vanish.

Edith is eliminated. Sex and other of life's familiar complications return. But Jack is not yet free; next he finds himself in the world of a "paranoid," where knives and objects really do take initiative and jump and cut and fall on people. Dangerous monsters do exist. Joan's house really *is alive*.

Next Jack finds himself in the world of a Marxist sloganeer. Workers attack capitalists who drive expensive cars through slums. Life is reduced to a new set of stereotypes. The class struggle takes place in an opulent gangster city with only riches or vice and crime. The rural areas are a world of lynchings and killings of Indians. The kids all drink and take dope while refugees from economic desperation file by. Everyone really is on one side or "the other. Issues really are black and white.

It is intolerable for Jack to have to be on one side or the other. The Marxist sloganeer is revealed and destroyed just in time, just as his revolution was about to kill Jack for his advocacy of the "cult of individualism." At last Jack is returned to the real, safe, world. Or is it merely his own world?

With artistic license in hand, Dick makes his point. We talk with someone, see a real physical being like ourselves, and hear words familiar to us. The words translate into meaningful actions that take place in the real world that is before our eyes and that both of us share. We say we only have different beliefs. It is hard to imagine that the person who looks, acts, and talks so much as we do may be living in a mental world that is not much like our own. But, the differences may well be vast.

On Being Vulnerable and Manipulated

These different individual and cultural views of reality all work for those involved. They are functionally valid in this sense. So what is the issue? Why the desire for truth and justice through art, science, philosophy, and the like? Why not leave well enough alone? Why hope to change anyone's thinking or our own? Why should not everyone think just as they like?

I will only touch on a few of the many answers to such questions. First of all, we are a species with a long and sordid history of exploiting one another. One who is superstitious or is living in a dream world may be particularly vulnerable to manipulation and exploitation from dozens of directions, of governmental, religious, economic, commercial, or even familial natures.[1] One may easily be so naive or so trusting of the conventional wisdom of their peers that they do not even know they are being manipulated. One so vulnerable has only the *illusion* of freedom and security. By mastering mental habits for weeding out nonsense, one can reduce the window of vulnerability and undo some damage. This is what Thomas Jefferson was referring to in his 1778 *Bill for the More General Diffusion of Knowledge*. Liberal education should be that which enables people "to know ambition under all its shapes," and prompts them "to exert their natural powers to defeat its purposes ... and ... guard the sacred deposit of the rights and liberties of their fellow citizens," especially "since experience hath shewn, that even under the best forms, those entrusted with power have, in time, and by slow operation, perverted it into tyranny."

Moreover, in many times and places people have been so certain of the truth of their own views and opinions that they have persecuted and killed those who saw the world differently. The religious wars and inquisitions of Catholicism, Protestantism, and Islam have taken millions of lives. The political absolutisms, particularly of the dictator ideologies, have taken millions more. Now, all of this carnage and destruction has been charmingly well meant. The killers were going to eliminate evils of one sort or another, so in the name of some "good" they burned, strangled, tortured, gassed, shot, bombed, and beat to death those whom they saw as evil (e.g., Hroch and Skýbová 1988). When critically thinking people saw the human mind fixated in such persuasions, some thought, as they listened to the screams of anguish and watched the loved ones struggle in despair, that it would be nice to really know the truth. A desire for justice and protection from folly is one factor recommending a search for truth and for criteria by which to detect nonsense.

Another question relates to the goal of democracy and self-rule. Self-rule requires the ability of people to agree. So all else being equal, why not agree on things that are more true than not? This, then, implies (but does not require) a search for truths. It also implies that we accept the burden of urging our fellow citizens to appreciate grounds for agreement.

History offers us a baton with which we must try to race ahead of ignorance, fanaticism, folly, denial, exploitation, and intolerance. But the baton that we accept from previous generations and pass on to the next is also a sharp double-edged sword that we must learn to handle carefully. It is not enough to be well meaning. History teaches the dangers of fanatical fighting for good, but it also teaches the dangers of not searching for truth.

But judging what is a worthwhile cause and what is not can be very difficult. Words like *good, truth, freedom*, get twisted about in Orwellian doublespeak, and it often happens that people end up fighting for a cause that they think is just, when in fact they are being psychologically manipulated, as Thomas Jefferson warned, by the rhetoric of some group or ideology. The inquisitors who tortured heretics to death were well-meaning seekers of truth and justice. They held trials that were careful by the standards, and in terms of the beliefs of, those days. Those were generally *not* scenes of drooling clergy, eyes glazed, running through the streets looking for victims. The horrible mass killings by Nazi and Stalinist movements were done under the banner of (misused) science and with the best of misguided intentions on the surface, to eliminate *in their view* sick mentalities and to consolidate power to create better worlds. The economic, social, and personal exploitations of colonialism and similar conquests have been the product of a wedding of drab greed with a glittering and well-meaning, if misguided, intention to extend "civilized values" to the world.

None of this emphasis on the well meaning is by any means to deny that thugs, sadists, the politically ambitious, and the greedy have involved themselves in all such crimes. But it would be shortsighted to ignore the power of dogma, ideology, and misused science to create states of mind in which well-meaning people can destroy other's lives wholesale and then congratulate themselves, as well as to permit sadistic murderers and cynical exploiters a cloak of social respectability.

Bridging Internal Realities

I have tried to approach the subject of individuality in perception from various points of view, but the main issue I want to return to is, given such individuality, how does one communicate a novel point of view? How can one help another person, with his or her own view of reality, to see the world in a new way *critically*?

Of course some people may have cultivated a capacity for empathy and open-mindedness, and in their case the problem is reduced. But how do radically new ideas get incorporated into the thinking of people who lack this sensitivity or who are even resistant to it? One should ponder what causes minds to be actively resistant to foreign ideas; but for the moment I only wish to discuss some general devices for bridging minds.

Art, painting, ritual, myth, theater, literature, films can all generate common experiences that can be the seeds from which empathy grows. Versions of these are found throughout human societies and it is likely that one of their profound functions is the sharing of individual realities and the facilitation of communication in social groups.

Science is a more recent invention than art, but it can serve the same function. Moreover, it has elements that allow the *collectivization* of critical thought to very powerful degrees. Logic is one part of science, and of some philosophy and religion as well, that lets us agree on rules for dealing with ideas. Logic and mathematics begin with simple ideas that can be agreed upon, rules of manipulation of symbols or items that can be shown to have validity based on simple demonstrations. Standards of measurement allow us to agree on size, warmness, heaviness, and so on.

Experiment is another common element of science. The word is used in several ways, but in one important sense an experiment is a way that one person can show another how to have the same revealing if counterintuitive experience. One of history's greatest experiments was also one of the most simple.

The theory of spontaneous generation, that organisms are generated by putrefaction, from rain and the like, was fought repeatedly, by Redi in the seventeenth century, Spallanzani in the eighteenth, and Pasteur in the nineteenth. The idea was difficult to defeat because so many people believed that they had actually seen life created thusly. Others believed correctly that they had seen animals only arise through what we would today call conventional reproductive means, like begetting like. How can one resolve two very different perceptions of the same happening? Should we give up and say that both are correct?

Redi acknowledged the position of his opposition, stated his own position, and then structured his own experiences and perceptions. He wrote carefully of his work so that anyone who cared to could repeat it. He began,

[A]lthough it be a matter of daily observation that infinite numbers of worms are produced in dead bodies and decayed plants, I feel, I say, inclined to believe that these worms are all generated by insemination and that the putrefied matter in which they are found has no other office than that of serving as a place, or suitable nest, where animals deposit their eggs at the breeding season, and in which they also find nourishment. . . .
(*Experiments on the Generation of Insects*, 1688)

Next he was exact about his observations. He had some snakes killed and soon saw worms (maggots) appear on them, and then eventually disappear. To find out if they had merely escaped, he obtained more dead snakes with maggots, but this time put them in a box from which they tried to escape but could not. In nineteen days the captured worms turned into egg-shaped pupae, which he placed in closed glass vessels, and in eight more days, a fly broke out of each pupa. Thus, he concluded the worms were the young of the flies. But had the worms been generated by other flies or created spontaneously from the putrefaction? Redi next put meat into glass vessels, some with paper covers. He saw flies enter the open vessels and in these were generated worms. Flies tried to enter the closed vessels but could not, and these did not generate worms. He even tried placing dead flies on the meat, which also did not generate worms.

> Hence great Homer, in the nineteenth book of the Iliad, has good reason to
> say that Achilles feared lest the flies would breed worms in the wounds of
> dead Patrocles, whilst he was preparing to take vengeance on Hector.

In these simple experiments Redi won his disagreement without blood-
shed. He appealed not to pure reason or to authority to make his points, but
to *the ability of the imperfect human mind and senses to deal with observation meaningfully when they are used in a disciplined and systematic
manner*. The later experiments of Spallanzani and Pasteur were versions of
Redi's and each was effective in his time. In each generation, people were so
sure of the evidence before their eyes for spontaneous generation that they
needed to deal with the issue in their own terms. And each time rather similar experiments were effective in helping people to share Redi's (and Homer's) perception of those phenomena and to help them question the observations on which their beliefs in spontaneous generation had been based.

Experiments do not necessarily prove things in an absolute sense. But
they can help to resolve differences in opinion by pinpointing the critical
points of difference between individual perceptions. They provide a formal
structure for perception, one that can be criticized and modified when necessary, and repeated by others.

They can also give the inventor of the experiment the benefit of useful
public criticism. Unless one can make one's thinking public, and in such a
way that really constructive feedback can be obtained, it is possible to get
trapped into ideas that are perhaps very logical but that are actually nonsense, usually because there is some flaw in basic commonsense assumptions that one has not thought to question.

Hence, like many scientists, when I publish the results of an experiment I
eagerly wait for others to do research to confirm or reject mine. In either
case I would learn something valuable to aid my own efforts and help me to
measure my progress in developing usefully disiplined thinking patterns.

Unlike many scientists, I see the deepest value of scientific research not so
much in the generation of useful technology or even of knowledge itself.
Rather, the social value is in its potential to contribute to critical ways of
thinking for anyone. When hard facts, or great accomplishments in medicine
or engineering, are the final remarkable results of this rational process, then
we can recommend some tools that may help one to stay clear of very attractive but seductive nonsense.

A practical problem comes in explaining just what scientific thought has
been, and we have usually done a very bad job of that. In part this is a problem because scientific thought is not one thing that can be memorized, and
much teaching is memorization. Also there is marketplace psychology. Our
society values material goods and power and sees memorized *facts* as useful

goods and *skills* as power that someone will hire them to apply. The student becomes the buyer, the teacher the seller of goods.

Various scientific methods have been successful and many scientists will agree that science is a mixture of formal method and the intuition of the trained mind. Moreover, as one philosophy of science or another comes into vogue it is at least fashionable, and perhaps compulsory, for scientists to write as though they are following that respectable method of thought, when in fact they may have come to their insights and conclusions differently. But some scientists are relatively open about their mental processes, and one can learn to recognize the more open individuals and learn from them, and to keep faddish methodological claims in perspective.

One can even learn some valuable truths from philosophical fads in science, if one takes them with a grain of salt. Bacon's ideal of gathering information open-mindedly before forming hypotheses is only partly realistic, but it can nevertheless be a valuable part of science. Popper's ideal that science proceeds by the formulation of falsifiable hypotheses, conjectures, and refutations is only partly realistic or correct, but it can be a valuable part of science. Kuhn's notions of paradigm revolutions, that truth is only what a group of scientists perceive as truth at a given time, is only partially realistic but it too provides valuable insight into the scientific process and underscores the need for scientists and others to remain open-minded.

There is no substitute for each of us turning to *several* of the great advances in the history of science, such as Redi's experiments, and thinking about what in each of them represented progress over competing ideas of the time. How did each serve to communicate, convince, resolve, refine, or reject?

Scientists too will overlook demonstrably real physical phenomena because of their particular views. Too often scientists behave as though they automatically become objective when they put on a white lab coat, read a few words about psychological bias, pick up a few words about one scientific method or another or receive an advanced degree. But psychological bias is not so easily brought under control. Indeed, there is always the very real danger that one's certification as a scientist, or *any* professional, or a well-read student, or a religionist, or an experienced old hand, is in and of itself sufficient to lead to *hubris* and its consequences. Indeed, most of us are too ready and even eager to assume that it is *the other person*, instead of ourselves, who is guilty of psychological bias, and thus we dismiss arguments that threaten our beliefs without really giving them a fair hearing. If we did not have this dangerous habit, it would not be so difficult to keep our minds open.

A primary goal of science is to construct so-called objective views of reality. This does *not* mean that there is only one possible objective view of reality or that scientists are intrinsically objective people. It does *not* refer to the immutable Truth of Plato, as some scientists in effect seem to claim, and that philosophers once thought science might have found. It certainly does *not* mean that the natural or socioeconomic worlds can only be explained in theories that reduce to special cases of the theories of chemistry and physics. No serious thinker any longer believes that. These untenable *nothing but* claims (that biology is nothing but a branch of chemistry, for example) *do*, though, have *a lot* of political and economic power behind them and *are* part of the inner reality or even the self-serving "ideological reductionism" of many vocal scientists and technocrats.[2] The extreme reductionist may emphasize only simple physical, chemical, genetic or cellular units—may even in the extreme see the mind as *nothing but* something that could be predicted from studies on the firing of nerve cells, or brain chemistry. The *holist* or *integrationist*, on the other hand, may emphasize complexity, interrelationships, organizational patterns, levels of organization, and principles that may *incorporate* physicochemical perspectives and language. Good reductionists and holists can look at cats with equally objective yet different methods and perspectives: so too good chemists, physicists, anatomists, behaviorists, ecologists, historians, and anthropologists.

Science (and much of scholarship—I refer to science here more akin to critical thought than to technology) begins with concepts that are careful and verifiable statements about the physical or relational properties of things. And it attempts to build on these toward more difficult questions. Science is a powerful, necessary, and complex *process*, not a body of absolute truths. Its body of facts are of quite heterogeneous reliability. Some are very firmly based indeed; others are quite tenuous. There is the issue of selective knowledge—of which areas of the natural world get investigated most intensively. What things, in effect, are the powers-that-be willing to pay for to hope to know well? How does this influence the general outlook and message of science? Does it create socioeconomic conflicts of interest that influence the objectivity of science as an institutional enterprise and the objectivity of discourse within it? Much of humanity has pinned its hopes and dreams upon this great enterprise and has a right to know what it does well and what it does not and why. Without a deep understanding of the heterogeneity of scientific reliability it is much too easy to underestimate or to overestimate the meaning of scientific facts, or laws, or thinking, and what is meant by scientific objectivity.

Objective views of reality means views of the world that can be communicated through common, robust, tested, if constantly evolving, languages that are open to thoughtful criticism and strive to be free of whimsy and self-serving or superstitious biases. The ideas are in a form so that they can stand or fall on their own merits—they are not merely supported by someone's inner experiences or beliefs. They are, it is hoped, the best opinions given the available data. Objectivity is not simply "true," however, for what is deemed objective at any time can change along with the community of interpreters or with history itself.

One person's subjective view of reality is real in the sense that it does exist as a phenomenon, but it is not necessarily objective. It is not necessarily based on impersonal calculation, repeatable observation, measurement, experiment, or systematic comparisons in which the tradition of science or scholarship as a whole has no permanent vested, ego, or emotional interest. This is not to say that scientists and scholars have not made mistakes and are not sometimes petty and narrow-minded. There is good science and bad. Antiscientific and antischolarship groups are adept at pointing to the bad science and scholarship in their attacks. But on the whole, scientists and scholars do try to be open-minded and have eagerly slapped each other's hands when some have strayed too far from the ideals of science. They have been relatively true to their cause and this has been relatively effective. It is simply difficult.

Science works quite well at the level of simple discovery, despite its failure to allow every technological breakthrough that we might want, and despite areas of ideological metaphysics. One would be foolish to hire just anyone off the street to build a bridge simply because that person has a strong, albeit personally real, vision of how the bridge should be built. To do so would be to tempt disaster. A professional engineer's opinions about how to build a bridge will usually be more reliable than those of a person not trained to use scientific information. Similarly, we are usually better off with "objectively" thinking physicians, judges, and historians, and would be wise to prefer their services to those of someone with merely strong views. Science is far from perfect, but it has been good enough to eliminate smallpox and give us satellite communications.

Since it has proved powerful over the centuries, diverse aspects of scientific thinking have been adopted in the thought habits of scholars and laypeople as well. This has occurred in large part because *individuals* have found good rational skills *personally* useful. Many scholars may have studied from delight or from idealism, but many of the public have taken advantage of their critical methods of analysis and their findings for other reasons. *Technical* thinking skills are important to the socioeconomic mechanism, and to simple discovery, and that is the argument that we hear when tax dollars are being spent on education. But *critical* thinking skills are also important from the individual's point of view, if only in terms of his or her relative ability to compete and survive. Knowledge and the ability to use it skillfully are power.

This is why members of ambitious tribes in New Guinea will pay well to have missionaries educate their children, and this is why wealthy people in the West pay generously to put their children through good schools. It is not *only* a matter of networking, status, or having a college diploma. Millionaires could live without the prestige of prep school and Ivy League diplomas, but their fortunes and the control of their domains of power might not survive poor judgment. The point is certainly not that millionaires *do* have excellent thinking, or that if one attends an Ivy League college they *will* acquire good judgment. National and world economic, political, managerial, and environmental practices are a mess. The point is only that when people who are in serious economic and political competition, whether in the jungles of New Guinea or the financial centers of New York, see the value of keen thinking skills, they will want to give their children an opportunity to have the best education that they believe is available. And of course, discouraging the acquisition of critical thinking skills can help perpetuate competitive inequity. Which is why liberal educations were long reserved for freemen but denied to slaves, women, Amerindians, etc. (In theory, a ruling class should want managerial and working classes to have at best largely technical educations as the mercantilists early understood.)

Perhaps the bottom line is that with each of us living within a *persuasive* subjective world and finding our being in restricted habits of perceiving the external world, we very much need ways to verify our perceptions and thinking. If we are to advance as individuals or as a society we need ways to reach outside of ourselves, to share and test our own perceptions, and to identify and verify the intelligent perceptions of others. Scientific thinking and scholarship are some ways to do this. They may not have the intensity of a shared religious, artistic, or sexual experience. But science (as critical thought, not ideological reductionism) is perhaps the most usefully empowering invention of humankind and should not be rejected or misused because of misunderstanding. Properly understood, science can augment individuality and does not smother it.

Cultural Pluralism and Utopias

The reader who shares with me the belief that cultural diversity enriches the human condition may have sensed a tension in this essay. On the one hand there have been many marvelous cultural experiments in the possibilities for humanity. There is much to be respected. There is much that cultures can learn from each other. Yet, given the brains we have, it is also too easy for any culture to become locked into its own dreams and perspectives and to become complacent of its own shortcomings. An appreciation of cultural diversity can provide a healthy challenge to perceptual provincialism.

On the other hand, I have recommended the weeding out of superstition for many individuals. But if such advice were widely followed, it could have political implications, it could tend to erode cultural diversity. I will not resolve this tension here. I emphasize that this is one of the great issues of our time: We have too often seen aggressive cultures seize upon the superstitions of vulnerable cultures, shame them, and then sell them a new set of self-serving superstitions packaged as truth and progress.

In evolutionary biology, one understands that species are broken up into small local populations that can be very different from each other. Natural selection assembles within each population combinations of genes that are advantageous for local conditions. These traits will disperse between populations and form new combinations that sometimes offer new evolutionary possibilities. In other words, the diversity of populations and their interchange allows features to be tested in different combinations, which can lead to biological "creativity" and "progress." I propose that culture is a lot like this, too. The United States got its ideas of democracy not simply from the English. Some ideas, such as the plans for a workable democratic federation, were modified from the Amerindians; other ideas were modified from the history of Rome, Florence, Athens. The Athenians in turn drew together important ideas from Egypt, India, and other civilizations. So it is that cultural pluralism has been critically important in historical development.

Popular models of utopia though, the utopias of literature, philosophy, and religion, tend to envision a single good society, an ideal culture, an ideal city, garden, or farm. From Gilgamesh and Plato to the behaviorist communities of Skinner and the technological cities of the modern architectural visionaries, there has been an emphasis on a single community and its buildings, fields, and technology, on the political management of people, on a comfortable, stable, peaceful pace of life. Inspiration comes from within. There does not seem to have been much aggressive thinking about the richness of cultural pluralism, about utopias in which the best of the range of human experience, discovery, and sentiment can coexist peacefully to challenge perspectives and facilitate continued exploration of the human potential.

Given this impoverished history of dreams for the future, I have some concern that when one speaks of weeding out nonsense the arguments could be misused for the hegemony of some single vision of science, the destruction of cultural diversity, the dehumanization and degradation of people who do not share the values of the powerful.

I can only challenge the writers, artists, and political thinkers to produce more realistic models of utopia and of pluralistic political futures, models that explore more realistically the human potential. (Though let them be also reminded that people divided are more easy to conquer.) Cultures will in any event change; mass communications and economics make that inevitable. But if the world tendency is toward homogenization, this underscores a failure of imagination and a failure to grasp and deal constructively with the forces of change and cohesion. If there can be change to become impoverished or oppressed, there might instead be change to encourage inventiveness and diverse exploration of the potentials for well-being, while avoiding isolation into vulnerable, divided groups.

NOTES

1. A remarkable development since only about 1983 has been the discovery of long-lasting emotional damage to the adult children of alcoholics (and of other compulsive personalities) and the understanding of its nature and dynamics. Much of this damage has resulted from subconscious manipulation of the family's perceptual habits and feelings by the alcoholics and codependents to maintain their own denial system. The phenomenon was described by therapists such as J. G. Woititz, C. Black, H. L. Gravitz, and J. D. Bowden in their books, and highly effective programs of self-improvement rapidly developed and began to spread, largely by word of mouth, among the 30 million or more Americans so affected. This example again illustrates again the enormous scale on which denial and manipulation can take place without being recognized. It also illustrates that accurate, "objective" knowledge of a situation can be the breakthrough that leads to more realistic and healthier perspectives and relationships.

The books of psychoanalyst Alice Miller have recently been important in revealing other ways in which parents may subconsciously abuse children and manipulate their perspectives, the social consequences of that, and how that awareness can lead to improved diagnosis and treatment of damage that manifests itself after maturity. Her titles are telling in themselves, such as, *Thou Shalt Not Be Aware* and *For Your Own Good: Hidden Cruelty in Child-Rearing and the Roots of Violence.*

2. The "nothing-but" game that is too often taught to us as students would be *often* valuable, and otherwise merely harmless intellectual play, exploration, and speculation if it did not have fierce ideological partisan players among the powerful and vocal, and if it did not have policy and social implications for our society. This litany of the faithful becomes dangerous when it is used to *delegitimize the points of view and searches for knowledge of others*, in the competition for grants, prestige, and influence. It is too often used to teach young scientists that they have nothing to learn from other areas of knowledge. We need, though, *balanced* approaches to the difficult challenges before us.

A related position is that at the end of reductionistic analysis, one will have necessarily discovered the basic essence of a phenomenon. This aspect of reductionism is close to the *essentialism* of Greek metaphysics (chapter 6).

There *once were serious thinkers*, from Newtonians to the Vienna positivists, who had faith that all understanding would one day be reduced to a single point of view. For example, human behavior reduced to "nothing but" biology, biology reduced to nothing but chemistry, chemistry reduced to nothing but physics. It just has not worked. The philosopher Karl Popper, for example, cogently argues in his book *The Open Universe* that the faith is unlikely ever to work. Einstein and Infeld (1966) describe that for centuries physicists held the faith that *all* of physics could be reduced to nothing but Newtonian mechanics. But it simply was not possible.

Most scientists today find it valuable to keep *some form* of reductionism in their tool kit. "Only by . . . shuttling back and forth between the worm's eye view of detail and the bird's eye view of the total scenery of science can the scientist gain and retain a sense of perspective and proportion," as Paul Weiss put it (Koestler and Smythies 1969). For me physics and chemistry are part of that rich world of detail. I am, in significant part, an enthusiastic *constitutive* and *methodological* reductionist. A good synthesist must be. The analytical traditions of *methodological* reductionism do work very well to generate that necessary detail at the level of *constitutive* materials and their organization. Its language can be extremely useful, too. (Though to *describe* something in the language of chemistry is *not* to have *explained* it or *predicted* it. Many vocal reductionists confuse reductionism with materialism.) But good synthesis, whatever language it uses, must also draw upon concepts from history, upon principles of the organization of higher level systems, and upon concepts from comparisons of several points of view. (See also Ayala and Dobzhansky 1974.)

After several hundred pages of defending reductionism against extreme critics (some philosophers and religionists) of its application to mind/brain issues, for example, Churchland (1988) can in fact conclude merely that "it is most unlikely that we can devise an adequate theory of the mind-brain without knowing in great detail about the structure and organization of nervous systems." The synthesist was prepared to agree at the outset with such extremely modest claims. One must underscore that there is no reason to believe that an adequate theory of the mind could emerge from a reductionist neurophysiological study of the brain *alone*. Thought may surely be a physical process, but it may be *about* nonphysical things (from rational numbers to centaurs and essences) and about physical systems that are directed by principles that emerge from higher levels of organization than mere physics or chemistry. These may have their own logic that can determine outputs that can in turn feed back and influence physical processes. We have *already* learned an enormous essential amount from anthropology, cognitive psychology, communication sciences, etc. The extreme reductionist may claim that the major contribution of my book has in fact been exactly to reduce philosophy and politics to neurophysiology. But close reflection would show that this is only superficially so. My thesis involves a synthesis, not merely a reduction.

Neurophysiology will help to *better understand* religious self-mortification of the flesh, for example. But it would be stretching the matter to claim that neurophysiology completely explains this or might one day have been able to completely predict the diverse patterns and manifestations of religious self-mortification.

In most scientific disciplines there is a recognition of *levels of complexity* in nature. At levels of higher complexity there may be *emergent properties*. Emergent properties may or may not require knowledge from the levels of less complexity, but they would not be *predicted solely* from knowledge at the lower levels.

Take for example what the evolutionary biologist calls the Darwinian *fitness* of an organism, which is a measure of its ability to survive and reproduce. The chemistry of the gene contributes to the fitness value, and fitness depends very much on DNA structure, but fitness is in no way merely a chemical property. It is a property of the product of a *system*, of interacting genes that help shape an organism and that in turn interacts with a given complex environment. Neither the gene nor the organism has any property that can be called Darwinian fitness, except in complex interactions that cannot be fully explained except to include organizational patterns with even historical legacies that would not have been predicted from physicochemical knowledge alone.

Similarly, *value* (purchasing power) describes a property of money that cannot be *predicted* strictly from the physicochemistry of coins and bills. Specific values at a given moment cannot even be *explained* (not merely *described*, much as one might use mathematical language to describe almost anything, however well) by physicochemistry. One would need concepts from history, psychology, game theory, politics, and even philosophy, among others. We were taught as students to continue to argue: that the issue is "nothing but" one of complexity and that game-theoretical, historical and psychological concepts will one day be reduced to physico-chemical systems, and then it will be clear that monetary value is nothing but a special case of physicochemical laws.

Well, sure. Maybe. But as a practical matter such claims eventually prove to be more metaphysics and faith than hard science. They can also become semantic gambits of the nothing-but game or chest thumping. No physics or chemistry text has chapters on history, economics, or politics; and typically the sciences do not even encourage their students to gain a solid knowledge of these subjects.

In *Beyond Reductionism*, edited by Arthur Koestler and J. R. Smythies (1969), Viktor Frankl points out that differences in perspectives can enrich knowledge, much as, in stereoscopic vision, two angles of sight allow us to see three dimensions instead of only flatness. In this sense perhaps specialization (championed by many reductionists) is not inherently bad, but only the habit of using one point of view. "What we have to deplore therefore is not so much that *scientists are specializing*, but rather that *specialists are generalizing*."

The integrationist will develop synthetic concepts that *incorporate* the knowledge and language generated by reductionism of the physicochemical constituents of a system. But the staunch reductionists will then falsely claim *success* for extreme ontological and epistemological beliefs; will claim that higher concepts *have thus been reduced* to special cases of the laws of chemistry and physics and were predicted or would have been predicted by them. *They confuse their reductionism for materialism.* On the contrary, general system theory and cybernetics, for example, were developed to deal with the fact that the principles of system organization are independent of any one particular physicochemical substrate. We can have a regulated system made out of metals, fiberoptics, or living tissue. Indeed, von Bertalanffy began as an embryologist and invented general system theory because reductionism in his field was providing interesting information, but no adequate conceptual framework for understanding the process of the organization of cells into differentiated organ systems. Wiener and others invented cybernetic (control) theory because it was clear that specific patterns of organization, and communication, are not specific physicochemical properties that arise merely as one adds more and diverse chemicals to a mixture. To fully understand the most interesting patterns, we needed new concepts. One would also need historical, ecological, or sociological knowledge, for example. How did the patterns become organized, how are they maintained, what are their functions at higher levels of ecological or sociological integration?

The scientist who has faith that all phenomena can be *explained* and *predicted* (not merely *described*) completely by the physics and chemistry of elementary forces or particles can be called a *philosophical explanatory reductionist*, in contrast to the *constitutive* and *methodological reductionist*. For example, a philosophical explanatory reductionist may argue that male and female behavior *will* one day be completely explained in terms of and predicted entirely by the chemical activity of hormones, while a constitutive/methodological reductionist would argue only that there may be *some great utility* in understanding hormone physiology for understanding gender behaviors. (Birke [1986] has an interesting chapter on feminism and reductionism.)

G. G. Simpson used the metaphor of a television set in his essay, "The Crisis in Biology" (1967). One could not predict the circuitry of the television set from the physics of transistors, though one may need to know this last *to fully understand* and describe the former (necessary but not sufficient). We can take this metaphor further. The content of the evening news will in large part determine the instantaneous active state of the transistor, and not merely its physics or even the circuitry. And one cannot entirely predict the content of the evening news from physics or chemistry. In most cases economics, politics, history, or psychology will be better predictors.

The extreme philosophical reductionist has faith that eventually the laws of physics and chemistry would *predict* all these things and that there is in principle no knowledge unique to more complex levels of organization. They see other attitudes in science as being "soft" or even "unscientific." They reduce their critics too, and lump them all together as, "nothing but" vitalists or mystics. But on close analysis their belief is counter to evidence and is probably untestable; it is merely a metaphysic claiming legitimacy by association with the accomplishments and linguistic utility of constitutive/methodological reductionism in partnership with synthesis.

An extreme form of philosophical reductionism is an ideology for many. Of course it (the nothing-but game, if one likes) *does work* to motivate and regulate idealistic young scientists. It is easy to slip uncritically into believing that ideas, however strange, must be True if they seem to work much of the time. It does work for fund raising by its partisans; it works quite *often* to see through complexities to generalities, especially in partnership with synthesis; it works to generate interesting data; and it works to generate useful data that can aid technology and industry. It can be a valuable professional and social *survival skill* for certain types of individuals in and near science. But it has aggressive partisans who claim that this is what science *is* and who have managed to sell their metaphysics widely. (See also Fuerst 1982.)

I use *ideology* much in the sense of Mannheim (1936). *Ideological reductionism* may promote as *truth* (or even Truth) a self-serving *faith* that one's particular reductionism should be given hegemony over other forms of thought. It may claim to be *the* objective point of view that will make a better world if the careers of its devotees and practitioners are favored and other groups of thinkers are discounted and perhaps even unemployed. Physics and chemistry have been extremely valuable to medicine, industry, and the military, and their "ideological reductionists" have powerful allies in international power structures despite the intellectual poverty of the metaphysics. Ideological reductionism survives because it works for a set of practitioners and it works for a powerful sector of society. Of course science and technology are value neutral and do not inevitably create progress (e.g., chapter 6). They have down sides as well as up sides.

There can be proper or extreme nonphysicochemical reductionisms. On the proper side, it can be valuable to reduce things to their patterns of *organization*, which can be quite independent of the particular physical or chemical materials and their special properties (e.g., Wiener 1954). *Structuralism* in the social sciences is a relative of systems theory and cybernetics that can, though, become extreme and co-opted by the nothing-but habit of mind. Sociobiology and economic determinism can also become ideological reductionism, though they do not claim to be physicochemical concepts. This point will be developed in later chapters.

Many scholars who are critical of "science" have *intellectual* concerns that actually apply to *philosophical explanatory reductionism* or to *ideological reductionism*, especially their "nothing-but-ness," but they are confused, because they have bought the preposterous claim that reductionism *is* science or *is* materialism. They may also have *social* criticisms of the cozy relationships that many scientists have with business and the military. Then when they see scientists with such powerful ties making an intellectually questionable claim to superior or even ultimate objectivity, it is no wonder that they become critical of science as a whole. Yet in theory, science can belong to everyone. Other aspects of this issue will be discussed in later chapters.

There are other ways to look at and divide reductionism than I have done here. Mayr (1982) distinguishes *constitutive, explanatory,* and *theory reductionism* for his purposes. He notes that there is no dispute among postvitalist scientists that the composition of all things will reduce to physicochemical systems (*constitutive reductionism*). This is simply materialism. It is false, though, that one can only understand a whole *in terms of* the molecular processes at the lowest level of organization (Mayr's *explanatory reductionism*, not exactly mine). One can understand the mechanics of joints without knowing the chemistry of the bones and cartilage.

Mayr cogently disputes that higher level theories can be reduced to special cases of the theories of more simple levels of organization (*theory reductionism*). Advocates confuse *processes* and *concepts*. Meiosis, gastrulation, selection, and predation involve physicochemical *processes*, but these are not *concepts* that emerge from chemistry or physics.

It has been wryly said that reductionism is *nothing but* the habit of claiming that something is *nothing but* something else. This charge would apply primarily to philosophical explanatory reductionism in my sense, or theory reductionism in Mayr's sense.

Mayr does not discuss the ideological aspects of reductionism. I emphasize the fact that reductionism can become *self-serving* ideology for networks of scientists. Marxists may argue that reductionism becomes a tool of the capitalist system and state ideology (e.g., Rose and Rose 1980, Graham 1987). (But to *what extent* can "the System" manipulate to its own self-interest the self-interest tendencies of the science establishment or of each individual scientist?) So this can be a larger subject for discussion and debate than may be at once obvious.

There is a close kinship between extreme forms of reductionism and *scientism*. Scientism is the attitude that the only valid knowledge or point of view is *scientific* knowledge or perspective. This can refer to cosmological knowledge, personal knowledge, or human values. Thus the insights of the poet, the midwife, the painter, the victim, the adventurer, and even the historian could be discounted *at the outset*, it is argued, simply because they were not gained by the techniques of, and phrased in the vocabulary of, the current vogues in science. Thus, according to scientism, a poor mother can *not* know valid things about starvation that a nutritionist does not know. Of course science has made many splendid contributions and folk knowledge has quite often been nonsense. One is better off to go to a hospital than to a shaman for curable cancer. But beyond selected examples, extreme reductionists may attempt to reduce the situation to a partisan matter of professionals who have knowledge and nonprofessionals who cannot possibly have valid views. In scientism's claims, a medical doctor should know more about any health matter than say, any midwife or massage therapist can. But this cannot be strictly so. It is obvious that an insurance broker who is a recovering alcoholic in a neighborhood AA program may know more about alcoholism and recovery in important and useful ways than a neurologist does. Clearly there are many valid forms of knowledge beyond those generated and controlled by the scientific establishment.

The even deeper issue is that *human values* not derived by the scientific establishment may be trivialized by scientism. Yet science is not even organized to establish solid values. So someone concerned that scientism can thus in the long run *only devalue* human life.

CHAPTER 5

Fragile Common Sense

"How many times have we lost?" she inquired — actually grinding her teeth in her excitement.

"We have lost 144 ten-gulden pieces," I replied. "I tell you, Madame, that zero may not turn up until nightfall." . . .

By the fifth round . . . the Grandmother was weary of the scheme.

"To the devil with that zero!" she exclaimed. Stake four thousand gulden upon the red." . . .

"Zero!" cried the croupier.

At first the old lady failed to understand the situation; but as soon as she saw the croupier raking in her four thousand gulden, together with everything else that happened to be lying on the table, and recognised that the zero which had been so long turning up, and on which we had lost nearly two hundred ten-gulden pieces, had at length, as though of set purpose, made a sudden reappearance — why, the poor old lady fell to cursing it, and to throwing herself about, and wailing and gesticulating at the company at large. Indeed, some people in our vicinity actually burst out laughing. "To think that that accursed zero should have turned up now!" she sobbed. "The accursed, accursed thing! And it is all your fault," she added, rounding upon me in a frenzy. "It was you who persuaded me to cease staking upon it."

"But, Madame, I only explained the game to you."

Fyodor Dostoyevsky, *The Gambler*

What Is Common Sense?

What is common sense? We usually define it as good, sound, practical judgment. Two and two makes four; dark clouds bring rain; a stitch in times saves nine; it doesn't take a weatherman to know which way the wind blows — and that sort of thing. We usually assume that each of us has it if we will only use it, and that it is then a very reliable tool. Parents teach their children to learn and use common sense in order to avoid serious mistakes in life — to learn to see and believe in the causal relationships between events and the values that successful people in the community have and believe in. Surveys show that citizens almost above all else want their leaders to have good common sense and to speak so that they can be easily understood.

If pressed, though, most would agree that really good common sense is not so easy to come by or even to define: possibly, "Common sense is as rare as genius," as Emerson put it. In fact, common sense seems to be an ideal that is seldom attained. Most of us can figure out that we may get burned if we play with fire, but beyond that sort of thing good common sense is hard to come by, and thus several score of proverbial suckers are born each minute.

A fool is a person whose common sense is a disadvantage. Magicians, con artists, unscrupulous political and religious leaders, and aggressively pragmatic businesspeople are among those who exploit the weakness in others' common sense to their own advantage. It has been said that the only difference between a fool and a wise person is that the fool will advertise having been duped. Be that as it may, most of us, perhaps all of us, believe implicitly that we have common sense, even though it is obvious that it often fails to serve us, and even though it is difficult to define. Is the power of common sense only a myth?

As Einstein put it, "Common sense is the collection of prejudices acquired by age eighteen." Common sense suggests that a ball keeps moving once it has been thrown because some force moves it (wrong). The earth sits still while the sun moves (wrong). Organisms are perfectly adapted to niches in nature (wrong). One cannot break boards with a blow of the bare hand, or walk across hot coals barefooted (wrong). People who talk fast are necessarily insightful thinkers (wrong).

Common sense is not entirely a myth. Under the right circumstances it may serve well enough. It can propel people to wealth and power. Most of our leaders and managers are people of good common sense who figured out the overt or covert rules of the local game and played it well. But it is fragile. This may be part of the reason why cultures often crumble and cannot adjust when changes come too fast. Common sense is mostly effective relative to locally usual circumstances. But it is real, and differences in commonsense beliefs and reason even allow us to distinguish cultures and subcultures. Thereby the members of a culture, class, or subculture are seen to have some common sets of ideas, experiences, reasonings, and rationalizations that work well for them in the lives that they have made for themselves, and that set their perceptions of life off from those of other groups.

Common Sense Is Culturally Shaped

I once did fieldwork among the Seri Indians of Mexico. As do some other hunting and gathering people, the Seri have what *we* call "speaking taboos," where fathers, sons, and brothers never speak with one another, once adult. They may live and hunt together, but they do not speak to one another. As far as I could learn, there is no fear that they will be somehow punished or sent to hell if they do speak, or that discord will set in, and *they* do not call it a taboo. It is simply good common sense to them not to do it or to expect each other to do it—dogs do not fly, men do not speak to their fathers. When questioned as to why the men do not speak (*why* questions are difficult to ask members of some other cultures), each will smile as though the question is silly, ponder the matter, and then try to guess at an answer, but the bottom line is that it is simply obvious that one *does not* speak with his father or brothers. We leave our bushes unguarded in our yards. It is common sense that no one will steal them. People usually don't do that. How would we react if asked why we don't steal bushes or expect people to steal ours? We don't fret that someone will serve us worms for dinner, because common sense tells us it would be foolish to expect that.

The Seri have always seen male family who speak Seri avoid talking with one another. Thus they were shocked to hear one of my Seri-speaking Anglo-American coworkers speak with his visiting father. They did not indicate that they thought he would be harmed or punished; it was simply extraordinary that something was happening that common sense said should not happen—like water flowing uphill.

When we see others follow alien rules, our impulse may be to say they are superstitious, stupid, or some-such. But they may in fact be very intelligent and merely be using good conventional wisdom much as we do.

Sometimes Westerners may similarly look mindless to foreigners, for example when they see us sacrifice ourselves for ideologies that may seem a bit arbitrary, when they see us eagerly jump to fads and mass advertising, or when they see us display little interest, understanding, or open-mindedness toward other cultures. Many of our food, sexual, and dress taboos and other superstitious customs do not flatter us in foreign eyes. We can often claim an advantage in that many of our silliest beliefs have been weeded out over the last few hundred years by science. Many of us have learned how to be especially critical of commonsense beliefs and customs and to weed out the worst. Still, we have quite a few erroneous commonsense ideas left in our society, so we should not congratulate ourselves and feel too superior to people in other cultures.

In talking with patriotic visiting scientists from the Soviet Union once, someone raised the subject of the persecution of dissidents, who were sent to psychiatric hospitals for treatment. I was corrected, told that there *is no persecution*, but that it is obvious that if these dissidents cannot see that the state is working towards a good end, then there is something wrong with their minds and they do need professional help. The Soviets thought we were concerned because obviously we had fallen for capitalistic propaganda. Similarly, they would call some of our imprisoned people who did not respect capitalist conventions of property rights, or draft resisters, etc., persecuted dissidents, while most of our citizens would see these as obvious criminals or crazies and would see the communist labels as obvious propaganda. Even if both their views and ours were biased by propaganda, the fact would remain that what was obvious and commonsense to party-line Soviets and to Americans was very different. And, what is obvious to the Seri is of a different nature yet. Then again, the world that is obvious to one scientist or another may be completely baffling to the average person.

The morning of this writing, as I walked over the Hennepin Avenue bridge to my office, some high school boys subdued one of their classmates and struggled to throw him over the railing and into the Mississippi River. He resisted with considerable energy but was clearly outnumbered.

Within a second I had decided that they were probably horsing around and would not harm him, and I moved on before the outcome was clear. But in retrospect, how could I and others nearby have been so certain and relaxed about the situation? There was nothing definitive in the straining boys' faces or voices that I could see. The only meaningful clue was context. Teenagers in my Minneapolis neighborhood are much more apt to be horsing around than killing each other. Reasonable common sense told me that the matter was innocent. A visitor from a much rougher city might have feared the incident was sinister, and he or she would also have been using reasonable common sense. Either commonsense view could be justified in the above case, but the only way to resolve the difference would have been to investigate the situation carefully and wait to see how things developed.

Commonsense responses and opinions may only be useful in particular social and physical environments. That much is obvious, and this is one reason why foreign situations make us uncomfortable. Our common sense may fail us and the new situation may seem illogical, silly, ugly, or even hostile. If

I had recently moved here from the Bronx, perhaps I would have been much more apprehensive about the struggle of the youths on the bridge. This could have created moral, ethical, and practical dilemmas for me: Should I interfere? Should I risk my own security or else leave, and cope with what could be called my cowardice? Social change, or attempts at normal living in a foreign culture, can be very threatening to some people, since commonsense beliefs, reasoning, and values may become tested and strained.

It might be one's first thought that what is good common sense and obvious is based only on ordinary workday experiences of relationships between events. But experience usually involves beliefs and customs. In fact, it may be impossible to have experiences that do not involve custom, belief, or imitation of others. We may experience a certain view of the city because we tend to conform to the customs of our peers in shopping in certain types of stores, and in keeping certain hours. A person who is a night owl may physically encounter and mentally note very different slices of life than a day-active person does. A policeman may experience a neighborhood differently from a social worker, and so on. Custom and belief can shape the actual nature of our physical encounters with others and the world, as well as what and how we see in these encounters, and then again they can shape how we emotionally or cognitively interpret each experience.

There is no reason why we should expect common sense to yield the same conclusions from persons of different backgrounds or experience. One's commonsense outlook is to a very large extent an accident of birth and the particular upbringing and way of life that has resulted. Some in different political parties may have different philosophies *because* their experience of the world has been very different. I recall a wealthy friend with a Princeton degree, having just flown back from a social lunch with (then Governor) Ronald Reagan in Palm Springs, sincerely announcing that there was no excuse for any unemployment in America. *He* could walk out and find a good job tomorrow and so could *any* jobless minority, he boasted. He did concede though, that he would have the advantage of various social connections. But "everyone knows someone, he insisted." Well, ... yes. At another extreme, I have known streetwise students who argue that nothing taught in college can be helpful in "real life." Well, ... read on.

When I read anthropological ethnographies as a younger man, I was left with the impression that people in other cultures are rather strange, that they may have little consciousness, or that they react emotionally in ways that are very different from most of the people I was used to. I gained the impression that so-called primitive peoples follow traditions and conform to stereotypes much as automatons or robots follow computer programs, more so than you or I do. One of the surprises and delights of travel has been to experience the fact that, for the most part, people with very different customs, manners, appearances, and tastes are nevertheless easy to relate to if one is open-minded. This is not to say that differences in cultural values, customs, and history are insignificant. Certainly they can lead to serious problems in professional or personal relationships. Yet, once we begin to grasp others' points of view, or they ours, it becomes clear that the differences in values and beliefs that often form barriers to congenial relationships tend to obscure the fact that we are mostly cut from the same cloth. The discussion in chapter 4 should be kept in this perspective. Fortunately, I have never encountered a culture at home or abroad where people are incapable of human warmth, humor, friendship, and that does not have bright, inquisitive, and perceptive persons with whom I can have interesting talks.

It is disappointing that little of this individuality comes across in anthropological writings. Perhaps some workers take it for granted. But others ignore individuality because it is nearly impossible to study scientifically on a cross-cultural basis. And, some individual anthropologists actually do relate to non-Western peoples as though the latter are cultural inferiors, and so go into their studies with intellectual blinders on. In any event it is difficult to glean adequate impressions of individuality, subjectivity, or intelligence from the anthropological literature, as several anthropologists have also complained (Bateson 1958; Herdt 1981; Lang, Ogan, Sarles, and others, personal communications).

When we see other peoples draw conclusions that we believe are false, we may tend to laugh at them and assume that they are not thinking very well. My experience with non-Westernized peoples has taught me that they may indeed be thinking clearly but are using assumptions that are unfamiliar to us. Moreover, when they do test or challenge their beliefs, and they do, they may do this in their own terms, not ours.

Masculinization through Semen Transfer

The sexual practices of one Melanesian people involves reasoning that is reasonable and acceptable enough as common sense, yet it leads to conclusions that are probably completely wrong. The published studies are adequate to allow us to trace the logic.

In most human societies there is a tension and suspicion between the sexes, on which one could speculate and expound endlessly. Even in our "civilized" culture, one might expect at worst to hear that:

women are not men's equals in anything except responsibility. We are not their inferiors, either, or even their superiors. We are quite simply different races. (Phyllis McGinley)

Yet, more extreme sentiments are not uncommon.

Marriage always demands the greatest understanding of the art of insincerity possible between two human beings. (Vicki Baum)

Disguise over bondage as we will,
'Tis woman, woman rules us still. (Thomas Moore)

Men are monopolists
of "stars, garters, buttons
and other shining baubles"—
unfit to be the guardians
of another person's happiness. (Marianne Moore)

The allurement that women hold out to men is precisely the allurement that Cape Hatteras holds out to sailors: they are enormously dangerous and hence enormously fascinating. (H. L. Menken)

This often unspoken fear-attraction of the opposite sex is handled in different ways in different cultures, and related customs appear to be a blend of resolutions to this as well as to other factors in the social, economic, and biological environment. Often there is essentially a competition between the sexes, with male powers and rites promoted and emphasized to appear to match the enormous power of females to give birth and raise and train children—thereby to alter the family and village economy and the directions of behavioral and emotional energies in the community. The significance of these customs may have been lost historically to the people themselves, just as we have forgotten the origin of many of the words in our own language. The behaviors that constitute the status quo are then performed because it seems *obvious* that one should behave in such-and-such a manner; such responses and thinking become a matter of common sense. However, they may be rationalized after the fact in terms of the local belief system.

In several parts of the world there have been, and still are, cultures that embrace homosexuality for typical, normally reproducing individuals. In some cultures it is expected of the men only, in others of both males and females. It may take many forms. The Sambia of New Guinea are of special interest to us in the context of this discussion. Their commonsense ideas about gender development are almost exactly the opposite of our own (Herdt 1981). This contrast is startling enough that it will allow me to illustrate several points.

Sambian men are fierce warriors who are deeply concerned about their power relative to female power—as in many societies. And they are concerned about giving females too much of their masculine strength and thus losing effectiveness as husbands, fathers, and warriors. "Ritual" homosexuality among the Sambia has the stated function in their culture of making a scrawny boy into a fierce, brave, muscular warrior and fertile, devoted husband and father. They also enjoy it.

Yet many people in my own culture might argue that ten years of enthusiastic homosexual intercourse would instead make a young man effeminate, cowardly, and reproductively incompetent. Moreover, many people might suppose that agreeably switching thereafter to heterosexual preferences might not typically be possible. Clearly, both assumptions are false.

How do the Sambia reason?

1. A man becomes weaker right after sexual intercourse, so he loses a bit of strength and power temporarily.
2. Body fluids have power and can cause growth—as milk does.
3. There is strength and power in semen that can be passed on to others, since it obviously causes birth.
4. Males mature more slowly than females, so obviously male maturation is more difficult and needs help.
5. The strength that men lose in the ejaculation of their semen may be passed on to young males to help them develop masculinity quickly. This sequence of events is always seen.

Around these simple notions and observations is a profoundly complex set of beliefs and rituals. Sambian homosexuality may serve a number of social functions (from a Western scientific point of view—in terms of what it produces materially and not merely aesthetically), from population control and a reinforcement of masculine loyalties among warriors, to a reduction of male aggression.[1] But discussion of these is not most critical to the point at hand, which has to do only with their common reasoning. I discuss only portions of this large subject here.

All nearly pubescent males are moved from their mothers (men and women live apart, as in many cultures) to the men's area to begin their masculization. Included in a series of initiations it is revealed to them that if they are to become men with strong bodies, courage, and pride; to become powerful warriors; to learn to manage a family and father children; then they should suck on the penises of young bachelors and childless young married men and eat as much semen as possible. They must strive to mature quickly or their promised brides may lose interest in them.

So the youths serve first as fellators and then as fellatees for some ten or more years, during which they are also learning the other essential aspects of manhood, including warfare. In some other cultures semen instead is smeared on the youths' bodies or rubbed into wounds to impart its power.

Interestingly, after their wives become fertile most Sambian men give up homosexuality.[2] For the minority who do not, it is thought odd, though non-judgmentally tolerated as individuality. Then, as in nearly all of Melanesia, there are postpartum sex taboos during which the men abstain from sex altogether for some two to five years after a wife has given birth. Alternative sexual outlets (as we might think would be necessary) are not usually sought during this period unless the man has several wives—chastity for men and women is the custom. This, as some Melanesian men have told me, is not difficult because it is simply a matter of "having good character."

Of course, from a cross-cultural perspective the matter is not at all so simple. Melanesians are not usually under the constant social pressure that we are to be interested in copulation and romance. Eroticism and talk of it have their proper time and place. They are not constantly exposed to romantic and erotic literature, television, movies, kidding, gossip, and our sexy mass advertising industry.

When they do become overly exposed to these influences or when missionaries shame them into giving up their postpartum sex taboos, as E. Ogan and others have studied, the reproductive rate may soar too high and cultural deterioration may follow the demographic changes—alcoholism, juvenile delinquency, a breakdown in organization.

So without the constant stimulation that we give each other, a couple of years of chastity might be much easier for a Melanesian of good character than for an American of good character. I do not wish to leave an exaggerated impression of the situation, though; there is a good deal of individual variability, and rape and adultery *are* known in Melanesia, as in our culture (at much lower rates). Still, the social norms do differ from ours to a degree that challenges our commonsense ideas of the fixity of sex drive and gender preference and that cannot be brushed aside.

Let us return to the logic involved in beliefs of a very personal nature. The Sambian custom of ritual semen eating is completely reasonable *from a commonsense point of view*. They always see young men mature as an apparent result of sucking penises and swallowing semen—cause and effect seem clear—and the phenomenon also fits with a myriad of their other observations and beliefs.

I do not *believe* that the Sambian practice is at all necessary for the development of muscularity and fierceness, but it might be effectively impossible to prove it to them. Without being able to experiment, it might be impossible to prove as a hard scientific *fact* even to any devil's advocates among my own compatriots. Semen was never part of my own youthful diet. But despite the fact that my draft board was quite satisfied with my body, the Sambia could still argue that I could not function as a powerful warrior in *their* world. They might argue that I could have done better. Or, perhaps I am some sort of exception to the rule, and so on.

Their beliefs also cannot be supported by current scientific knowledge. But it would be difficult to convince them that their common sense is wrong and that my judgments are right. They may not understand or respect scientific thinking, just as many people in our culture do not. And that would make my scientific evaluation no better than a personal opinion in their eyes. If I were to tell the Sambia that such customs are not practiced in my culture and yet youths do mature, they have a variety of possible responses available to argue back with me and it would be difficult to defeat them in debate.

1. Perhaps we whites (and other Melanesian peoples) actually do practice such homosexual acts in secret and must conceal the fact by our customs. They could certainly muster some evidence for this.
2. Perhaps only our most muscular and manly warriors have practiced the secret custom. We could not say no for sure.
3. Perhaps we have found some substitute for semen eating (so the relationship still holds for them at least). The breakfast of champions?

Commonsense ideas may be wrong, but then *they are not simply neutral mistakes.* Often they are based on fairly good reasoning within the context of a given world view. People everywhere tend to defend such ideas through intellectual arguments or psychological rationalizations. This is an important point that is too often overlooked.

One might naively assume that truth should have an immediate appeal. No one likes lies or claims to believe in them. We are all *eager* to learn truth. However history shows that often the opposite is true. If I have chosen my example well, then even as they read this many readers will be trying to find reasons why the Sambian experiences do not really challenge some of our own flimsy commonsense beliefs about human nature, sex drive, gender preference, and stability. Perhaps they typically only pretend to enjoy such sex when young. Or, perhaps they typically cheat and do not observe the postpartum sex taboos later in life. The Sambia, for their part, may do much the same about our customs.

Common Sense May Dismiss Evidence

History shows that incorrect commonsense ideas are notoriously difficult to shake. It took hundreds of years for most people to accept that the sun does not move, or that putrefaction does not generate life. My own experiences as a scientist working in the field, particularly with rural people or people who are not keen on scientific thinking, has been that they are remarkably resistant to facts that should challenge their commonsense beliefs.

When I have insisted, for example, that a certain animal is not poisonous, to locals in the United States or elsewhere who strongly believe that it is deadly, words will not convince them. I have often even let the lizard or other animal bite me to demonstrate its harmlessness. But the locals may have suspected a trick or else quickly reasoned that I am immune. When a brave local youngster has trusted me and let the animal bite him and does not become ill or dead, then the people have insisted that I have passed on some power temporarily to him (or that it is a trick). One cannot shake their beliefs so easily. Sometimes this is because commonsense beliefs are a part of a larger world view, and such views themselves are relatively stable because they have arguments built into them that resist challenges to their logic. In any event, those were thought-provoking incidents for an idealistic young scientist. On reflection, though, they were very similar in nature to forms of intellectual conservativism in my own society, and even in my own profession. This resistance to new ideas retards the progress of science and scholarship and retards the ambitions of our civilization to grasp a more accurate image of reality. Each of us, like Oedipus, is vulnerable to it. What exactly is this mentality? We shall touch on this question again and again.

The biologist Alfred R. Wallace in 1870 accepted a now-famous bet with a devoted believer in a flat earth; Wallace could not prove that the earth was spherical. Wallace, accustomed to dealing with reasonable people and certain that he could prove his point, to his surprise got nowhere. The flat-earther refused to look through his instruments, refused even to accept the premise of the experiment. Finally the bet-holder did award the bet to Wallace, but the flat-earther was infuriated and he libeled and actually threatened Wallace, and even spent some time in jail (three times) as a result (Schadewald 1981). This example is amusing but not surprising. There are even today many flat-earthers in the United States who cling to their senses and intuitions and do not accept orbiting satellites and space photographs as evidence of a spherical earth. The governments are lying to us again, they charge. *No evidence* can shake a completely closed mind. Most of us are not guiltless, but are somewhere between open-and closed-minded.

One is reminded of the anecdote about the man who is convinced that he is dead (Vogt and Hyman 1959). A psychiatrist argues with him but cannot shake his basic assumption. Then the psychiatrist thinks of a fine new tactic.

PSYCHIATRIST: Tell me, do dead men bleed?

PATIENT (after carefully considering the question): No, dead men do not bleed.

PSYCHIATRIST (takes a pin and pricks the patient's finger, whereupon blood gushes forth): Well, now what have you to say for yourself?

PATIENT (contemplating the bleeding digit): Well, by golly! I guess dead men do bleed after all!

The implication is not necessarily that closed-minded people are organically damaged. But they have programmed themselves so that if they are out of step with reality, there may be no way to penetrate their reasoning or, indeed, for them to exit voluntarily from their thought systems. Since the precise nature of reality has proven to be slow to locate, it is extremely dangerous to fall into a closed-minded mentality; one may drift or be pushed into thinking that is ever farther from reality as new facts are confronted. The example of the Sambia might lead most readers to be more cautious about commonsense wisdom concerning sexuality. But I am painfully aware that a closed-minded person may only be pushed further into an elaborate belief system. For example, one could think, "My golly, the Devil has gone to such great lengths to confuse us that he has even created this elaborate system of beliefs and set of immoral customs for us decent people to find while we were looking for gold and riches high in the mountains on the other side of the world! He is capable of such devious strategies. What next?"

There was an article in the newspaper about a "multimillion-dollar nationwide" lobby of a group organized to promote their belief that the Nazi death camps and the Jewish Holocaust are a myth. As in the case of the flat-earther above, this group offered a reward ($50,000) to anyone who could prove that even a single Jew was murdered in Nazi gas chambers. A Mr. Mermelstein, who lost a mother and two sisters at Auschwitz, submitted evidence, but the group refused to pay and so Mr. Mermelstein sued them. He won. The lobby group protested. Does this sound familiar? If the group takes the position that concentration camp deaths were all made up by Jews in a conspiracy including Joe Stalin, Winston Churchill, Franklin Roosevelt, Dwight Eisenhower, and on down to my insignificant refugee neighbors, then it may indeed be impossible to prove any murders to them. They may claim that the news films were faked, the Nazis' own tons of records were faked, and all the witnesses have been lying. As with the flat-earthers, and in the spirit of the generic philosophical argument, they can set the standards for proof so high that of course no one can prove to them what they do not want to believe.

Since the government has lied and deceived in the past, such groups can use this as evidence that the government has faked the evidence of moon landings, space-satellite photos, or of the murders of millions of people. I read another article about a group that supposedly controls *both* the communist and the capitalist nations, in which the author, the then leader of a well-known right-wing political society, argued that the fact that these "Jacobians" have kept their identities secret for generations is *proof* of what clever and devious people have been involved. This general idea was not original to that author, but it was one version I am certain was intended to be taken seriously and not as diversional reading.

Water witching is an ancient custom that has been long established in much of the United States. It is familiar to most of us and some substantial proportion of our population believes that it works. The water witch, or diviner, handles a branch that supposedly leads her or him to water and even violently pulls her or him toward the spot to dig or drill. Since at least the nineteenth century, tests where no visual clues to water are available have shown water-witching holes to be no more effective than holes dug at random. This does not prove that the practice is always nonsense, but it does indicate that logic and critical judgment of evidence have little to do with the custom's survival over large geographic areas. There is some correlation between the difficulty in finding water in an area and the *number* of water witches, as we might expect—supply follows demand. Yet in some areas where one can dig *anywhere* and find water, water witches are still believed to be necessary and effective in finding water. Psychological comfort seems to be provided by the water-witching custom (Vogt and Hyman 1959).

> Contrast the clear-cut indications of the rod with the vague suggestions of the scientific geologist. The rod's message is decisive and unambiguous. The diviner says, "Dig here," and pinpoints the site precisely. The geologist supplies general information but leaves the pinpointing to the consumer. The scientist supplies guidance alone; the diviner relieves the consumer of all responsibility of choice.

> The action of the rod also provides reassurance at a time when the anxious seeker for water most needs it. The scientist, ever honest and ever aware of the fallibility of his method, qualifies his judgment and does not guarantee success. The water diviner goes about his business with the certitude that comes from blind faith.

But psychological comfort notwithstanding, the question of why commonsense beliefs have been so resistant to change remains, overall, a profound and complex one.

Perhaps there is no system of thought that does not exist alongside information that could challenge it. One might expect any living, dynamic thought system to owe its stability to arguments, right or wrong, that protect its basic assumptions, or to habits of thought that make it acceptable and even seemingly prudent simply to ignore or dismiss threatening evidence. I sometimes think of this mechanism as being functionally like the skin, with its ability to heal wounds and form protective scar tissue.

Control theory experts have in fact referred to "self-sealing" conjectures that are "irrefutable," in contrast to proper scientific hypotheses that invite critical tests of their logic, assumptions, and predictions, and that invite refutation (Watzlawick 1977). Puncture a self-sealing conjecture with evidence or logic, and its proponents will have a good defense waiting to heal the wound. *Good* scientists, on the other hand, search for valid ways to test and reject their conjectures. Scientific concepts are thus being continually modified, in details, at least, and this apparent instability can look very unappealing to many laypeople, and even to some scientists who are impatient for truth or certainty or who are defensive of their own beliefs.

Common Sense and Stable World Views

Common sense world views are not simply collections of facts, beliefs, superstitions, and opinions that accumulate like miscellaneous magazines in a dentist's office. We see systems of thought that have survived in competition with other systems of thought because they have aquired degrees of organic wholeness. They may be elegantly integrated like flesh-and-blood creatures are. They may have evolved defenses that protect them against violation from outside ideas. Some may have evolved the capacity to partially assimilate new information and thus adapt to new information environments while preserving their essentials. They may demand proselytization and thus have evolved high reproductive and dispersal rates that propel them from human mind to human mind. Such thought-system organisms are found all the way across the political spectrum, from the fanatical right to the fanatical left, and including vigorously closed-minded factions in the middle. Some cybernetic thought-system organisms may be benign and leave us room to retain or reject them by reason or experience, some may be malignant and take complete control of us. We can *never* live free of such creatures, we must learn their natures and how to live with them, retaining some reason and mastery in our own households where possible.

It may be as unrealistic, given the status of education, for the idealistic scientist or scholar to suppose that everyone will one day completely embrace critical thinking—as unrealistic as it is for a given religious group to believe that one day everyone will come to see things exactly as the group wishes. We should only expect partial acculturation. Such partial acculturation and reserved skepticism are very much what tribal people in New Guinea or elsewhere do when they encounter Western religious and scientific beliefs. It is very human to combine the new beliefs with old ones in some mix that seems to give the maximum power and explanatory potential. They may remain skeptical of much of the new, and cautious of being taken advantage of. The Westerner may be surprised to find that friendly Christianized and educated members of such societies still secretly hold traditional magical beliefs. Indeed, highly trained scientists and scholars are not able to consistently apply critical thought and open-mindedness.

Even in Western society many beliefs from pagan times survived determined offensives from both Christianity and scientific thinking: the little people, the notion that the heart is the seat of passion, some species of ghosts (see also chapter 6). If faith in science or in theological programs within religion begins to falter, some people will turn back to their own common sense and superstitions and may even make the more confusing systems such as science into adversaries to the extent that they dare. Is any of this very different from a traditional situation in which one New Guinea tribe will buy magic from a neighboring tribe and then get angry and try to get its payment back if the magic seems not to have worked?

It is often implied that common sense is nothing more than the habit of blinding oneself to evidence.[3] But that is true only in part. Common sense involves many things including *beliefs, perceptual habits*, and *systems of rationalization*, which are themselves part of a world view. The moon and the sun do indeed appear to be similar in size, and our senses, unaided by careful measurement, calculation, and insight, would never tell us that they are vastly different in mass. Our senses alone would never tell us that a pint of oil and a pint of water at the same temperature contain different amounts of heat. Common sense thrives in large part because it is consistent with what one sees, and it works. Each form of it has resulted in workday survival, success, and wealth for whole groups and classes of people.

Modern religious "fundamentalism," or literalism, whether Judaic, Islamic, Protestant, or Catholic, so much involves commonsense thinking that it should be mentioned. Creationists, as a familiar example to people in the United States, appeal to common sense in their attacks on evolution and natural selection. For example, they argue that *chance* cannot result in *order*: the universe should run to disorder without the intervention of God, and so the laws of thermodynamics reveal that natural selection can't work. The probability that random mutation will result in apple trees or parrots is more remote than the probability that a hurricane will assemble a lot of junk into a jet airplane.

Of course such things never happen. Common sense tells us that chance cannot create order, and so many people have accepted the creationist argument. The argument has vivid commonsense appeal, but in order for it to glitter many critical facts must be ignored, distorted, or dismissed.

1. Random mutation is only *part* of genetical dynamics, much as randomness is only a *part* of certain games of skill. Genetics is not wholly a matter of random mutation any more than poker, bridge, Monopoly, or Scrabble are games of pure chance. DNA replication mechanisms and natural selection pressures can be *shown* to order the structure of genes, much as a very skilled player can be seen to control a game of poker or Scrabble even though there is a random mixing of the cards or letters. Of course a player has a *mind* and natural selection does not, so this still may not make common sense to them.

2. Even the most simple atoms and molecules *do* in fact order themselves. It is not true that systems always run to disorder without intelligent intervention. Crystallization is a conspicuous everyday example. The molecules in a solution come together into highly organized patterns from their own properties. So too with the atoms themselves. The hydrogen and oxygen in the vast oceans of the world are not randomly arranged, but everywhere form molecules of H_2O. It is completely out of the question that chance alone could order all of the trillions of water molecules. But it is not chance alone that is involved. Molecular forces constrain the interactions of hydrogen and oxygen and produce structure and pattern. More complex molecules also arrange themselves as a result of simple forces.

3. Of course a room will not move from disorder to order itself neatly, and a hurricane will not build a jet airplane from junk. Moreover, mutation alone will not build an apple tree. Nor will crystallization. But no one ever claimed they will. Evolution is a rather slow process of change and adaptation, involving heredity, recombination, mutation, and competition for survival among the varients. It is not the sudden emergence of complex creatures by sudden bursts of mutation, analogous to a hurricane, or even by slow rates of mutation alone.

When confronted with some of the above, and surely by more detailed information in open discussions, I have seen leading Creationists simply brush the difficulties aside; one said at a meeting of the American Association for the Advancement of Science, "Will you just tell me, in common-sense language, how *chance* can ever result in *order*?" Apparently they feel that they have a logical proposition and that any information that might complicate the seemingly obvious or suggest that the words and concepts are being used improperly are confusions and temptations that they must not yield to. Crystals are crystals, not apple trees—how can there be a design without a designer?

I know many scientists who insist that Creationists must be knowingly dishonest and are dangerous manipulators of their audiences, since they repeatedly misrepresent data and ideas and constantly repeat discredited arguments. But I wonder. It is possible that they see any complications to their commonsense positions as being temptations that in a sense are not real and that they must not fall prey to.

For example, one established (nonbiologist) scientist who is also a Creationist agreed at that same meeting that there is much evidence to contradict Creationism, but that this apparent evidence must be interpreted in light of an epic struggle in the universe between good and evil. Satan, "the father of lies," planted the fossils. Certainly, some creationist literature, such as *The Twilight of Evolution*, by H. Morris, even explicitly calls evolution Satanic. Then perhaps they do not accept and pass on to their followers the weaknesses that scientists have found in their arguments because, given such a world view, they would be accepting illusions, apparent realities only, and doing the work of Satan. I cannot be certain, but the matter does seem to be much more complicated than that they are all simply cynical manipulators of people's emotions. The more involved interpretation would be consistent with many other examples of elaborate reasonings that people will go through to maintain their various commonsense realities.

Not all doggedly defended commonsense views by any means involve religion. And, indeed, not all religion is firmly tied to commonsense thinking. Theology often is an effort that with logic and scholarship attempts to move belief beyond common sense and intuition, and to reconcile the various contradictions in the Bible. Indeed, within religion there is always a political struggle between so-called fundamentalism, or literalism, and theology, between common sense and religious scholarship. (For theological analysis of Creationists and literalism see Heyers 1984. For secular discussions see Futuyma 1982, Kitcher 1982, Ruse 1982.)

As I write, we live in times of economic, political, social, and ideological uncertainty, and literalists of *all* the biblical religions have been taking more power. In times of general uncertainty, at least some shift in the direction of the common sense of each culture would be expected even without the vast financial resources and powerful psychological and organizational techniques being used, as detailed, for example, by Conway and Siegelman (1982).

But common sense, whether secular or religious, whether fear and Satan are brought in or not, can involve deeply and strongly held beliefs and habits of thought about reality and about one's proper path of participation in the accepted reality. Moreover, imbedded in it may be persuasive reasons for dismissing any challenges to it. World views, even if informal and implicit, may involve so many interdependent elements of thought, perception, values, and habit, that it can be very difficult to change any major piece without offsetting the others. We depend on these for routine social functioning and survival.

In an entertaining science fiction book, *The Mind Parasites*, Colin Wilson discusses the importance of our mental habits by using a vivid metaphor.

> I looked into his brain, and what I saw horrified me. It was like a town whose inhabitants have all been massacred and replaced by soldiers. There were no parasites present; they were unnecessary. . . . They had entered his brain and had taken over all the habit circuits. When these were broken, he was virtually helpless, for every act now had to be performed with immense effort, through free will. We do most of our living through the habit circuits: breathing, eating, digesting, reading, responding to other people. . . . To destroy a man's habit circuits . . . is to strip him of everything, to make life as impossible for him as if you had stripped him of his skin. The parasites had done this, then quickly replaced the old habit circuits with new ones. Certain circuits were restored: breathing, speaking, mannerisms (for these were essential to convincing people that he was the same person and in full possession of his senses). But certain habits were completely eliminated—the habit of thinking deeply, for example. And a new series of responses had been installed. We were "the enemy," and we aroused in him boundless hatred and disgust. He felt this of his own free will, in a sense; but if he had not chosen to feel it half his circuits would have gone dead again. In other words, having surrendered to the parasites, he remained a "free man" in the sense that he was alive and could choose his actions. But it was consciousness on their terms—either that or no consciousness at all. He was as completely a slave as a man with a gun pressed to his head.

It may be very difficult to know, for any given individual, how complex habits of thought intermesh and to what extent a small perturbation may have widespread implications. An idea or notion that may seem small or in-cidental to one person may be tightly attached to the very foundations of being in another. I am hardly aware of what guns I own, what type of car I drive, or what is popular on television, whereas for some people at the other end of the spectrum guns, automobiles, or television series are so much a part of their outlook and life-style that these may become virtually an ex-tension of their personal identity. This is a modest example but may help to make the point that the arrangement of each individual's inner world may have an interconnectedness that is difficult to anticipate.

I think that academics on the whole tend to minimize the interconnect-edness of facts and ideas for individuals. Often this is because it is their habit to carve off facts from the whole and treat them as isolated units for research or as items to be taught or memorized. The result of a lack of the habit of holistic thinking may be an insensitivity to how a new fact or idea will fit or not fit into a commonsense scheme. As a researcher and teacher I am constantly surprised to find resistance to apparently innocuous new small facts and ideas. The resistance usually traces to some larger set of assumptions about life. Education theorists have found that such things present major barriers to learning. But schools are not typically set up to deal with such personal dynamics. A student can look stupid but may be crippled merely by various prior assumptions. Or a student who appears to have learned something may not *actually* grasp the issue, because of prior assumptions. (As in Pyramid Film's *A Private Universe.*) In the course of over twenty years of research, I have found that my own greatest contributions have often come when I have been able to root out and overcome commonsense assumptions.

Last, let us not neglect the political, power, implications of common sense. One's control over oneself, as well as one's self-confidence, involve decision-making ability, which can in turn involve common sense. Threats to common sense can be threats to self-confidence, autonomy, and personal power.[4] Social position and power in the family or larger social unit may be based on respect for good common sense. Thus, challenges to the commonsense belief system can be seen as threats to individuals in the family, friendship network, etc. — threats to the spheres of power of familial, community, political, and religious authorities. The predictability of social interactions may rest on common beliefs and ways of thought. Threats to commonsense popular beliefs may be seen as socially destabilizing to individuals in pure democracies and to rulers in the more centrally controlled societies.

In some form these principles have been long appreciated. For example, consider Machiavelli's (1469–1527) still-followed instruction manual for politicians, *The Discourses*. In the section, "The importance of giving religion a prominent influence in a state, and how Italy was ruined because she failed in this respect through the conduct of the Church of Rome," Machiavelli goes on to argue that popular beliefs should be supported and worked with, even if they are seen as false, for this makes the people more stable and easy to govern and lead.

> It is therefore the duty of princes and heads of republics to uphold the foundations of the religion of their countries, for then it is easy to keep their people religious and consequently well conducted and united. And therefore everything that tends to favor religion (even though it were believed to be false) should be received and availed of to strengthen it. . . . Such was, in fact, the practice observed by sagacious men, which has given rise to the belief in the miracles that are celebrated in religions, however false they may be. For the sagacious rulers have given the miracles increased importance, no matter whence or how they originated; and their authority afterwards gave them credence with the people.

Institutions, from the family and business to science, church, and state, have as a priority their own stability and tend to have strong effects on the common sense of their members. They tend to first "gain legitimacy by distinctive grounding in nature and reason," as Mary Douglas (1986) sees it.

> Any institution then starts to control the memory of its members; it causes them to forget experiences incompatible with its righteous image, and it brings to their minds events which sustain the view of nature that is complementary to itself. It provides the categories of their thought, sets the terms for self knowledge, and fixes identifies.

Scientists and Scholars Are Also Vulnerable

Scientists and other academics certainly also have their own commonsense patterns of thought. Scholars often consciously try to improve their beliefs, to remain open-minded, and to brace themselves for revolutions in thought. Indeed, scholars may even too often prematurely announce that they have made revolutionary new findings or try to shoot down some giant in their field of study, since this would make them look bright and get them professional advances.

But while scientific and scholarly common sense is usually orders of magnitude better than lay common sense in *given arenas*, even in these it may nevertheless have some of the same difficulties. In a particular field of study there are assumptions that are commonly taken for granted and that go unexamined. There may be convenient arguments ready to dismiss good criticism without really weighing it. More correct ideas may be quite available, but will go largely unappreciated if they cannot be grasped in terms of prevailing canonical beliefs. The community may only slowly *improve* its perceptions, to be able eventually to fit more correct ideas in with its interests, perspectives, and career system. Sometimes a great finding will cause an *immediate* revolution, but even then it may have been necessary for a *mistrust* of the prevailing common sense to make it possible for the finding to be seen and appreciated. Mendel's genetic studies, overlooked for decades, were rediscovered by several workers only after those who studied heredity were ready to begin looking at particulate inheritance and abandon popular assumptions and statistical views that had concealed the phenomenon.

Good new ideas and data may not get published and known so easily if they do not fit within the canons of common sense in the discipline. There are legendary examples of this. Einstein's early papers were not taken seriously except by a small group. Kreb's classic paper on the citric acid cycle was rejected by several journals before it was published. Lindeman's important studies on energy flow through ecosystems were only published after long battles with editors. Wegener's excellent study of continental drift went through decades of ridicule before it was salvaged by paleomagnetic studies in the mid-1960s. Beveridge cites other such examples in his *Art of Scientific Investigation* (1950), and every scientist will have less famous examples. Yet, critics might make too much of such examples, serious and instructive as they are.

Scientists must function with human brains and in a context of social and economic pressures. Fads, for example, are rampant in science, though usually these are relatively short-lived and are in the turmoil of the cutting edge of new and controversial ideas where the public is least apt to be misled by them for more than a couple of decades each.

Science has its share of "old-boy-networks" of young and old men and women who promote one another and who put their political alliances and ambitions ahead of quality and the objectivity that may be necessary to transcend common sense. My experience is that most scientists dislike this, while a large and perhaps growing minority take it as a fact of life and may even thrive on it. Such difficulties notwithstanding, the scientific community on the whole has been ultimately grateful when specific mistakes have been corrected, and this is to the credit of open-mindedness in scientific common sense. Nevertheless, the same factors that make lay common sense fragile still contaminate science and academics to a nontrivial degree that is worse than annoying.

The main points I wanted to make in this chapter are that common sense may allow some people to prosper, but it is nevertheless relative. It is dangerous in that it is fragile and may prevent one from adapting to changing circumstances through the inertia with which it burdens individual minds and systems of authority and management. Science already has profoundly changed and can further change and improve our common sense.[5] Modern common sense seems to be based relatively more on truths and open-mindedness, and is more adaptable thereby, than species of common sense that have incorporated scientific thinking less. But the speed with which science and scholarship might do this is restricted, since the glue that holds commonsense world views together is stronger than we often appreciate. Moreover, science has its *own* brands of common sense that prevent it from advancing as rapidly as it might otherwise. Furthermore, new myths are being generated constantly to support selfish interests.

NOTES

1. Homosexuality has become a cultural commonplace, or even obligation, in an astonishing variety of ways among cultures, and for a great variety of apparent social (in addition to personal) reasons. There is no "essence" or single cause of homosexuality. The large variety defies brief classification, though I will provide at least a brief context for the Sambian system, since common wisdom in our culture about homosexuality is rarely informed by scholarship.

By way of contrast to the Sambia, least relevant here are those societies where men have been politically powerful and free to use both women and boys as sexual playthings.

Then there is a category of societies where same-sex behavior is tolerated or encouraged among the young unmarried people, probably to keep them from interfering with those in the marriage system. There may be some elements of this in the Sambian pattern. Other societies, such as the Romans in their late empire, merely have broad social tolerance for homosexual and bisexual behavior and relationships and, unlike in the Sambian culture, these are common *at all ages.*

Then there is a large category of cultures in which mature adults are expected to be bisexual and typically take on specific moral obligations for the care and training of a young person of the same sex whom they also become sexually involved with, and may live with, and in some cultures even marry temporarily. This set of models is well developed for males in warrior or monastic societies, or among shamans, where a great deal of training and loyalty may be necessary to hold the system together. When women have had such systems, there has been less study and the reasons are less clear. The preclassic Greeks, the Japanese Samuri, the Big Nambas of Melanesia, and the warrior class within the Azande of the Sudan would fit here (Adam 1986).

Among the Sambia, homosexual behavior may seem a bit more promiscuous than this last, but *it observes strict incest restrictions* according to kinship rules. Homosexual obligations are also well structured in that they *exclude the mature men* (over twenty-five), who hold political power.

Possibly most important, in terms of political stability, it should take place between *potential* enemies (though *not* actual or deadly enemies) outside the formal system of close-kin obligations, *not* between close friends, relatives, or older sponsors as among the Greeks, etc. Intratribal warfare has long been intense, and suspicion and mistrust are serious obstacles to peace. Thus, *as with the marriage pattern*, same-sex behavior may have one "political" function of building goodwill, thus easing the incidence or level of conflict once the males mature (Herdt 1984).

Thus, while intense male bonding ultimately may serve small warrior societies well, by augmenting strong loyalties and a system of apprenticeship, it would perhaps not make sense in terms of our notoriously bloody European war styles. European wars for hundreds of years have had disgustingly high death rates. It may be easier on morale if male bonds are kept at a less intense level. Moreover, the European system and high turnover rates of troops imply that men be quickly drafted from farm or city life and trained rapidly.

None of this is to question that the *immediate* main motivations of the Sambia are as stated to make the males as manly and as competent as warriors and husbands as is possible, to enjoy themselves, and to maintain a certain "masculine spirituality." Similarly, we give Christmas presents because we enjoy giving and getting, some for the spirituality, but it is *also* vital for our economy. Our system prefers bought presents to inexpensive ones that we make ourselves. Enduring customs serve multiple personal and social functions.

2. I should comment briefly to preclude a misunderstanding. Common sense might suggest to us that the sexual transition of most Sambian men to exclusive heterosexuality after ten years of exclusive homosexuality is evidence of an "essentialism" or a spontaneous return of "natural drives." But the situation is in fact very complex and I do not wish to leave the reader with the impression that a strong genetic determinism for heterosexuality specifically rather

than eroticism generally is necessarily in evidence here. Determinisms of various sorts, such as astrological, genetical, historical, or divine, have a simple appeal and many of us turn much too easily to them without much critical evidence. Sambian homosexuality is strongly directed toward the production of masculinity that includes heterosexuality. Thus, throughout the ten years of fellatio, each young man is constantly reminded of the purpose of the acts and of the heterosexual transition that he will one day make. Moreover, Sambian men fear women greatly and believe that without proper mental and physical preparation sexual contact with women may weaken and even kill them. But the result is strong *attraction*.

This fear-resulting-in-strong-attraction may seem at first to be a contradiction. But I suggest that it may be easily understood in our own terms. Sambian men are brave warriors and look forward to sex and love with women as much as to victory in battle. Their masculinization is a triumph over their fear of death in both battle and sex, and they are proud of their skill in dealing with both "perils." Just as it is macho in our society to love death-defying mountain climbing or fast cars, might not heterosexuality and war be similar to Sambian men?

Typical Sambian homosexual and heterosexual desires are both strongly influenced by moral values, imitation of peers, the dictates of political authority, social stigmatization of alternative sexual outlets, and culturally outlined individual aspirations. In these and other ways Sambian popular beliefs and values may channel young men through complex courses of gender preference, and there is little if anything that is spontaneous about it at any stage, except the passion itself in the adjusted male.

Gender preference also forms under complex conditioning and social pressures in our own society; so the conclusion does not necessarily follow that if it is not usually or largely a biological imperative, then it is arbitrary, and therefore adjustments to or from one form to another could be simple, as one might have incorrectly assumed should follow from this discussion. (Mangan and Walvin [1987] are also of interest here.)

None of this is to rule out a *large* component of genetic determinism in gender preference for a *small* fraction of people. Interpretation is open to question, but it is interesting that a small percentage of Americans withstand enormous socialization pressures and even persecution, to heterosexualize, and about the same small percentage of Sambians resist enormous pressures to homosexualize (though nonconformity is not persecuted).

3. "Men fear thought as they fear nothing else on earth—more than ruin, more even than death. Thought is subversive and revolutionary, destructive and terrible; thought is merciless to privilege, established institutions, and comfortable habits; thought is anarchic and lawless, indifferent to authority, careless of the well-tried wisdom of the ages. Thought looks into the pit of hell and is not afraid. Thought is great and swift and free, the light of the world, and the chief glory of man. But if thought is to become the possession of many, not the privilege of the few, we must have done with fear. It is fear that holds men back—fear lest their cherished beliefs should prove delusions, fear lest the institutions by which they live should prove harmful, fear lest they themselves should prove less worthy of respect than they have supposed themselves to be" (Bertrand Russell).

4. It may be said that persons who represent and support the status quo would be threatened by change because it may cost them their *positions*. There are also factors that go deeper than any conscious or subconscious desire to hold on to power and position. Persons with good common sense may have high predictive abilities and rise to authority during unchanging circumstances. When changes come, they may be lost without common sense as an intellectual tool. They may not be able to understand the reasons why things are out of control or know how to correct them. As a trivial example, suppose that one day you found that you had to relearn to write or perform some skill using the normally subordinate hand and you did not know how to teach yourself. Most of us would avoid such an unpleasant situation if possible. Think how difficult it can be for some of us to change smoking, diet, or posture habits. Persons

faced with ever-more-difficult change could become resentful and might conjure up all sorts of possibilities for why things are not going as predicted: from seeing elaborate plots to finding simple neighborhood scapegoats. So change may threaten much more than position in society. It may threaten a whole mental strategy for dealing with reality.

Similarly, we can see some reasons why historically it has been appealing when the alleged possibility to return to commonsense values has been held out to the public during times of social or economic uncertainty—this traditionalism would appear as an offer to reintroduce simple, supposedly time-tested, personal strategies for survival and day-to-day decisions.

There are many reasons why many people have been uneasy about and even hostile to science and scholarship. One important reason is relevant in this context. The history of science has been in large part a history of confrontations with common sense and of victories over it. The sun and the moon seem to be about the same size, but science has convinced most literate Westerners that this is an illusion, and so on. Science has often been an enemy of blind faith in intuition, the senses, revelation, authority, and all that popular beliefs have been traditionally based on.

To good scientists, science is a tentative and self-critical *process* and cannot afford to be authoritarian. It is the opposite of faith. But, for the person who does not understand good science very well it may in fact be another faith, and one that holds out little intuitive appeal or emotional comfort. One may well see science asking people to believe astonishing things that their common sense says are impossible and ridiculous. One may accept parts of it that one seems to grasp, or that make machines work and bones mend, but remain skeptical of other parts even without understanding the technical issues involved. Given such a position, that is a reasonable thing to do. The survival impulse would be expected to make us very cautious of handing over all of our personal power to some new someone, or something, that we do not feel we completely understand.

One could argue, though, that people do seem to give themselves over easily to leaders, and one could cite examples such as the mass suicides of the People's Church at Jonestown. But that was a select group of intensely faithful who believed themselves to have no spiritual escape from a corrupt larger society. They were a select group who *thought that they understood* the situation. (The leader was *convincing*.) Even at that, many did doubt, and refused to kill themselves, and they escaped or were murdered.

5. When science is taught as facts to be memorized, and the nature of scientific thinking is given only secondary or no emphasis, then the stage is set for intellectual confusion. One is asking the student to take much on faith, and one may unintentionally be sending the false message that even the best of science is little different from various types of faith. The student is denied his or her heritage as a member of a civilization that has sought improved powers of judgment for much of twenty-five centuries. (See also ideological reductionism, chapter 4, note 2.)

CHAPTER 6

Philosophy as Perceptual Template: Readings of Nature

Each beast, each insect, happy in its own:
Is Heav'n unkind to man, and man alone?

. . . .

All Nature is but Art unknown to thee;
All chance direction, which thou canst not see;
All discord, harmony not understood;
All partial evil, universal good;
And spite of Pride, in erring Reason's spite,
One truth is clear, Whatever is, is right.

. . . .

Here then we rest: — "The Universal Cause
Acts to one end, but acts by various laws."

. . . .

Nothing is foreign; parts relate to whole;
One all-extending, all-preserving, soul
Connects each being, greatest with the least;
Made beast in aid of man, and man of beast;
All serv'd, all serving: nothing stands alone;
The chain holds on, and where it ends unknown.

. . . .

Order is Heav'ns first law; and, this confest,
Some are and must be greater than the rest,
More rich, more wise: but who infers from hence
That such are happier, shocks all common sense.

. . . .

Condition, circumstance, is not the thing;
Bliss is the same in subject or in king,

. . . .

> *Heav'n breathes thro' every member of the whole*
> *One common blessing, as one common soul.*
> Alexander Pope, *An Essay on Man* (1732–1734)

The Intimate Politics of Nature

Circumcision, no matter how compulsively practiced by some societies, is distinctly unnatural, and today even its hygienic value is doubted. The Hebrews explained that they circumcised their sons because in Genesis God ordered Abraham to circumcise every man child of his own, or born in his house, and even the servants, in order to begin the everlasting covenant. In their eyes even patriarchy and the burden of motherhood were not necessarily natural, but were also ordered by God: "In sorrow thou shalt bring forth children; and thy desire *shall be* to thy husband, and he shall rule over thee." In many cultures deep beliefs about right and wrong are based on instructions supposedly laid down by supernatural personalities or by wise ancestors. Strong convictions may also be associated with "social contracts." Yet other strong convictions among Western people (and in some schools of Chinese thought, for example) also are based on notions of the "laws of nature" and of "purpose." This chapter will explore some old metaphysical beliefs that tend to persist covertly and typically unexamined even in contemporary Western secular thought.

An imaginary conversation:

SHE: I think we should share responsibility for taking care of the baby.

HE: But child rearing by fathers is unnatural. Among animals, among apes, the male is supposed to hunt and protect and the female to nurture.

SHE: So what if it's unnatural? So are chopsticks, and half of humanity has learned to handle them. Grab a diaper. You can learn.

Such a conversation might commonly take a different turn.

SHE: Well, is it really unnatural? What about birds where the father takes care of the young? Even some male primates do.

HE: Yeah, but not *most* primates. The *basic, typical* primate nature is for the male to defend and lead, and the female nurtures.

In both cases the father is arguing an issue of family politics from a moral position, that what is *natural is right*. But a third form of the argument is possible based on the faith that what is natural is *wise*.

HE: Fathers aren't born to know how to take care of children. Mothers are. Do we want our kid to become neurotic?

SHE: What's your evidence that fathers *can't* learn and mothers *don't have* to?

HE: Important things like that could not be left up to chance. I have faith in nature, whether it is the result of God or evolution.

SHE: But language is important and we learn it. Learning is not chance. Just look around you. Where did you get that crazy idea? What kids need is love and this can help to bring the baby and you closer together.

If one listens, it is remarkable how often Westerners argue about right and wrong based on notions of what is "natural," "basic," or the "wisdom of nature." Such beliefs have had great political implications and have been used to attempt to control the terms of public debate on slavery, competition, homosexuality, contraception, feminism, conscription, collectivism, individualism, criminal rehabilitation, class inequity, the preservation or exploitation of nature, science policy, and even nutrition.

I have long loved to learn about the diversity of nature. As my knowledge of plants and animals grew, it became evident that one can find creatures to prove just about anything is natural, but also counterexamples. I also learned that the idea that *anything* has a simple "essence," let alone the diverse human species, has been notoriously misleading. Most species of plants and animals are in fact highly variable within and between populations over their geographic ranges. Biology in the nineteenth century had to abandon the old notion of the "type" with its implications of basic essences (*typological thinking*). Ethical arguments based on laws of nature and typological thinking have tended to be largely contests of will and cunning. One may largely end up simply rationalizing, by selecting examples from nature, and blinding oneself to socioeconomic or political mechanisms that underlie the *status quo*, or desires for change.

Eventually I learned that Kant, Voltaire, and others had started to figure this out a couple of hundred years ahead of me. The moral arguments from the design of nature, the arguments from the wisdom of nature, ideas about basic essences all trace back to habits of perception and terms of discourse that the most influential Greek philosophers shared. Under Christian rule these took a backseat to *divine law* based on the Bible for nearly a millennium.

Then trade began to revive, cities grew, social relations and legal interactions became more complex in Europe. The effort to base law evermore on reason was especially strong in the thirteenth century, and combined with a rediscovery of Aristotle's works. People increasingly had to *reason* to try to link the words of the Bible to changing and complex circumstances. They cast about for further guidance and came to see nature as a second text of God, in which was written His Laws (e.g., Gierke 1987). This was natural law and *naturalistic ethics.* A relative of this would be *natural theology.* Social laws based on reason would often return, as Aristotle had, to an examination of the apparent *design of nature*—presumed knowledge of order in the world and of how this is preserved.

Enlightenment thinkers, like Voltaire, Hume, Kant, Rousseau, Ricardo, Malthus, and later Darwin, though, next planted the seeds of the destruction of design-of-nature thinking in its formal terms. Kant and Voltaire had a faith that they were actually only cleaning up "design" arguments, for example, by sweeping out the weak and silly arguments. But later influential thinkers decided that there was not much of interest left after housecleaning.

As housecleaning goes, though, dust and disorder settle back soon enough. Scientists, for example, do remember that the words teleology, design, essentialism, and various brands of naturalistic ethics represent thoroughly discredited ideas, but they tend not to know what the words mean or to recognize the ideas when they influence their thinking today.

A biochemist told me, for example, "I don't believe in evolution, because evolution says that women should resign themselves to social roles that are inferior to men's, since men are evolved to be the leaders. Evolution can be shown to be false, since it makes crazy predictions."

"Hold on," I replied. "First of all, evolutionary theory does not say that. These are the ideas of certain sociobiologists, and of social Darwinists (Spencerians), who would like to believe that their social views are supported by the authority of science. They are merely rationalizing. Other evolutionists have rebutted these views. You are right to be concerned about the political misuse of evolution, but don't throw out the baby with the bath. Second, the big problem here is with ideas such as teleology, essentialism, and naturalistic ethics. It is such metaphysical beliefs, which are well discredited but still popular in covert forms, that allow people to use select or twisted facts about nature to rationalize their politics. The problem is with the belief that *is-ness* means *ought-ness,* that things have come to be because of some purpose, that there are immutable essences, and that if we can know these *true* essences and purposes of things, then we can know what is right and wrong. Historically, Darwinism has been in fact the most significant single blow against ideas of the plan of nature, the wisdom of nature, essentialism, purposefulness, superiority and inferiority, and supposed moral imperatives from nature."

The biochemist paused and reflected. "Perhaps. But can we deal with that? No one I know has ever even heard of teleology. Right or wrong, the idea that what is natural is right is too widely believed to attack. Evolution is visible, it means something to people, it can be a useful target."

The example illustrates that there exist legitimate concerns about the political misuse of science, but without sufficient understanding of the mechanics of that misuse. Without sufficient understanding it may be difficult or impossible to understand the threat, properly defend oneself, or on the other hand to take effective countermeasures. One could even end up undermining one's best potential arguments.

Thinking in terms of teleology, essentialism, and naturalistic ethics has been used to promote slavery, sexism, colonialism, social immobility, and other injustices and to deprive groups of potent education, since long before Darwin was born. It will be instructive next to reconstruct an ancient world view in which all these philosophical beliefs came together in overt form and took on a strong mutual interdependency.

The Teleological World of Classical Greece

The teleological world of Aristotle is in disrepute among conceptually oriented scientists and historians—it has been quite soundly discredited as a formal doctrine—and yet in covert and sometimes overt forms it is still a key part of the fabric of beliefs that so many Western people live for, rejoice in, die for, and suffer in, even today. For over 2,000 years Western languages, schools, arts, religions, literatures, philosophies, and commonsense wisdom have simmered in this seductive sauce and have preserved its strong flavor. Intellectual tools were forged in its fires. Since Western languages and common sense were built upon it, unconscious effects on thinking and the perception of reality persist even today. It touches judgment and values, religions, political ideologies, and perceptions of right and wrong from capitalism to Marxism, and it is scarcely possible for anyone to escape its effects, even though most seldom hear or use terms such as *teleology*. It is not enough to declare that one is against it. Perhaps we should try to understand it.

What is teleology? In my dictionary it sits innocently between *telencephalon* and *teleost* and has such muffled definitions as: 1. "The study of final causes;" 2. "The fact or quality of being directed toward a definite end or of having an ultimate purpose, especially as attributed to natural processes;" and 3. "The philosophical study of manifestations of design or purpose in natural processes or occurrences, under the belief that natural processes are not determined by mechanisms but rather by their utility in an overall natural design."

Our vivid perceptions can blind us to truths. Our common sense can blind us to truths. In this chapter, by this example of ideas closely allied to teleological thought and natural theology, we will see that even sophisticated philosophical systems have long blinded people to truths and retarded intellectual and ethical aspirations in spite of the best intentions.

Let us begin about a century after the births of Buddha and Confucius, with the birth of Socrates in 469 B.C. Nature was teleological, designed, and purposeful to these Greeks who sensed pattern in nature but who obviously did not have modern physical and biological theory to understand it. Theirs was a world of gods, souls, and spirits and they sought to use their reason to understand how such metaphysical forces could produce the changes in nature and the apparent order that they saw. In doing so they tended to be anthropomorphic and dualistic: if ethical and aesthetic values could be the ultimate causes of human works, then why not also of animals and objects and their interactions? Order they saw as design. And as works of art are judged good or bad, some animals or persons may be superior and inferior in the scheme of things.

The world was seen to be like a tapestry, fundamentally a series of perfect and eternal essences woven into a harmonious design that was good and beautiful (though this may be hard for human eyes to see in its faded and frayed material manifestation). Things had their natural places as higher or lower in the design, and things were placed purposefully and wisely for the purpose of providing harmony and virtue for all. Thus teleology becomes the key unifying factor in the essentialism, design, and ethics of Plato and Aristotle. If the philosopher could see the "true" purposeful design beneath the fray and discoloration, he could understand his place in the design and know when he was following a virtuous path or when his eye was distracted from the true design by some loose threads and discolorations.

It has only been since about the 1700s that important Western thinkers have begun to see the apparent reason and harmony among organisms as being in significant part the result of the disposition of the human mind to select organized patterns out of fields of objects and happenings that are, like ink blots, in fact more ambiguous and value-neutral—rather than the result of God's plan or of reason and *ethical values in nature itself.* Until then, design thinking could be good common sense, good philosophy, and good science. For many it still is.

While one usually associates teleology with Aristotle, it was certainly present in various forms in the earlier Greek philosophies and in Greek religion. The following passages from Plato's *Phaedrus* give the flavor of certain thoughts entertained in segments of the Greek society of the day. Some Greeks, such as Socrates and the Orphics, detailed a belief in reincarnation that recalls that of the East Indians (e.g., Reale 1987). Socrates speaks of soul. He defines it as that which is moved by itself and is not born and does not die. He likens it to the winged horses that pull a god's chariot and he goes on to describe how it may fall into earthly forms and then, through reincarnations, has the opportunity to return to the gods and to the True Forms. Socrates was not simply spinning tales, keep in mind; his aim was to define proper moral conduct and attitudes, to discover Right and Wrong, and how life should be lived. Perhaps humans could discover the Good Life and the Ideal State.

All soul has the care of all that is inanimate, and traverses the whole universe, though in ever-changing forms. . . .

Whatsoever soul has followed in the train of a god, and discerned something of truth, shall be kept from sorrow until a new revolution shall begin; and if she can do this always, she shall remain always free from hurt. But when she is not able so to follow, and sees none of it, but meeting with some mischance comes to be burdened with a load of forgetfulness and wrongdoing, and because of that burden sheds her wings and falls to the earth, then thus runs the law: in her first birth she shall not be planted in any brute beast, but the soul that hath seen the most of Being shall enter into the human babe that shall grow into a seeker after wisdom or beauty, a follower of the Muses and a lover; the next, having seen less, shall dwell in a king that abides by law, or a warrior and ruler; the third in a statesman, a man of business or a trader; the fourth in an athlete, or physical trainer or physician; the fifth shall have the life of a prophet or a mystery-priest; to the sixth that of a poet or other imitative artist shall be fittingly given; the seventh shall live in an artisan or farmer, the eighth in a sophist or demagogue, the ninth in a tyrant. . . .

. . . in all these incarnations he who lives righteously has a better lot for his portion, and he who lives unrighteously a worse. For a soul does not return to the place whence she came for ten thousand years, since in no lesser time can she regain her wings, save only his soul who has sought after wisdom unfeignedly, or has conjoined his passion for a loved one with that seeking. Such a soul, if with three revolutions of a thousand years she has thrice chosen this philosophic life, regains thereby her wings, and speeds away after three thousand years; but the rest, when they have accomplished their first life, are brought to judgment, and after the judgment some are taken to be punished in places of chastisement beneath the earth, while others are borne aloft by Justice to a certain region of the heavens, there to live in such manner as is merited by their past life in the flesh. And after a thousand years these and those alike come to the allotment and choice of their second life, each choosing according to her will; then does the soul that has beheld truth may enter into this our human form: . . . passing from a plurality of perceptions to a unity gathered together by reasoning; . . . looking down upon the things which now we suppose to be, and gazing up to that which truly is.

These passages from Socrates/Plato are useful in illustrating several points that are modified and formalized in other sections of Plato and in Aristotle, and they give a hint of some general themes within classical Greek metaphysics and its modern derivatives.

1. Form, motion, and change, even of inanimate objects, are associated with "essences," True Forms, or souls, which are *eternal* and *underlie all material things* (essentialism).
2. True Forms (the ideal chair, the ideal stream) are not of this earth, and souls will try to return to their "natural" place to join the Truth that they had once seen and can still remember to a degree. Nature is in motion with *purposefulness* (teleology).
3. There is a natural hierarchy with some forms *higher* and actually *superior* to others, and more noble, since some forms or souls are closer to Truth than are others. This is part of a general plan that reveals the wisdom of natural order (design).[1] Philosophers, their disciples and lovers are near the top, and sophists and tyrants rank down near the livestock (as in the Great Chain of Being).
4. There is a moral cast to all this. Uncorrupted motion is toward the good. There is *absolute moral Truth*. There is moral virtue in nature and if we can discover the wisdom in nature it can teach us what is Right (naturalistic ethics).

One might try to picture a purposeful world of both living and inanimate objects that move when the souls that have taken up residence in them, or move them, strive to imitate or return to the True Forms that they dimly remember having once glimpsed. Along such lines, Aristotle would later explain, for example, that the acceleration of a falling body is caused because things move quickest when they are near the natural or the right state.

Socrates believed that all objects can be ranked on a scale, as good-to-miserable copies of their particular True Form. Plato's True Forms existed off somewhere of unearthly geography. The average traveler could never hope to visit the land of ideal birds, chairs, houses, trees, and men. One could only see poor earthly copies, and contemplate the *immutable "essences"* through reason (rationalism).

Aristotle objected to this dualism in his teacher's philosophy. He doubted the existence of extraphysical souls that move in and out of objects, and he saw the essences and motive forces as being properties of objects. Aristotle brought substantial observation and study of the material world back into philosophy (empiricism). This is why Aristotle gave much attention to the study of plants and animals: their matching of form to function seemed his best guide to the teleological nature of the world (as in *The Parts of Animals, The Movement of Animals, The Progression of Animals, The Generation of Animals, The History of Animals*).

Aristotle was a great natural scientist, and we cannot help even today but admire his intellect and ambition. We can see in hindsight that his work has been both a gift and a curse. He invented formal logic and set a model for attempting to make critical empirical observations motivated by hypotheses. At the same time though, in many if not most cases we now know he was wrong. How can we explain the profound influence of this man on all of us? For one thing, the genuine brilliance and astonishing breadth and ambition of his work remain awesome even from today's perspective. Also, it cannot have hurt to have had as a student Alexander the Great, the man who conquered the known world and spread the ideas of his teachers.

Much of Aristotle's natural history involves explorations of his ideas on causation and order in the universe, and without appreciating this, it can be boring and difficult to read his work. In order to understand Plato's and Aristotle's approaches to discovering the order in nature, we must recall their views on causation. First Plato, briefly.

1. The *Demiurge*, or God, molded the world from preexistent matter.
2. The *Pattern* exists in the world of True Forms and was the model that the Demiurge worked from.
3. The *Receptacle* is the material, limiting, distorting, disordering aspect of the world—the poor, coarse, lumpy clay, if one will.
4. The *Good* is the Form that provides purpose and direction, the source of the teleological organization of the world.

For brevity I will not deal with both Plato and Aristotle, but will attempt to focus on the latter, since his direct influence has possibly been greatest on science and its interface with religion and Western common sense, especially since his "rediscovery" in the thirteenth century.

Aristotle concluded that there are four causes for everything being and changing. He used the artistic creation as a model for his scheme.

1. *Material Cause*—physical substance such as clay or stone
2. *Formal Cause*—the form, design, blueprint, or artist's working sketch
3. *Efficient (Moving) Cause*—the shaping agent, such as the artist or craftsman
4. *Final Cause*—the purpose or function of the object, such as beauty or pleasure in the case of a work of art

He placed much emphasis on the determination of Final Causes, or purposes, because they would reveal the natural and ethical order of the universe. *Beauty, pleasure, harmony, goodness, can cause things to exist and happen.* But he also defined other causes, particularly the Efficient Cause, which is somewhat like the concept of *proximal causation* in modern science. Thus, some of his approach will look more or less familiar to a modern reader. Final Causes remain familiar as a controversial subject in the *semantics* of evolution and behavior.[2]

Aristotle viewed life in terms of opposites, dualisms, as did most Greeks. Chance, or coincidence, was seen as an opposition to purpose. By demonstrating in his natural history studies that objects seemed usually to have functions, he was able to reject chance as being major in nature and so he felt he was able to claim purposefulness as the operative force in the world.

> So if . . . things must be the result either of coincidence or of purpose, and not be the former, they must be the latter. But . . . things are admittedly natural; so there must be purposiveness in what happens and is by nature. . . . Art partly completes and partly imitates the work of nature; therefore if art is purposive, so is nature. . . . The purposiveness of nature is most manifest in the lower animals, such as spiders and ants, which without art, inquiry, or deliberation do things which look like the work of reason. Tracing this tendency downwards, we find adaptation to ends even in the growth of plants. Evidently, then, a final cause is at work in nature. . . . If ever there was only rest then motion in objects would have originated as follows: (1) As Anaxagoras says, . . . reason introduced motion and order into them, or (2) as Empedocles says, viz that things move when Love is making one out of many or Strife many out of one, and rest in the intervals. (*Physics*)

His vision is of a world where there *must* be harmony of all things and one can supposedly find out what is *Good* and *Right* by finding out what is *natural* (*somewhat* like Taoism at roughly this same time in China).

Among the Socratics, especially Aristotle, the ideas of essence, design, purpose, and naturalistic ethics became so closely woven together that it is academic to separate them. At other points in history they have been more separate ideas. Yet, the potent logic of Aristotle and Plato seems virtually to have magnetized them so that they have a tendency to reclump (if covertly) in Western thought, even in contemporary thought.

With the advance of modern science and Galileo, Descartes, and Newton, the idea of Nature as a manifestation of God's *ongoing* will or wisdom (*theism*) gave way in degrees to a mechanical clockwork world operating according to God's laws, without his ongoing intervention (*deism*). This was one of the major shifts in all of Western thought. Various forms of deism became the basis for natural theology, which was popular among scholars in England and Scotland into Darwin's time, and which we will see influenced the economic philosophers such as Adam Smith.

Teleological Design Overtly Crumbles, Covertly Survives

It is against this background that one can best understand the enormously pervasive impact of the Darwinian revolution starting in the late nineteenth century. It was certainly a move beyond theism, even beyond the prevailing deism. It did not require atheism, but God and nature could never again confidently be seen in the old ways.[3] *Darwinism was not simply a threat to the idea that man is "superior to the beasts," but it provided powerful ammunition against the whole teleological design scientific and common-sense view of the world*: Darwin proposed that form matches function in organic nature not because preexisting Reason, purposefulness, and design prevail to produce an ethical universe, but because of a simple mechanical process, "the survival of the fit." If things exist for their own sake and not merely to serve ultimate ends, then "all discord" is *not* "harmony not understood," as Pope claimed. The reality of ethical and harmonious design, moral hierarchy, and inherent purpose in nature were thrown into serious question on the basis of material evidence from the decades of empirical research that followed throughout Europe and America.[4] Darwinism in its correct form (unalloyed to popular progressivism and social Darwinism, that is Spencerism) denied the ranking of creatures and races of mankind into higher and lower forms on any absolute *scala naturae*. Worms are as "fit" for their mode of survival as humans are for theirs, and the human would be inferior if it tried to survive on the resources available to an earthworm.

Knowing all this helps to understand why it is that the so-called fundamentalist, or literalist, religious attack on evolution is tied, at the level of its intellectual leaders, to a moral crusade that is a widespread attack on, and criticism of, the entire Enlightenment (e.g., references in Conway and Siegelman 1982). The Age of Reason is regarded by some as having been a failure that has led to *moral decay and a denial of God's plan*. Some of them fear that if children are taught that humans are animals, then they will think it is *natural* and *right* to behave badly, as they believe animals do. Whether society was more moral before the spread of scientific thinking is plainly doubtful, and is not something that should be taken at the critics' word.

The Enlightenment constituted the most profound revolution in Western thought since the origins of Greek philosophy and the spread of Christianity. Indeed, much of the Enlightenment developed into a dramatic, if incomplete rethinking of teleological design perspectives in the common sense, philosophy, and religion of Europe.

A passage from Kant's *Critique of Teleological Judgment* (1790) gives a feeling for the thinking that he and others were struggling with.

> In cold countries the snow protects the seeds from the frost. It facilitates
> human intercourse — through the use of sleighs. The Laplander finds animals
> in these regions, namely reindeer, to bring about this intercourse. The latter

> find sufficient food to live on in a dry moss which they have to scrape out for themselves from under the snow, yet they submit to being tamed without difficulty, and readily allow themselves to be deprived of the freedom in which they could quite well have supported themselves. For other dwellers in these ice-bound lands the sea is rich in its supply of animals that afford them fuel for heating their huts; in addition to which there are the food and clothing that these animals provide and the wood which the sea itself, as it were, washes in for them as material for their homes. Now here we have a truly marvellous assemblage of many relations of nature to an end—the end being the Greenlanders, Laplanders, Samoyedes, Jakutes, and the like. But we do not see why men should live in these places at all ... for nothing but the greatest want of social unity in mankind could have dispersed men into such inhospitable regions.

So, the "end" being served would not exist in this region of the earth at all if there had not once been discord, or a lack of harmony, in some other area. Life is not harmonious design.

Europe in the 1600s and 1700s smarted badly from religious wars and persecutions. Sects were proliferating, each claiming to have the Truth. Science on the other hand, seemed to be creating intellectual and economic progress. There was a hope for humanity and stability if rationality based on scientific thinking could discover God's natural laws and offer an alternative to the sorry state of affairs that authority, revelation, and inspiration had proliferated. Some Protestant ministers eagerly turned to science even as a weapon to argue the improbability of the miracles that Catholics had used to claim theological authority.

Newton, Locke, Descartes held out the promise that a universal clocklike mechanical world of nature and society would be discovered. In the early Enlightenment, Voltaire in France, for example, argued that with reason one should sniff out and destroy all the myths, superstitions, and other nonsense of the traditional order and free up "True" natural laws of social conduct to operate. The "hidden hand" of the natural laws would insure that a virtuous social order would result.

A notion of social *progress* had also been emerging in Europe. It was based on a faith that some mysterious mechanism would be found to show that not merely was *change* taking place, but that society was moving forward to a better end. The idea of automatic progress had elements of design and purposefulness to it and reflected the optimism of the design teleologist.[5]

The deists, though, did not realize how much teleological design remained covert in their faith that the ethical properties of some hidden hand would ensure harmony. Deistic ideas became adapted by the rising middle class who increasingly were in economic and social competition with the old order represented by the aristocrats and clergy. It was convenient for them to believe that if they were free to pursue their private privilege, aggressive self-seeking, and competitive egoism this would eventually produce a virtuous society by action of the hidden hand. Scientific thinking had helped commerce and now supposedly it would show what fools all those old fellows were who were preaching divine rights of kings and of rich men not being able to get into heaven.

Quite prior to Darwin, such largely deistic critics of overt "design" as Voltaire (1694–1778) had poked fun. For example:

> The bees may be regarded as superior to the human race in this, that from their own substance they produce another which is useful; while, of all our secretions, there is not one good for anything; nay, there is not one which does not render mankind disagreeable. (*Philosophical Dictionary*)

Of course, the confirmed design teleologist would not be crushed by such jabs, as Voltaire well knew. In *Candide*, his great satire of conventional design thinking, in which that which is natural is for the good (and so this is "the best of all possible worlds") Voltaire writes:

> "Well! my dear Pangloss, said Candide, "when you were hanged, dissected, stunned with blows and made to row in the galleys, did you always think that everything was for the best in this world?" "I am still of my first opinion," replied Pangloss, "for after all I am a philosopher; and it would be unbecoming for me to recant, since Leibnitz could not be in the wrong and preestablished harmony is the finest thing imaginable, like the plenum and subtle matter."

So mockingly, if this is the best of all possible worlds then it seems to be filled to the brim with "necessary evils."

It was with the scientific approaches of da Vinci (1452–1519), Copernicus (1473–1543), Galileo (1564–1641), Kepler (1571–1630), and Newton (1642–1727) in mind that overt teleology—fundamental to Western thought as it stood at the time—was coming under most formal and effective attack. We may recall Francis Bacon in his 1623 *De Argumentis Scientiarum*, "Inquiry into final causes is sterile, and, like a virgin consecrated to God, produces nothing."

Among intellectuals, while theistic teleological design was becoming replaced by the more subtle deistic design in the early Enlightenment, by the late Enlightenment even this last was starting to crack and peel. Rousseau began as a deist but turned his critical rational analysis upon deism itself. By 1750 he had the insight that "mechanical" rules of social conduct did *not* produce harmony and virtue. They produced conformity, compromise of compassion, hypocrisy, empty self-esteem based on false values and shallow rewards, jealousy, the multiplication of needs and problems. He saw alienation of the inner self from its full potential for happiness and contentment in innocent love, deep friendship, honesty with oneself and others. (In the construction of the inner self, the baby first begins experience with the potential for love, trust, openness, open-eyed curiosity, without shame for its feelings.) The "progress" of civilization might produce the elimination of ancient religious superstitions, yes, but people evermore smothered in personal myths and out of touch with themselves and evermore subject to manipulation from without. He became a critic of the faith of the *philosophes* and the middle class, that simply replacing revealed faith with reason and science would automatically result in progress to harmony and virtue. The political liberation of the middle class alone would not make a better world for all. Knowledge can be used for good or bad, to nourish or to exploit. Values and thoughtful intervention are necessary. In effect he became a critic of the hidden-hand notion.

In this, Rousseau holds as a philosopher a parallel position to the economists Ricardo and Malthus in England, who later demonstrated that the free market of Professor of Moral Philosophy Adam Smith or the rational anarchy of William Godwin were not going to result in greater prosperity for all due to the presumed laws of some "Invisible Hand." Horrors! Malthus realized. Left to themselves, the laws of natural reproduction would eventually smother the planet with people and increasing want. Uncompromised competition, Ricardo realized, and unchecked economic expansion, would lead eventually to domination by landlords and not to freedom from subsistence for everyone.

These insights that natural laws left to themselves would fail to produce harmony and virtue for all were in effect criticisms of deism, a covert form of design, now including progress, that often included elements of teleology and naturalistic ethics. They suggest that more sponsorship and popularization of philosophy and science will not automatically produce a better world. These can only be tools, and while they may be *essential* tools, some sets of values must be constructed and ways found to implement them. There remains even today this division between those progressives who believe, in effect, that more science will somehow automatically lead to progress, and those who believe that intervention and the broad assertion of realistic values based on rational analysis is essential.

Despite the profound influence of Rousseau and Darwin on their times, it was in some cases on restricted issues and in others on restricted if important audiences. Aristotelian thinking found covert ways to survive into our time. In the end, Rousseau became recycled into a father of middle-class romanticism and democracy, selective responses to the larger issues that he raised. He was eventually misleadingly identified with the phrase "Noble Savage" and stereotyped as a back-to-nature advocate. When Darwinism came along, no problem. His model was twisted about wrongly into a mechanism for biological design and progress. Others misused it so that they could feel comfortable that rotten behavior and conditions are natural and inevitable, progress or not.

Thus we refer to "social Darwinism." More properly, we should associate such social thinking with earlier writers such as Herbert Spencer, Thomas Hobbes, Bernard Mandeville, and lesser known mercantilist and laissez-faire writers.[6] Then, the selfish competition of the upwardly mobile middle class could still be seen as good, because supposedly science shows that it will lead to improvement and progress for all. Somewhat as social success was evidence of predetermined grace to Calvinists (Weber 1958), social success was evidence of being inherently "most fit" to prototypical social Darwinists (poverty then is *dis*grace and/or biological *un*fitness). Self-interest and selfishness still will make a better world (or are at least inevitable). Covert deism embraced a mispainted Darwin and certain middle-class values remained unfazed at their foundation.

Design, naturalistic ethics, and other Aristotelian arguments persist in popular thought today, and are useful in the architecture of some ideologies, but they are at the core intellectually completely vulnerable, because science and critical scholarship have shown them to be useless at best, and usually misleading. There is no longer any reason to believe that flowers were created to give mankind pleasure through their beauty, or beasts created to pull wagons and provide meat. It has not proven useful to believe that falling objects move faster because they are approaching their "natural state." It is more accurate to know that the speed of a falling object accelerates a certain amount each second, and that the rate of movement of one body toward another is related precisely to the mass of each.

Ecologists once did believe they saw strongly regular patterns and interactions in nature and, as in the Clementsian school, they thought of communities of organisms as virtual superorganisms, coevolved for mutual benefit and stability, a virtual deistic clockwork mechanism (though not directed to human needs). Decades of detailed research have recently shown that much of this is illusion. One was tending to treat negative data as exceptions, the classical self-deception of typological thinking that allows stereotypes to seem real. Patterns there are, but actually quite irregular and fragile ones, and they have changed through even recent geological time. There is no evidence to support a "coevolved superorganism" sort of model. Since the 1960s ecologists speak more of *assembly properties*, by which some degree of patterned interaction consolidates as species are independently shuffled about by shifts in climate over tens of thousands of years

(Simberloff 1982, Regal 1985). These ad hoc patterns are certainly not goal directed.

None of this is necessarily a death blow to morality or religion. A belief in gods or in morality certainly does not logically or in any absolute way depend on Aristotle's teleological argument for a "Prime Mover," or on evidence of purposefulness or of any overall plan in nature. There are many belief systems in the world that are unthreatened by demonstrations of the falsity of such thought, and many modern Christian and Jewish scholars, for example, have certainly been able to reconcile beliefs and moral codes with scientific knowledge. Social stability does not even seem to require belief in *absolute moral truth*. Indeed, ethical absolutism has been relatively rare among the world's cultures. All around the world, peoples in hundreds of cultures have had their own sense of right and wrong based on love, or fear, or loyalty, or tradition, or imitation, or good sense and social contract, or strong notions of decency, or the Golden Rule, and combinations of these.

Incidentally, various speculations have been made as to why Darwin waited for so many decades to publish his theory of natural selection in 1859. In evaluating this question it may be of interest to keep a time frame in mind. Copernicus published in 1543 and was not tolerated by Western religions over nearly three centuries. Darwin studied and wrote at a time when *universities were still controlled by the clergy*. (In fact it was Darwinism that ultimately was to play perhaps the major role in the development of academic freedom in universities; e.g., Metzger 1955.) We may recall that Giordano Bruno and many nonscientists had been persecuted or killed and tortured for their heretical beliefs by the Inquisition. Decades later, Galileo fared better than Bruno. In an extremely complex interplay of personalities, politics, and doctrines, Europe's greatest scientist was arrested but allowed to recant his "heresies" and to live out his last years in the relative comfort of house arrest. His reputation also survived organized attempts at character assassination. The episode was not a bright spot in history.

> Wretched human beings. ... You have been spoken to a hundred times of the insolent absurdity with which you condemned Galileo, and I speak to you for the hundred and first, and I hope that you will keep the anniversary of that event forever. Would that there might be graced on the door of your Holy Office: "Here seven cardinals, assisted by minor brethren, had the finest thinker in Italy thrown into prison at the age of seventy; made him fast on bread and water because he instructed the human race, and because they were ignorant." (Voltaire, *Philosophical Dictionary*)

Interestingly, the case in time became an enormous embarrassment (since the earth did in fact turn out to have a "double motion") and the Church attempted to absolve itself of blame. But Napoleon captured the secret records of the trial and related affairs from the Vatican, and the French government returned them to Rome (1846) only on the condition that they be published—eventually (1867) most were. Anti-Roman Catholics traditionally ridiculed the Church over the whole affair, but historian A. D. White calls this "unjust" and recalls Protestant persecutions of scholars over the same and related issues (White 1896).

Darwin's ideas were to offer an even more pervasive challenge to the Plan of Nature doctrines and common sense of Europe than had the works of the physical scientists; for recall that Aristotle's best evidence for the purposeful rationality of nature and human life had come from zoology. A Christian *physicist* such as Newton had been able to retain faith in design (even though Hume and others severely criticized the reasoning by which he supported such arguments, e.g., Hurlbutt 1965).

So teleology, design, essentialism, and naturalistic ethics have stayed around, often clustered in the Platonic or Aristotelian fashions, because they have been an unconscious part of Western common sense and language for millenia and because there have necessarily been economic, semantic, religious, and even political pressures to keep them in ideologies and popular thought. It can be difficult even to talk or think without them. Surely they will keep reemerging even in critical discourse in one form or another, overt or covert, for many many years to come. Each of us must find ways to recognize and live with them.

The Perception of Inferiority and Exclusion

Beliefs about what is natural, progressive, superior, inferior, essential, primitive, and advanced—superficial stereotypes of what is "the true nature" or is "typical" of a class of people—have been used to justify slavery, colonization, racism, sexism, and other miscellaneous religious and secular exploitations, exclusions, persecutions, and oppressions even in our time. The tactics work well since essentialism, design, naturalistic ethics, and teleological reasoning remain imbedded in Western common sense and in some formal doctrines. Aristotle justified slavery, for example, arguing that it is a *natural* institution, and that non-Greeks are *inferior* to Greeks, since he believed they were further from the Good in terms of philosophy, art, etc.

The role of Aristotelian teleology in social attitudes and policy is a complex matter. For example, there has been asymmetry in racial and social relations in cultures that were not exposed to Aristotle either directly or indirectly, and the economic reasons for this have been obvious. Yet the involvement of such metaphysics in much of the tragedy of slavery and colonialism cannot be ignored on this basis. History would have had a different character if the dialogue in Western society had been different. There has not been one essential cause, such as economics, that shaped institutions of slavery.

Hanke (1970), in *Aristotle and the American Indians: A Study in Race Prejudice in the Modern World*, recalls the complex disputes within the Spanish church/state over the morality of enslaving the New World Indians. There were enormous economic and political pressures and efforts to enslave the Amerindians and the arguments "for" involved Aristotle's doctrine of natural slavery. Policy vacillated on and off between two views of morality. Aristotelian teleology determined the ground rules for the debate. On the other side were those who argued that the Amerindians did have a "high" culture and so slavery would be effectively "unnatural" for them. African slaves were imported at times when Amerindians were thus unavailable as slaves, and the world saw one of the great and vile displacements of human beings in all of history. Apparently it seemed acceptable to enslave Africans and haul them across the oceans because the Moslems had had black slaves for so long that it seemed natural. Here we see some presumption of an "essential" nature of black people, of a grand "design" in which they are inferior, of some "purposefulness" in which the subordination of one class of people is for the good of the whole, and some notion of "naturalistic ethics," that supposedly makes it right.

So while slavery existed in various times and places apart from any exposure to Aristotle, the particular impact that the slave trade had on the lives of many Africans and New World Indians was very much affected by decisions that pivoted on design teleological considerations.

Colonization by Protestants was sometimes less charitable to the colonized peoples than in the case of Roman Catholics and so an uneven blame should not be placed on the Spanish Catholics. Much of the Amerindian population in what is now the United States was not merely exploited but killed, and not simply to protect wagon trains, as portrayed in older films. While many whites did respect Indians, the view that Indians were *basically* "inferior primitives," and of the white man's "destiny," generally prevailed. (There was also the Judeo-Christian notion of a "chosen people" among the whites.) Sophisticated sociopolitical systems, and even Western education and industry among ethnic groups such as the Cherokee, did not spare them from persecution and near genocide (Honour 1975, Drinnon 1980, Jennings 1975, Kuper 1988, Sheehan 1980). So in both the Anglo-Protestant and Latin-Catholic colonies, life, death, and liberty often rested on notions about the "natural course of events," the "essential" or "basic" nature of peoples, the "destiny" or purposefulness of history, on fabricated stereotypes backed by "essentialistic" or "typological" thinking, and on notions of a people's being "advanced" or "primitive," "superior" or "inferior."

The above systems of classification and ranking are, in their general form, very deep elements in European thinking, and while they are often associated today with nineteenth-century social Darwinism, they represent a revitalization of something *much* older and more pervasive in Western metaphysics. Indeed, the construction of social Darwinism was essentially an effort by nonbiologists to defend stark competitive values and perceptions. The evidence from organic evolution has always contradicted their schemes. Contrary to popular belief, alleged virtues of hierarchy, overt competition and compassionlessness are not biological truths. Competent evolutionists try hard to clarify the context in which terms such as *basic nature* and *primitive* and *advanced* might be used, and they try to avoid blanket categorizations of particular organisms as being overall superior or inferior. Competition can lead in any direction: to cooperation or extinction, to more vulnerable as well as to more sturdy organismal structures.

Implicit appeals to latent beliefs in naturalistic ethics and design are used today to sell things from soaps to ideologies. We live in a society where "natural" foods and even shampoos with "natural" ingredients sell briskly. (The eating of human placentas was even reported among some tiny groups of middle-class Americans who decided it is "natural.") On the other hand our society is diverse, and some of us at times are attracted by words such as *new, synthetic, modern-wonder invention,* and *scientific breakthrough.* Artificial is good when it is part of the supposedly inevitable march of "progress."

A wide variety of supposedly unnatural sex acts are illegal in many areas. "Artificial" birth control is a sin for many Westerners. Beliefs such as these, in their most extreme forms, contribute directly to the population explosion in some countries and cause many millions of people to have to cope with poverty, anxiety, or guilt. Moreover, one may believe so absolutely in the rightness of so-called naturalness that when they see others reject this idea they can become very upset and concerned that society is going to the devil. For some, social changes such as women leaving their traditional, supposedly "natural," roles may be as much cause for concern as rising crime rates.

While it is openly antimetaphysical, Marxism too, has been accused of being covertly infected with teleology and related beliefs in several respects. Is there the implication of moral progress and final causes in the assertion that history moves towards a goal? Does the assertion that ends justify and are superior to means imply necessary evil? Does the assertion that some states are more advanced than others imply essentialism and natural hierarchy (see also Collins 1967)? Some forms of Marxism claim that they follow a "logic" of nature and they use this to claim a moral superiority and to justify acts that would be condemned if they were done by non-Marxists. Of course Marxism traces back through European common sense through deism and theism to Plato and Aristotle, just as do other Western philosophies and religions.

A Few Metaphysical Elements in Modern Science

It is clear that ancient philosophy can still affect our judgment, laws, morals, opinions of ourselves, and treatment of others. But does it still influence modern science as it did in the past? It may in too many general and particular ways to mention here. Most scientists enter the profession with unlabeled metaphysical preconceptions, and it is very hard to learn to see and correct these. A scientist may correct his or her thinking only after years of struggling with a concept and realize, "Darn, I was being teleological!" or "Gosh, I was thinking typologically!"

It is still not uncommon for scientists to calculate a statistical average and treat this value not simply as a statistical measure, but as the representative of the entire category, containing the essence of the category. Whereas of course one can average anything. The average size of objects in my living room does not represent anything typical or essential about the objects in the room. (See also Beatty 1989.)

There remain metaphysics surrounding common views of science as a *necessarily* progressive social enterprise. Yet for major thinkers since Rousseau, the discoveries of science and the developments of technology seemed to undermine the early Enlightenment faith that more science would *automatically* result in a better, more virtuous world for all. The findings of Ricardo, Malthus, and Darwin seemed to agree that leaving life up to the "laws of nature" could more easily result in disharmony than in harmony and virtue. Science and the laws of nature are morally neutral. This is not to say that a vast fraction of the middle class and many scientists became disillusioned. They largely saw Darwin as a form of social Darwinism and a justification for inevitable progressivism, and Rousseau became misconstrued by many as a "back-to-nature" romantic.

However, the massive and senseless slaughter of World War I and the worldwide depression did progressively cause more people to question the *inevitability* of scientific/technological utopia. Whereas *Faust, Frankenstein*, and *The Food of the Gods* had earlier portrayed *individual* scholars and scientists erring, Capek proclaimed in his 1921 play, *R.U.R.* (where robots first appeared), that scientists are naive to think that humankind will be able to retain control of technological eras. Aldous Huxley wrote of total technological dystopia in 1931 in *Brave New World*.

Yet natural and social scientists mostly continued a faith in the *inevitability* of progress until the atomic bomb and Carson's *The Silent Spring*. The specters of an arms race and of industrial competition, of supposedly rationally and scientifically organized corporate dynamics and the corporate mentality, leading to widespread pollution, economic traps, or even mind control or conflagration led even many scientists to doubt that more science would *automatically* lead to progress. Some scientists themselves even began the *Bulletin of the Atomic Scientists* to discuss their concerns.

Still, much metaphysics remains today covertly in science and in technological thought in deistic and in more secular versions. The idea that more knowledge together with economic competition, perhaps corporate organization, will *automatically* push society along some track toward a better world for all still sells to a very large constituency. This is one reason that reductionism can so easily transform from a wonderful methodological utility into a naive ideological imperative (chapter 4, note 2). In caricature, the ideological reductionist would insist that the only "true" science is that which investigates the basic essences of nature in the language of mathematics, the forces of physics, and the materials of chemistry. All knowledge can be reduced to these basic truths, and other forms of science and scholarship are "primitive" and "inferior." Any knowledge unique to higher levels of complexity is illusion. More funding of this "true scientific" knowledge will automatically result in better technology and a more virtuous world. Calls for thoughtful caution and careful planning are nay-saying. Much of what passes for materialism today drips with Greek metaphysics and the secularized, self-serving vestiges of deistic natural theology.

The notion of *inevitable* progress comes to us ultimately from design thinking, and even scientists often assume uncritically that what is new is necessarily best, rather than a decline, an interesting or uninteresting detour, simply more knowledge, or yet another fad. Certainly we have made meaty progress in many areas. But, for example, scientists may tend to ignore the older, supposedly "antiquated," "obsolete," literature and read only the new. Their teaching may reflect this same habit, and the students may be left with peculiar ideas of how the latest interests in a field can help them to grasp life intelligently. I have seen courses in introductory biology that are primarily biochemistry, with perhaps a few dreary sections on the memorization of taxonomy. I have seen introductory courses in physics that never would prepare the students to evaluate soundly the place of physics in human thought. The conceptual dimension may be passed off as "historical stuff" or "philosophical stuff" by the "practical-minded" scientist who is focused on keeping up and staying respectable in terms of the fashions and trends in his or her field.

Yet almost any leading scientist will offer the opinion that perhaps even most of the contemporary literature is not very interesting or important, and at the same time he or she will have a few stories about turning to some older works and finding them to be a wealth of detail, thoroughness, and even fresh and worthwhile ideas. When I studied at the Brain Research Institute, for example, it was a standing joke that someone would conduct a neuroanatomical study only to go to the library and find that the great Cajal had already made the finding. Examples are easy to give. I well appreciate that there are several factors involved in why we tend to overlook much of the older literature, but the point is that one reason involves the common uncritical assumption that we are heirs to progress and that the latest will be best—despite many conspicuous examples to the contrary. We will even laugh at ourselves for behaving this way, shake our heads and go right on, some of us not quite understanding why we do it, and uncomfortably offering diverse rationalizations when we get into discussions about it.

Scientists may also slip into preaching naturalistic ethics. In a society where many people unconsciously or even consciously base at least some of their ethics or morals on feelings and opinions about what is natural, scientists can come under various external and internal pressures and enticements to satisfy an audience interested in the question of human nature or the "way of nature." Many people want scientists to tell them what is natural so that they can try to figure out what is right and wrong, how to live their lives. Scientists can be swept up by such attention and begin to feel unrealistically important as individuals. They may start to claim that they have simple "scientific" answers to difficult social questions.

Ideally the scientist would *try* to approach research with complete objectivity. Some growing number, though, argue that objectivity can never be approached, and so instead we should adopt an adversary approach much as is used in the courts of the United States: polarize an issue, let the sides fight it out, and assume that the victor is right. There are deistic underpinnings here again. There is the old middle-class European notion from the 1700s that competition is the natural law that produces a supposed organization in nature and human affairs and that creates moral progress. This is not too unlike Adam Smith's Invisible Hand in spirit.

While there is a bit to at least seriously *consider* in their view, my observation is that in practice, as often as not, the adversarial system is used to conceal dishonesty, incompetence, sloppiness, laziness, insincerity, and mediocre intelligence and is responsible for much of the venality that the public notices among scientists. I am not convinced that this is a productive trend to endorse.

Moreover, one recalls lessons from life in general. I recall a German film by Reinhard Hauff, *Knife in the Head*, in which a geneticist is injured by the police, who paint him as a criminal. His friends paint him as a martyred victim of police brutality. His brain is damaged and he despairs over which side to believe. The truth turns out to be simple, but not at all what either side had envisioned. Both sides can easily be wrong, as we all know. Adversarial systems such as courts may not be a good model for approaching the truth. As Justice Holmes once said, "This is a court of law, young man, not a court of justice." In legal cases, decisions must be made within a certain time frame and perhaps an adversary system is not a bad way to get the job done. But in science the object is truth, not to get the matter resolved and to move quickly on to the next case. Making quick decisions is bowing to external requirements of reputation, grantsmanship, and promotion. Of course one cannot live apart from these. But neither should we embrace them, with motivated inattention to their potential costs.

This discussion has public-policy as well as individual implications. Our universities and federal and other granting agencies have policies that directly and indirectly promote competition and polarization among individual scholars and departments. What is the effect of these policies? Some see healthy consequences and others see quite the opposite. In science, *ideas* should compete, but on the other hand *individuals* need to cooperate in many ways, and it is much too simplistic to suppose that there is a tidy relationship here that can be easily managed from above. Moreover, competition does not necessarily promote the free expression of creative energies. It can smother one by forcing participation in faddism. It can even encourage dishonesty.[7] I will deal with this subject again in other chapters. But for the present, let me say that the damage done is easy to see and the presumed benefits seem mostly hypothetical and wishful. I believe that there would be quite too much competition and polarization even without institutional pressures.

While we have gone a bit to the edge of the subject of Greek metaphysics, the discussion has taken us deeper into the serious problem of hidden biases in judgment and values. They are deeply rooted and not so easy to pinpoint and excise. They are like a banana peel or ice on which, in a bad dream, we seem destined to keep slipping, after each fall regaining false confidence and rushing blindly ahead. For now let me suggest that perhaps the best the thoughtful individual can do is to try to understand some sources of our biases and how they may affect us, and try to explore personal strategies for negotiating around them.

NOTES

1. Similar thinking superficially becomes codified in classifications of plants and animals and human societies into primitive (inferior) and advanced (superior).

Early biologists held a complex variety of progressivist views (Bowler 1976) and even some modern biologists who have not studied the matter, along with most of the lay public, continue to see creatures on a scale of progressive advancement toward *greater fitness and complexity*. The facts belie this. Parasites and cave animals are conspicuous examples of creatures that have become more *simple*, and in some sense more fragile, in their evolution. Even in the evolution of vertebrates there is overall simplification of the skeleton and musculature. *Some* features may become more complex in the evolution of vertebrates and other groups (for example, the musculature of snails, or some frog tongues), but there is not a uniform trend toward greater complexity. Even at that, complex forms are not more fit (than, say, bacteria), nor protected from extinction. But *if* there were a uniform trend to produce complexity, would it salvage Aristotelian metaphysics? Is complex uranium a goal, or *better* than simple hydrogen?

2. Modern philosophical and scientific treatments of teleology usually have a different flavor than I present here. Academic philosophers in the recent era have been concerned with logic, ethics, and semantics, and they have tended to view history and other disciplines in terms of these current issues. Today they are very much concerned with the *logic* of teleological arguments (e.g., Woodfield 1976, and essays in Canfield 1966.)

In evolutionary biology there are actually reasons for using words like *purpose* or *ultimate* function. In biology one cannot agree to stop inquiry once there are mathematical descriptions, as is often possible in physics and chemistry. One must deal very carefully with the tricky questions of *ultimate* causes, the primary functions and survival values of the traits of organisms. Ideas of community organization, or "the balance of nature" in ecology, and of "motivation" or "drive" in behavior, for example, have origins in Aristotelian teleological thought and are tricky; the semanticists still struggle with this legacy.

It is well for the reader to keep in mind that this published dialogue about teleology in modern philosophy and biology focuses most intensely *not* on the overall theory, and how subtly the notions of purpose can interact with notions of essence and design, and even naturalistic ethics, including questions of superiority and inferiority, but primarily on semantic and logical problems in *describing* organic form, function, and purpose. For example, natural selection involves selection of features in a sense "for" particular survival functions or "purposes." One proximal function of the bird's wing is to add weight to the bird and to get caught in thorns, but this cannot be its function of origin, which was flight; ergo (one might argue), can one not logically *say* that ultimate purposes *do* actually exist in nature? If, though, a penguin uses the wing only to swim, then was "the purpose" of "the" avian wing still flight?

It would be most correct to say that, of a structure's *many* physical, physiological, or behavioral functions, it is those that have survival value to which we might first look in trying to understand how the structure evolved by natural selection. Talking of "the" (essential) function or "purpose" of a feature is mostly sloppy shop-talk that will not hold up to scrutiny and that has little value aside from allowing rapid conversation. Describing evolutionary *causes* can allow us to *imagine purposes* or *goals*, but so what? This does not have serious philosophical or metaphysical implications and the matter is largely one of propriety in semantics. What do we gain by saying that "the purpose" of a wing is flight, rather than simply to understand that flight has survival value for birds of most species and that the wing has evolved largely in connection with this function and by an opportunistic process? Little, unless we are trying to salvage the word *purpose* with its metaphysical nostalgia.

3. One can correctly argue that there was *no* true Darwinian revolution at all, since so many scholars, even biologists, misunderstood Darwin and used his name to perpetuate old incorrect progressivist views in new forms (Bowler 1988). But Darwin did introduce powerful new materialistic perspectives that the most serious thinkers would have to deal with over the decades. He had a profound effect even if *revolution* overstates a complex historical phenomenon.

4. Natural theologians argued that the patterns of relationships that were revealed between species by comparative anatomical studies do not necessarily reveal genetic relatedness and evolution over time, but are instead God's design.

But it was *departure* from pattern that in the end helped to win over the opponents of evolution. Limb vestiges, for example, can be found in the adults and embryos of some limbless vertebrates such as snakes. However, studies showed that some species retain such rudiments and others do not. This was evidence of disorder rather than harmonious design. Extinctions: why would a wise and virtuous God create forms and then destroy them?

5. The worlds of Plato and Aristotle *as a whole* were static but allowed the optional earning of *moral progress*. By about the sixteenth century Europeans had been strongly confronted with the reality of social and geological change, and by the late 1700s the idea of social change had become associated with the old notions of *moral progress*, though produced by the natural course of things. Utopias were being placed in the *future*, rather than *elsewhere*. I would support the idea of the *possibility* of humanly directed progress. But I am critical of the notion that progress will *automatically* and conveniently be produced by natural, social, or economic forces without intelligent intervention.

6. Social Darwinism is a complex issue and much in need of a fresh and careful reexamination. I dislike the term, which is not only a misnomer but protean and imprecise as well. Yet it is established in common usage. In some versions advocates are promoting ruthless *individual* competition, for example, and in other versions they are promoting the rightness of the competitive supremacy of one *interest group, class,* or *state* over another. The exploitation, for example, of so-called primitives by moderns.

Bannister (1988) details these and other complexities (also Bowler 1988). He also details that specialist *scholars* such as Sumner were not so unsubtle as history often painted them. Hofstadter's (1955) landmark review tended to paint social Darwinists in poster colors. Historians have not identified satisfactorily an essence of social Darwinism. Bannister's often interesting review, though, can be misleading since its own special aims can too easily be overlooked. He is concerned that left-wingers have taken to uncritical label slinging. The reader, though, can come away doubting that, whatever things have been meant by social Darwinism, it has had any reality or weight at all as ideology in popular thought, public policy, or scholarship. Bannister may go too far in rehabilitating the reputed social Darwinists.

Social Darwinism in *popular thought*, as social psychology if not as formal ethical theory, has been quite pervasive, though. It is no mere myth and it is easy to encounter today. As a biologist I probably notice a great deal of it and may ponder and analyze its appeal and nuances, since it stands out as such superficial biology used with such complete conviction. It is disturbingly common to find not only the public but even mature scientists using essentialism and naturalistic ethics and calling this objective materialism in this and other instances. I repeatedly have to deal with this with biology students, too, who expect that they will be able to use evolution to fortify their particular preconceptions about the essential brutality of human nature and the "rational rightness" of associated social norms. Indeed, social preconceptions about what evolution is and how it works have *historically* always been major barriers to the correct understanding of evolution, even by professional biologists (e.g., Bowler 1988).

Social Darwinism is most clearly understood as the crude biologization of the effort from the end of the Middle Ages to rationalize a set of "non-Christian" or "noncompassionate" sentiments and behaviors as *basic* and *natural* ("biological"), and usually too to condone them as therefore *ethically right*. The need and effort to develop such an ideology can be traced back to well before genetics or even Herbert Spencer, and had to do with the need for the middle class and government to try to defend individual, class, and national selfish competition and compassionlessness as competing mercantile states developed, and got into the habit of needing wealth to support large armies and navies. In this sense specific details of the ideology, such as the use of genetical language, are less significant than its functional status, to meet the social and emotional needs to reconcile the (still problematic) ethical conflict between the values of new competitive, wealth-oriented social and economic systems in Europe, against the values of the unambiguous *traditional* Christian *idealism* of the founding era and Middle Ages, that society should encourage cooperation, meekness, charity, compassion, and brotherhood, *not* self-love, pride, competition, compassionless indifference to the misfortunate, and the accumulation of wealth and luxury. Darwin got drafted in the ongoing cross fire because in a crude form he was useful to the ongoing effort to ground the rationalizations in "human nature" and biology.

Before going on, let us get one thing out of the way and not let it become a red herring. Even the ancients long appreciated that humans *are quite often* hypocritical and vicious and will deceive themselves. Whoever seriously doubted that? Even Rousseau, contrary to popular impressions, did not claim that "noble savages" have really existed. This was for him a theoretical abstraction or potential, like Galileo's frictionless plane.

But what of an *essentialist* badness or goodness? Behind the strife and meanness, are people "basically" good? Defenders of social Darwinism often claim that essential goodness, or that human behavior is completely plastic (as some Skinnerians and Lysenkoists *do* seem to advocate), constitute the *only* alternatives to their beliefs. But this is just cheap debate to make their position look more reasonable than the alleged alternatives. These "sole alternatives" are straw men in this context. A *few* religious writers and *some* believers in inevitable progress did argue that humans are "naturally good," especially in the seventeenth through nineteenth centuries (Lovejoy 1961). But there are not only two cartoonish positions available (basically good or completely plastic), as alternatives to social Darwinism. End of digression.

Mercantile states and new ways of economic life developed in Europe with the eclipse of the Middle Ages. People tried to understand their own behavior. Why were professed Christians behaving as they were and was it allright? Some sought to condone the new social values using naturalistic ethics and the language of science.

Thomas Hobbes (1588–1679) argued not merely that humans have the biological capacity to be vicious and that some conditional factors may elevate the level of such behavior; but he took an essentialist position and linked it to naturalistic ethics. In his eyes, a human being is not even a social animal, but in essence a ferocious animal (not original with Hobbes), who has a natural right to compete and strive for power and profit. It is *naturally* just for one person to invade another's rights, and in the state of nature there is naturally a war against all (*bellum omnium contra omnes*). But, in the end it is *rational* for humankind to give over some of their natural rights to the state or they will destroy themselves (e.g., Thilly and Wood 1957).

Today scientific evidence suggests that this sort of view is quite simplistic. Even aside from the wealth of information on the potency of socialization processes, and the high degree of gentleness in many smaller human societies, there are newer comparative studies. For example, chimpanzees will sometimes squabble and fight over food and other things *when these are limited*. And they will defend group teritories even quite violently against *foreign* bands (e.g., Goodall 1986). Yet *within a community*:

> chimpanzee males are surprisingly generous when it comes to material things. . . .
> Their control rests on giving. They give protection to anyone who is threatened and
> receive respect and support in return. . . . The two basic rules are "one good turn
> deserves another" and "an eye for an eye and a tooth for a tooth." . . . The rules are
> not always obeyed and flagrant disobedience may be punished. (de Waal 1982)

Indeed the stability of that social system seems to be based on patterns of *sharing*. If one insists on using anthropocentric terms, then our nearest relatives show a lot of charity and ethical behavior, with or without conscious reason, as do, say, hunters and gatherers. Even studies at playgrounds suggest that dominant children tend more than others to protect and share. No noble savages here, but no "basically" ferocious animals either.

Bernard Mandeville (1687–1733) was one of the most widely read and influential writers of his century. Like Hobbes he took an essentialist position, and the essence was "bad," though he went beyond Hobbes. To fortify his essentialist argument, his strategy was to locate the element of self-interest in just about every behavior, including supposedly altruistic behaviors, and to claim that he had reduced them all to equivalency. Since self-interest was a vice in his day, he argued that all private behavior was vice. He even reduced compassion to vice. "The weakest Minds have generally the greatest share of it, for which Reason none are more Compassionate than Women and Children." This was evidently before "bleeding-heart liberals" was a marketable stereotype to label people who, allegedly only to selfishly soothe their squeamishness, or to look generous, supposedly mess up the natural course of things.

It may have been an intellectual eye-opener for many Europeans in those days to reveal that there can be a large component of self-interest even in conventional sanctimonious rhetoric and posturing. And *surely* there *can be*. Mandeville saw some very real weaknesses, indeed real dangers, in the positions of the zealous Societies for the Reformation of Manners. Moreover, he might have been the first to detail that people are not merely deceptive, but that they are extensively deceived and *self-deceptive* when it is in the socioeconomic interest. In *functionalism*, behaviors exist to support the social system no matter what the actors may think their motives are or what the net effect of their values and behavior patterns are. He should be credited here as a pioneer. (See also false consciousness, chapter 9.) He was a physician (who specialized in nervous disorders), and in his mocking social writings, supporting his satire, his thinking superficially was empirical, insightful, and scientifically analytical. *As a physician* he was actually an excellent empiricist for his times.

But Mandeville went badly overboard in trying to *equate* all forms of self-interest. He also went on and argued questionably that all vice (*not* major crime, though) should be tolerated since in his observation it is natural and it produces the wealth and power of the State. Man's "vilest and most hateful Qualities are the most necessary Accomplishments to fit him for the largest, and, according to the World, the happiest and most flourishing Societies." Government should not step in to stop these but should allow "natural" behaviors to play out. Man is "an extraordinarily selfish and headstrong, as well as cunning Animal," and only flatters himself to have any moral virtues when he has been manipulated by "the skillful Management of wary Politicians." Pride, selfishness, greed, vanity, cunning, lust for luxury thus are not only the *essence* of Man, but it is rational to condone these (*contra* Hobbes) since they produce the glory, wealth, power, and industry of the happy state, he argued in *The Fable of the Bees* (1714; quotations here from the composite edition of 1924 by Kaye). Thus he linked the much older essentialist notion that humans are basically "bad" to design and teleological thinking—he saw a social design and final causes that he took as a given were good.

Like many of the social Darwinists who would follow in his footsteps, Mandeville argued that humans have an essential human nature and that it is selfish, covetous, greedy, vain, competitive, cunning; it lusts for status symbols and luxury; and it is rational that this should be tolerated or only manipulated, but basically not stopped because it is *biologically determined*. And it produces good.

Mandeville did reject some versions of deism and teleology, and he stopped short of claiming that this is the best of all possible worlds (Gunn 1975). But the underlying logic of his argument is consistent with the familiar logic of the design teleology, defenses of necessary evil, essentialism, and naturalistic ethics that prevailed in his age. He was a little more than kin and less than kind of his critics Adam Smith and Alexander Pope, for example (see also Lovejoy 1961).

With parochialism, he took the goodness of mercantile societies as a given (since they seemed to embody the values of his own class where one admired, desired, and respected power, wealth, and glory). He did not weigh well that (the "plan" or "design" of) his whole system, public as well as private, might have been flawed, for example, or even rotten. By defending all allegedly "necessary" evils he cut himself off from the option of weighing a full range of alternatives and systemic improvements. He did not weigh well the extent to which the way of life in societies such as his own may have actually exaggerated traits to produce the behaviors that he saw as basic human nature. Indeed, his reductionistic and essentialistic leveling strategy seems to have left him not well able or inclined to help the reader to distinguish low-level traces of capacities for envy, ruthless competitiveness, or lack of compassion from conditional high-level exaggerations. To what extent can a way of life, covert notions that good and love are in limited supply, the ways we treat children, and the role models we provide for them, bring out "the worst" in people? (A *lot*, we now know. See Alice Miller's books, for example, and Montagu 1978.) In such ways he lacked the analytical depth and synthetic power of other Enlightenment essayists on hypocrisy, such as Rousseau.

Mandeville made the common mistake of using his view of his own peers and class as a paradigm for "basic" human nature. Thus, for example, he claimed that high levels of greed, envy, competition, and lust for luxury are natural and inevitably drive people to work hard to make lots of money. No point in trying to stop this. It is "biologically determined." But he slipped, as Horne (1978) noted. The poor (most of the population), Mandeville argued, are lazy and must be kept poor and their education limited to keep them motivated and capable of being manipulated to work hard.

One should keep in mind that after brandishing a crude, attention-getting view of the world, he would, especially in subsequent editions of *Bees*, qualify his views. So for scholars, there is a Mandeville that holds up a little better to critical examination than does the cruder bombastic Mandeville that so captured the public imagination and implanted his perspectives in popular psychology. Scholars ask, which is the real Mandeville? For present purposes, it does not matter.

Mandeville's books were much opposed, yet the crude versions of his sentiments and logic have endured and have been enormously influential in popular thought. He outlined a superficially coherent world view and metaphysics useful to those who had been developing extreme mercantilist economic policy and sentiments and to those who would develop laissez-faire (technically debatable but influential wealth-oriented economic philosophies that have had some renewed influence since the Nixon and Thatcher/Reagan administrations) in details and in the attention that they demanded by being stated so provocatively and entertainingly (Horne 1978).

But as phrased so moralistically and nakedly, his words were an embarrassment and he has been quite rejected in name. He seems to have had only one *confessed* follower among writers. Mandeville claimed that it is human nature to be greedy and selfish and otherwise "non-Christian" and that this is indeed vice. He argued that vice in society be named as such and *condoned*, (though, unconvincingly, not condoned in oneself) because vice (not necessarily crime) creates the wealth and power on which is based the admired state. Scandalous in his day, somewhat like deSade, he led the reader down a course of theological mischief.

Nevertheless, he was read exceptionally widely and his basic mind-set provided a popular metaphysic and perspective that one sees reflected in popular social Darwinism: a) The world is *inevitably* a dog-eat-dog operation in its essence (essentialism); b) ruthless competition and indifference to suffering should be condoned, brotherhood and the eye of the needle or not (naturalistic ethics); because these are c) allegedly natural behaviors and sentiments (biological determinism); and d) social wealth and good come of them; the end is good even if produced by much "necessary" evil (elements here of design and teleology); e) government should keep out of each person's economic dealings and natural venality or should skillfully manipulate these to increase their benefits (wealth, power, glory) to society; f) there is natural hierarchy, and supposedly the greedy nature of the rich is superior to the lazy nature of the poor (design); and it is rational and right (naturalistic ethics) to favor the one over the other.

In Mandeville, workers must also be kept poor and naive for the progress of the the happiness (wealth, power, glory) of the hive (state), as a reliable source of cheap, manageable labor. In Spencerian Mandevillism, though, they are kept so because they supposedly (there is no good evidence) have inferior heredity and it is natural and right that the "most fit" should prevail and exploit others as resources in the "inevitable" competition for survival, supposedly leading to social progress, and to a purification of the race. Another sometimes distinction: mercantilists have long favored population explosion because it ensures cheap labor (Teitelbaum 1988). But I have never heard a *biologically sophisticated* social Darwinist advocate population explosion. With such qualifications one might instead rename social Darwinism, or Spencerism, "geneticized Mandevillism." It is a little more than a recent chapter in an older effort to write a new pseudosecular scripture.

Mandeville's superficially empirical metaphysics (in alliance with the older mercantilist and the newer laissez-faire policies) had been insinuating themselves into European ethical and economic social thought for almost 150 years before *The Origin*, or even Spencer's *Social Statics*. Many apologists for callousness, selfish privilege, or ruthless competition were more willing to use the name and language of a respected if controversial natural scientist like Darwin to authorize their sentiments and round out their arguments than the name of a taunting, scandalous, satirical, if entertaining, essayist and pub celebrity.

In correct Darwinism *there is no essentialism*, no inherently superior genetic type for a population (also see Dobzhansky 1960). The fitness contribution of genes changes with, is relative to, circumstances, and these are often shifting. Fitness is *not* an absolute, essential quality or value. Moreover, only in *some* cases is fitness manifested in strength, cunning, aggressiveness, or in other common images of competitiveness. This too is conditional. Progress too is a relative term in biology (e.g., note 1).

There is no "law of nature" to support popular images of competition, and naturalistic ethics. Natural selection and self-interest have also produced a variety of *cooperative* systems ranging from loose sociality even to multi-individual colonies such as corals; not merely the dramatic competitive situations that social Darwinists would like us to focus on.

Nature and the principles by which it operates are so diverse that if one attempts to use naturalistic ethics, one will find evidence for just about any moral position one cares to argue. Natural diversity simply contradicts essentialism and presents enormous complications for naturalistic ethics.

The claim that it is *right* to compete to spread as many genes as possible, is not *necessarily* what is usually meant by social Darwinism, but it deserves mention here since some enthusiasts of sociobiology like to extend the technical theory on which this is based to ask us to wink at nepotism, racism, rape, etc. But even the basic technical theory is complicated. It is not simply optimal to have lots of children. *How* should one invest one's genes in terms of the specific biology of mates, for example? To what extent should one inbreed or outbreed? Even crude theory for this is not worked out. (For a critical technical discussion of other aspects of sociobiology *specifically*, see Kaye 1986, Kitcher 1985, and references therein.)

It may be that to be rational or right by such thinking a white American should travel and mate with African blacks or Indian Hindus to take advantage of heterosis and high reproductive rates in those populations or others. Since most white Americans do not do this, their reproductive behavior would be irrational or wrong by such logic. Or, maybe they should instead mate incestuously with those who have genes most like their own. But the theory has been so superficial and tentative that it has been easy to argue post hoc that whatever they are doing on average is right, whatever that is.

The co-option of Darwin's name and of some fuzzy genetic thinking by Mandevillians and Spencerians has had one special, important impact other than merely to confuse historians. By linking genetics to essentialism and naturalistic ethics, and sometimes design and teleology, the way was paved for extreme forms of eugenics. This contributed to new and especially terrible, if bizarre, chapters in the history of racism. It appears that recent historians of social Darwinism have tended to overlook such "pre-Darwin-and-Spencer social Darwinists" as Mandeville because they have followed Hofstadter (1955), who wrote in 1940–42, when it may understandably have seemed, as he noted, that eugenics had "proved to be the most enduring aspect of social Darwinism."

Historically, there was a powerful consequence of the dog-eat-dog paradigm on the side of *caring,* too. Many argued that if science does show that we and life are actually so brutal, then we must work especially hard, it is in our self-interest, to try to think out good values, to make a better world, and to contain the dangers that natural forces evidentally offer. Some forms of this response, then, recall Hobbes's position: it is rational to curb "natural" ferociousness. But other forms would not necessarily involve an esentialist view of "human nature."

7. Here, in assuming that honesty is important to establishing truth, I may be accused of bourgeois sentimentality. Such criticisms come from the more general argument that objectivity and honesty are myths. There may be faith instead in intellectual progress caused by a dialectical logic that involves constant clash. Issues of this sort will be explored and criticized in chapter 9. Some critics of science are vocal that capitalist science serves and has sold out its ideals for career ladders manipulated by money interests, and so it is inherently dishonest. This is a very different preoccupation than the issue at hand, as will become more clear in chapters 9 and 11.

CHAPTER 7

Language and the Construction of Reality

The reason for a fish trap is to catch fish, but when the
fish are got, the trap is forgotten.
The reason for a rabbit snare is to catch rabbits. When
the rabbits are got, the snare is forgotten.
The reason for words is to convey ideas. When the ideas
are grasped, the words are forgotten.
Where is the man who has forgotten words? Eagerly
I would talk with him.
Chuang Tzu, "Means and Ends" (fourth or third century B.C.)

Language Informs and Misinforms

Clear thinking is not simply logical thinking. Illogical thinking can be an enormous handicap, but a failure to appreciate the nature of words and language, their power to create illusion, their psychological and political dimensions, is a common and major aspect of muddy or manipulated thinking.

Language may *seem* merely to be a vehicle for objective description of reality. As late as 1473, teachers at the University of Paris were actually bound by oath to teach the *realism* of words (in reaction to the *nominalism* that threatened to emerge and disturb their Platonic/Christian realism from throughout the Middle Ages), that words represent eternal Realities. But in fact, concealed in language are schemes to organize life's complexity.[1]

Language directs how one breaks up the complex world into categories, selects items and events to pay attention to, and puts values on these. It helps *construct* one's reality and does not merely *describe* it. How are we to grasp that by which we grasp? The problem is a bit like the difficulty of trying to taste one's tongue.

An effective, working, conceptual grasp of the hidden aspects of language is far more difficult to master than a conceptual grasp of logic. That is the main point of this chapter. "The word is not the thing," "The map is not the territory," are phrases that have been used by semanticists and logicians for generations, but still they require restatement. Chuang Tzu made a similar point thousands of years ago in "Means and Ends," and still we stumble on this issue.

The symbols in a logical sequence must be absolutely fixed for the conclusions to be valid, as we fix symbols in mathematical manipulations. But words are not fixed symbols by any means. Generally each has serveral distinctly different possible meanings and also different subtle shadings of meaning in different contexts and to different persons. One may put words into a logical manipulation and not notice that the context at the end of the manipulation has changed subtly in such a way that different nuances of the words' meaning come into play and the conclusion reached logically is in fact misleading and invalid. Logic alone could never provide reliable thinking, which is a major reason why academic philosophers have expanded their focus from symbolic logic to include semantics and the like. Language and meaning have become highly technical subjects, but for our purposes it is sufficient to keep in mind that the issue is, as has been said of Mozart's music, too simple for children and too difficult for the virtuoso. When we think we have mastered it we find we have not.

It must be said at the outset that whatever terms I may use for brevity, even verbal language is more than words and grammar. It matters enormously in what *context* these are used and seen. One might assume very different intentions if told the words "I love you" by a parent, a date, a friend, a minister, or a complete stranger on a dark and deserted street. Words have great power, but one cannot simply change behavior by changing words, as one might assume if the following discussion were merely read superficially.

Convenience Redefines Words

Voltaire observed that, "It is forbidden to kill: therefore all murderers are punished who kill not in large companies and to the sound of trumpets."

Combat and killing, armament, capital punishment, and such present difficult moral conflicts, and particularly for those who accept the Ten Commandments as the word of God. The words of the commandment of Judaism, Islam, and Christianity are plain and unambiguous—"Thou shalt not kill"—period. There are other contrary instructions from God in the Old

Testament, even in the same book of Exodus; for example, "He that curseth his father or his mother shall be surely put to death." Should God or the Hebrews do such killings? We may presume God, since the people were shortly instructed in unambiguous terms not to kill. Moreoever, any such instructions *to* kill may be entirely suspect, since the reports indicate that those *not* to kill were written out on stone, and with thunder and lightning and the earth trembling and trumpets and clouds of smoke, and with God descending in fire and in many other ways making a very special occasion of the presentation of the Ten Commandments.

The Hebrew definition of killing may not translate well, but the moral issue for Christians, at least, was made very clear. Jesus took a pacifist position and repudiated even defensive armed support even as he was taken to his execution, insisting, "All they that take up the sword shall perish with the sword." That should settle the matter for Christians at least. Indeed, those closest to Jesus, *all* Christians for the first several generations, were conscientious objectors, until about the time of Constantine, in the third and fourth centuries. Yet the blood has and will run—believer against nonbeliever, brother against brother, faith against faith, nation against nation. All religions, even the vegetarian Jainists of India, who will not usually kill any life including insects, eventually find justifications for the killing of their fellow humans, as John Ferguson, for example, details in *War and Peace in the World's Religions.*

Of course many followers of the biblical religions would never kill anyone under *any* circumstances. There are those who would only kill if there were a *clear and immediate* danger to their own lives or those of loved ones. But even official and routine killing for more abstract reasons is sanctioned—for example, to destroy a presumed enemy, punishment, the theory that execution discourages crime, warfare to show strength, "to protect national interests," or "honor." It seems that somehow *to kill* does not always mean *to kill.*

Once it is agreed that it is all right to kill in self-defense, then the definition of *self-defense* will be stretched until it becomes convenient. People seeking to promote an arguable case will commonly try to get a eulogistic label, with positive connotations, for their cause, rather than a dyslogistic label, with negative connotations, that would make the task harder for them. George Orwell's explorations of the politics of language are splendid reading (in *1984* and elsewhere). Bureaucrats generate some of the most outlandish examples in order to promote and defend their programs, but the practice is quite general and this is one reason that language evolves. Our Department of War is now a Department of Defense. Some "euthanize" animals in laboratories whereas they were once killed. The words *democracy, individuality*, and *freedom* have incredibly diverse usages these days. And so on.

In the minds of some Western religionists there is no conflict between unambiguous statements of morality in the Ten Commandments and social desires and needs. Perhaps God's words were not chosen carefully. Perhaps God intended to command only that socially unsanctioned killing is wrong. The common sense of some expedient, aggressive, or indeed militant, Christians tells them that Jesus, the Prince of Peace, could not have been literally serious when he said that the meek shall inherit the earth (or that one can be meek and yet aggressive?). This would be convenient indeed for those who believe that the Bible is the word of God and not primarily a record of Hebrew and early Christian history and beliefs.

The earliest Christian leaders took seriously the doctrines of love and nonaggression, for example in Romans, "Bless them which persecute you: bless and curse not." Or, "Dearly beloved, avenge not yourselves, but rather give place unto wrath: for it is written, Vengeance is mine; I will repay, saith the Lord." As a child I was raised to believe such things deeply and so I quickly became very sensitive to the fact that many Christians I met in our expedient society skip over such central messages or find convenient ways to rationalize them away so that they can behave as they feel society and common sense demand. Most groups, religious or secular, define words and turn ideas to their own convenience. It is the rare group that does not.

It is sobering to reflect that *no* great religious figure, Moses, Jesus, Mohammed, Buddha, has been able to state a doctrine so clearly and logically that people have not changed it radically and then institutionalized their own version in that figure's name. There are contending sects in *every* major religious tradition.

Religion provides good examples of the adaptive plasticity and evolution of words, because one might think that lightning and thunder and the fear of eternal damnation would override other considerations and fix meaning. They do not because the issue is a perceptual one and not a logical one. The dynamics are largely unconscious.

People will modify for their convenience the *boundaries* of categories that words supposedly describe. "Bacon is *not* pork," a Jewish friend once insisted. "My family *is* kosher and *we* eat bacon. Bacon is different from pork."

The power to define can be the power to destroy. A major way to mistreat or even kill people if one wants to do so and still sleep well is to tell oneself that one's enemies or subordinates are off in *another* category, some version of "subhuman." Secular people may not need to visualize the subhumanized ones as evil or morally inferior, but only as trivial and discredited on the basis of their lack of adherence to "civilized" standards. This has been a popular habit over the centuries.

It was done to slaves, exslaves, women, poor people, homosexuals, and colonials. Margarita Levin (1988) argues that some modern radical feminist scholars are now prone, ironically, to trivialize and discredit a wide range of individuals, groups, knowledge, science, and culture that they are opposed to with the label *masculine.*[2] Extreme economic determinists among Marxists and capitalists alike will dehumanize others by using labels that strip away individuality and reduce them to historical and ideological forces, and to creatures having only false consciousness or brute concerns. Whatever truth there is to either position is all too often lost in exaggeration and cruel abuse.

Language and Power Relationships

We usually distinguish between illegal killing (*murder*) and legal/moral killing (*execution*). But what unit of society can make the distinction for us and thereby sanction killing? What are the power relationships in language? Who, what group, gets to define right and wrong? For believers in the Bible the issue is who gets to define God? Much religious disagreement and crusading boils down to the attempts of factions to "own" the definition of God.

What is *society?*—our circle of friends or the whole human race? If a group of cronies decides to harass or kill someone, we tend to disapprove and label them a gang or mob, and lynchings and kangaroo courts have unpleasant connotations for most of us even though they may be socially sanctioned killings. If the governing body of our larger society (the ultimate power structure to which we must answer and bow?) agrees to executions and bombings, then we may feel differently. Most feel that killing is *morally* justified if it is *legally* proper. Yet the Nazis killed many German citizens "legally," and we can agree that this was wrong. The Nazis saw it as proper to kill people who they saw as enemies of society and the state in terms of politics, beliefs, or genetics.

The rightness or wrongness of other examples of socially sanctioned killing are more difficult to agree on. In the undeclared Vietnamese war many powerful words were used to talk about the killing—self-defense, enemies, patriotism, duty—and yet the American public remained confused and divided over which labels to use. Was the United States "defending" itself or its "values" and thus justified in killing? There was debate over whether or not the Viet Cong were to begin with really our "enemies" and a "threat" to the United States, though America had become involved in the war because initially the opinion makers bought the government's use of those words.

Moreover, who kills a man or boy if he is drafted, placed in a highly dangerous situation, and then is blown up? My grandfather was killed by our own bombs in World War II. The bombs were intended to kill, but not our own people. Was it not murder? This is not an issue for me as an adult, though I did have to think it through when I was young and grappling with the semantics of moral issues. Can one human through considerable effort kill and not be accountable? In our culture, particularly since the Christian Platonic and Calvinistic traditions, *motives* weigh more heavily morally than the *effects* of actions do. In some other cultures the rule is more apt to be an eye for an eye and a tooth for a tooth.

This is not, again, a theological discussion. These are rhetorical questions. I raise them because they underscore that a simple act like killing can have a wide range of emotional connotations depending on the label and boundaries that we place on it. Words can cause our hearts to pound and bring us to our feet. Words such as "I love you" can create a physiological reaction in us whether the statement is "true" or not. As the song goes,

> How can you please me?
> Then torture and tease me?
> All in the name of love.

We reward soldiers and encourage them to kill for us and risk their own lives by labeling them patriotic, heroic, dutiful. Words like *killer* or *murderer* might be accurate in many cases but would hardly be rewarding, since they may carry serious connotations of disapproval as well as their technical meanings. Medals and honors say brave *hero*, not brave *killer*.

Such control generates an enormous amount of power when a person, group, or class can capture the definition of certain superficially innocuous words. Whose definition do we adopt of what is *basic* or *natural* or *essential* or *progress* or *scientific*? Such things set agendas.

Suppose reporters want to investigate the workings of a government department and the editor tells them, "That is not *news*. I want you to cover *events* as they *break*." Then the editor has defined the nature of what the most honest reporter looks for and reports or does not report. That will in turn influence what viewers or readers see or do not see, what we may think is important about the world, and what we think we know. When we uncritically accept the *newsworthiness* of items in papers or in programming on television or accept a definition of *news*, then editors have extended control over us in a very deep way. Unflattering photographs are in "poor taste" when a personality is popular, but become "revealing" and are published if they become involved in scandal. It is irrelevant whether or not editors are conscious of their power as they "impose" particular values. Their values have come into control whether they try to control us or not.

In fairness, news people at conferences often insist that it is critical for citizens to understand the nature and limitations of media coverage, and that the main threats to democracy and truth come when citizens do not understand such things and do not seek independent information sources. Let the buyer beware.

Media and reporters are loosely controlled by owners and editors, who form a sort of class or society of people sharing similar social values, including ideas of what is and is not news. This is a primary reason why the major networks tend to cover the same few stories each day out of the simply enormous number of dramatic, interesting, and significant events that are happening around the world. There is even a strong tendency to cover these stories at a similar level of depth. One becomes particularly aware of how narrow national media are when one travels abroad for extended periods and is exposed to different ideas of what is interesting and/or newsworthy.

Hayakawa (1950) pointed out that the Soviet establishment honored Pablo Picasso's communistic political views but did not display his "degenerate" paintings. The U.S. news establishment, on the other hand, honored his paintings but did not display his political views. In effect each establishment censored part of what Picasso wanted to tell the world. (Here, people in either nation might resent my use of the word *censored* as applied to their selectivity and exclusions.)

Behavior similarly is influenced very much by what the tastemakers have decided is "good" and what is "tacky" and one sees people change their clothing, behavior, furnishings, language, and so on, when the "right" people announce that something is "tacky." When I was young the term was *not-cool* for kids, while adults responded to *no-class*. Yet tastes go in and out of style, revealing that a thing or a behavior is not "tacky" in any essentialist sense.

At various times in our history we have seen struggles over which group gets to control the definition of *patriotism*. In the late 1950s I saw competing groups of conservative "little old ladies" in San Diego literally attack each other with canes and umbrellas over this issue of who the "true Americans" were. A few years earlier, the McCarthy era, in major part cold-war mass hysteria, may also be viewed as a national political struggle to control the definition of *loyalty*.

Labels help to identify and keep the ranks of each opposing faction in step. They may determine victory or defeat if public opinion buys one label instead of another. Control of words is control over the perception and thinking of others. Control of definitions is critical for the aims of political, religious, and other groups, and so the subtle and sometimes not so subtle, unconscious and sometimes not so unconscious, struggle for control is an ongoing process.

Individual Word Games

Often, changes in meaning may not be linked to philosophical or larger political agendas. People may misuse words that they hear floating around in common usage, which they do not really understand, because they sense in the public pulse that the words "sound" important. (Some behave as though this practice gives them the benefit of the collective judgment of the herd. It should take only a bit of reflection to realize what a dangerous attitude this is. It is only that one's personal thoughtlessness is inconspicuous, except from a distance.) The misuse may catch on and compete with or replace the original. As a small example, since experiments are high status in modern science, almost any research operation that involves data collection or laboratory equipment may be called an experiment by students and technicians, and sometimes even by senior scientists. "Buzz-words" certainly plague science too, however hard scientists try to standardize language.

At a meeting in 1874, Herbert Spencer, who promoted a wrong but influential version of Darwinism (social Darwinism) to the social sciences, was complaining because the physicists were insisting on a precise and restricted meaning, a too-limited and definite use of the word *force* for his purposes. The physicist Clerk Maxwell responded, and cuttingly noted of Spencer, "He had himself always been careful to preserve that largeness of meaning . . . by using the word sometimes in one sense and sometimes in another and in this way he trusted that he had made the word occupy a sufficiently large field of thought" (Lindsay and Margenau 1936).

In order to hold or gain power people may cite a famous concept or big name as supporting a certain position (often a commonsense position) when they are actually only vaguely familiar with the ideas of that person or concept. They parasitize prestige and respect consciously or unconsciously. The result may be that the person's work becomes generally misunderstood. (How many absurdities and horrors have been committed in the names of the respected, like Jesus, Mohammed, and so on? And why so few in the name of Bernard Mandeville or Jim Nobody?) This seeking of authority from respected authority is probably why Perry's respected Harvard undergraduate study (below) is misrepresented by many educators. Hypothesis testing in the name of the respected Karl Popper as *the* way that science must be done is militantly in vogue in many branches of natural and social sciences, yet among many strong advocates I have met probably no one who has read Popper or his commentators even moderately. They may be surprised or unbelieving to hear what Popper's position actually is (see also chapter 11). Einstein's theory of relativity and Heisenberg's uncertainty principle are often cited in support of all sorts of odd beliefs.

Scientists may try very consciously to define words precisely and to standardize their meanings, but they often fail. Science has made very much progress in this regard, but any superiority of scientists is relative, not absolute. They have often had no more luck than God did on Mount Sinai, with the exception of long-since resolved issues where in hydrogen is hydrogen and a gram is a gram.

People can hardly be stopped from redefining words if they want to badly enough. It is no wonder that an awesome profusion of languages and dialects have been among the most plentiful creations of our species.

When the meaning of a word becomes unstable in science or scholarship, it may be due to misunderstanding, carelessness, or intent. Sometimes it is indeed desirable to limit, expand, or modify the meaning of a word because research reveals that a phenomenon is actually more complicated than early hypotheses had envisioned. But if not all specialists then agree on the change, different usages come into style.

Such genuine problems set the stage for conscious or unconscious abuse. If a scholar can change the meaning of a word and get others to follow him or her then, as with nonscientists as well, some degree of leadership and recognition can be established. And language changes another small bit.

There are many excellent books, essays, and even technical journals on words and semantics, and I do not wish to try the impossible task of abstracting or even outlining them here. Their titles give a flavor for the range of implications for this subject. *Science and Sanity* by Alfred Korzibski, *The Tyranny of Words* by Stuart Chase, *Symbol, Status, and Personality* and *Language in Thought and Action* by S. I. Hayakawa, *Language and the Discovery of Reality* by Joseph Church, *Language and Human Nature* by Harvey Sarles, *The Human Use of Human Beings* by Norbert Wiener, *Communication and Social Order* by Hugh Duncan, and *Rules and Meanings* by Mary Douglas are relevant here. The important point is that although much of this becomes obvious once one starts to discuss it, we need constantly to be reminded of it in particular cases. It seems to be an even trickier problem than for a person with bad posture to visualize that and then stand and walk straight for more than a few minutes.

Style Is Not Content

Our kind is enormously vulnerable to mistaking style for reliability of content. Even white males with deep voices and a steady gaze may spend precious time and money to improve their style and image in the world of money and power. They may take public speaking courses, and this may indeed pay off. They may hire image consultants to tell them how, in order to win trust and respect, they should smile or joke, look serious or angered, at what angles to hold their heads, how to move their hands and bodies, what sorts of clothes to wear, and where their families should stand when the cameras or public are around. Machines are used to speed up voices in the media without changing the pitch because it has been confirmed that one responds more favorably to fast talkers. We associate being able to talk fast with thinking fast, intelligence, good judgment, and reliability. We *know* consciously that this is not necessarily so. Yet we continue over and over to fall for style, and one repeatedly sees particular styles win public confidence.

De Bono (1986) lists seven elements of what he calls the "intelligence trap," which causes intelligent people to assume they are thinking well when they are not. One is especially relevant here: Verbal fluency and quickness are so commonly mistaken for good thinking that bright people are often *encouraged* by parents and teachers to develop these qualities more than critical thought.

Any number of books advise that it is essential for communication to write plainly and never to use a ten-dollar word when a twenty-five-cent word will do. Yet even well-educated readers generally do *think more highly* of writings with a lot of ten-dollar words, and even convoluted styles, than they do of simply stated writing. Salesmanship and communication are not necessarily the same. People will borrow the style of a higher-status convention of discourse (register) to garner symbolic authority. Even scholars and scientists may do this to satisfy and indeed impress their peers, and to give force to their arguments. I (somewhat whimsically) lament that a substantive part of our literature, and inclusive of that circulated among scholars or purporting to be profound and original, seems to evince, notwithstanding concurrence on the utility of unembellished composition in effective discourse, the author's stake in validating an image of erudition and intellectual adroitness, whereas more prosaic formulations would be less efficacious in eliciting a subverbal mentation translating in effect to, "Hey, this guy must really be smart, I have to put some honest work into figuring out what he is saying." Or, "This guy's been to good schools, so it is worth my time to read him." Or, "People will certainly be impressed that I have read this and can have an opinion." A prestige automobile may have high-performance capacity that is sometimes needed; but it also says that the driver has money—prestige language similarly is sometimes necessary for exactness, but sometimes used to impress and persuade. On the other hand, other people may respect plainspoken, "commonsense" styles to a degree that is anti-intellectual.

One factor *restricting* effective communication between groups and disciplines is related to the above. Different groups often have different cues that subverbally confer and confirm authority and reliability. Certain styles and paces of speaking, a use of particular words, mannerisms, dress, degrees, institutional affiliation, race and even sex—particular combinations of these—signal respectability with a given group's elite. The elite and followers of one group may not even begin to examine carefully the words of a wise person from another who lacks what is in their eyes the right style and intellectual profile. Here again one may be misled from truth by taking image too seriously, by confusing social respectability and style for intellectual reliability.

Language and Manipulation

One of my fondest memories is of a wonderful little antique railroad train that winds its way patiently past waterfalls and lush forests, up steep and spectacular tropical mountainsides, from Puerto Limon on the coast of Costa Rica to the plateau where San Jose is located, at over five thousand feet. It is no doubt an engineering marvel and a benefit to the country. In 1914, Frederick Upton Adams, in *Conquest of the Tropics: The Story of the Creative Enterprises Conducted by the United Fruit Company*, wrote that one should pause and reflect that the trip "was made possible by men who did not hesitate to risk and sacrifice their lives in a work that will ever stand as a monument to the constructive genius of American citizenship." But my admiration is more bittersweet than Adams's heroic language insists. The men who died building it were mostly imported, "humble Jamaican negroes," for "nothing could induce the average native to enter the deadly zone of the *tierras calientes*, the dreaded hot lands of the Caribbean coastal region." No less than four thousand workers died.

Did the thousands of "humble Jamaican negroes" and the few white engineers really understand the hardships and risks beyond the rhetoric when they signed on? Did they sign on and stay on in light of other reasonable economic options? How could anyone say with a straight face that four thousand men "did not hesitate" to sacrifice their lives to build a monument to constructive genius? The natives who *knew* the area *did hesitate*.

People will say such things with a straight face because: 1) they may really believe the picture that the words paint; 2) in *denial*, as in the denial of extreme compulsive people such as gamblers, alcoholics, chronic achievers, procrastinators, etc., deep inside one may grasp the truth but deny it at the conscious level to avoid facing disturbing implications; or 3) one *may* grasp the truth consciously but wear a mask of innocence a) for economic interests, b) to present a sanitary image of one's ideology, or c) for fear of nonconformity.

Adams wrote his book to argue that the United States had "neglected" her tropical neighbors while the flags of "Great Britain, Germany, France and other progressive nations" floated over hundreds of millions of "tropical natives." So the United States should move to "stabilize" Central American governments so that its capital and enterprise might move evermore confidently into Central America. But one man's "stabilized" government becomes another's "banana republic." I was among those who once believed words such as Adams's, that U.S. forces had brought purely progress and freedom to Guatemala, El Salvador, Honduras, and Nicaragua, until I drove there in 1964 to do biological field studies and discovered these nations of poor Indians living daily intimidated at gunpoint, and, to an astonishing degree, not happy about it.

Adams was dazzled by the supposed mechanical efficiency of the corporation concept that had been developing in the United States. "It had no useless parts. It made no useless motions. It made no mistakes. . . . It was immeasurably the greatest achievement of Man."

One knows how important it is to find ways to keep to one's budget, yet Adams's miracle of low-priced fruits seems tarnished when one has learned that too many tropical native laborers have lived in too much fear to request higher wages. What I knew of Mexico before the Revolution of 1910–20 seemed still alive in upper Central America (not Costa Rica) in 1964.

Adams wrote that in Mexico prior to the U.S. developers, "Mexico—well Mexico was in its normal condition. Porfirio Diaz had not yet clubbed the semi-savage factions into a coma of temporary peace and prosperity." Such words paint a picture and reinforce stereotypes that conceal that Diaz was long hated by poor (most) Mexicans because of his killings of political dissidents, the fact that he encouraged foreign companies to practically enslave Indians and appropriate their lands and to operate and extract natural resources and pay almost no taxes. Mexico's superficial prosperity did not trickle down to the masses. Profits flowed *out* of the country. In 1911 Diaz fled to Paris. Adams's words conceal the murder of the quite civilized President Madero, and the torture-death of his brother, by the Diaz followers. They conceal the reasons why the revolution of 1910–20 was not merely a palace revolution of the sort that Latin America has seen hundreds of, but became a unique and vast social revolution (Bailey and Nasatir 1960).

If one does like Adams's style and buys his language, then if one later learns about the brutality of Diaz, one is predisposed to think "Oh, those Indians he had clubbed and killed must be the semisavages I've heard about who were standing in the way of peace and prosperity."

The issue here is not whether a railroad brought benefits to Costa Rica or whether corporations kept fruit prices low; it surely did and they surely did. The issue is how words can paint pictures that can shape a great nation's foreign policy, make unwitting people march willingly into the jaws of death, on the one hand, or basically decent people approvingly applaud while others drop like flies, on the other hand. So, for example, in the following, what or who has been reduced to Adams's "American" or his "frontier?" "It was not solely a desire for profits which caused these men to combat the seen and invisible dangers of the tropical vastness. They did it in response to that instinctive spirit which ever has urged the American to face and conquer the frontier."

In addition to wishing to avoid bad thinking of our own making, it is important to understand language because powerful forces have long understood it particularly well. A manipulation of language has been used extensively to manipulate the thinking and behavior of groups for political and/or economic and/or religious and/or ideological motives. Call it propaganda, rhetoric, brainwashing, effective marketing strategy, public relations, divine lies, official mendacity, psychology, disinformation, or a dozen other terms—it is a subtle and noncoercive (nonphysically coercive, at least), though extraordinarily powerful, form of manipulation and control over each of us by both impersonal traditions and interests, and by conscious agents.

Tacit Philosophical Indoctrination

Of course this is what language largely is in early life anyway: thought and behavior control. The child learns to see the world as its social group does and to behave as they do largely through language, as well as through imitation, physical and emotional rewards, and punishment. There is much literature on this from both Western and Eastern points of view and the areas of agreement are clear and fundamental. Alan Watts, for example, noted, in an introduction to Zen (1957):

> Thus the task of education is to make children fit to live in a society by
> persuading them to accept its codes—the rules and conventions of
> communication whereby the society holds itself together. ... We have no
> difficulty in understanding that the word tree is a matter of convention.
> What is much less obvious is that convention also governs the delineation
> of the thing to which the word is assigned. For the child has to be taught
> not only what words are to stand for but also the way in which his culture
> has tacitly agreed to divide things from each other, to mark out the
> boundaries within our daily experience. Thus scientific convention decides
> whether an eel shall be called a fish or a snake, and grammatical
> convention determines what experiences shall be called objects and what
> shall be called events or actions. How arbitrary such conventions may be

can be seen from the question, "What happens to my fist [noun-object] when I open my hand?" The object miraculously vanishes because an action was disguised by a part of speech usually assigned to a thing!

We could say that the hand is fisting, much as we say we are sitting. We do not say we are a separate noun object, a "sit," when we make the action of sitting. Instead of calling a person an adolescent, a category of being, we could simply drop any specific noun designation and say they are leaving childness or becoming adultness. But that would be awkward for us in our language because we do not think that way, although people in other cultures may.

Karl Heider (1979) called the Grand Valley Dani of New Guinea "peaceful warriors" to draw attention to the fact that their constant warring turns out on close study to be very different from our own image of war. It also comes to mind here that the Dani language is predominantly verbs, as are various other non-European languages. Many peoples see the world much in terms of phases of happening rather than largely first in terms of distinct entities that are acted upon and that change their state. We tend to see the world first in terms of items or things with fixed identities or essences that are moved about by forces (much as the pieces in a chess game), causes "acting on," and effects "happening to" "things." This works well for us, but other systems have worked well for non-Western peoples.[3]

The Dani language is also interesting in that it has only four number words: one, two, three, and many. They are more apt to care about and to notice "other, richer things" than quantity.

West of the Dani are the Kapauku. Pospisil (1958, 1963) discusses their number system, which in striking contrast is a "decimal system that stops at 60 and starts over again, having as higher units 600 and 3600. The people impressed me as having an obsession with counting. They may not remember the name of the parties to a transaction but they never fail to recall the exact number of shells or beads paid for the merchandise." The writer would reward his best informants, "by giving them permission to count thousands of his small trade beads. . . . The emphasis upon quantity assumes forms which come as a shock to a Western observer. My informants when confronted with a magazine picture of a smiling girl failed completely to react to her beauty. Instead they started to count her teeth." The Dani and the Kapauku have very different ways of paying attention to and concerning themselves with the world.

Other New Guinean peoples of the Papuan language group to which the Dani belong may have number systems based on 27 or 28. This may seem strange to us but obvious to them. It is based on a series of points on the body, starting at the little finger, moving up the arm and over the top of the head. Most of us can at once visualize our fingers as dividing naturally into ten entities, and that is the basis for our number system. But we would have to stop and think about which parts of the arms and head we would want to regard as points providing an "obvious" basis for counting. Why, for that matter, is it any more obvious to count on our fingers than to use all the joints in the hands, for example, or all the bones between the joints, or both (Ifrah 1985)? Similarly, how should we divide up the heavens? We long ago decided that combinations of stars make obvious patterns and we named these the constellations. But in other cultures other constructions seem obvious.

Questions of this nature have varied dimensions and one is akin to the different reactions evoked by saying that a glass is half empty rather than half full: a pessimistic versus an optimistic description of the same exact thing. Words carry a lot of instructions for thinking and feeling, for dividing the world up into categories and for *reacting* to them, and are not simply uncomplicated objective descriptions. They tell us what to value.

I was astonished in Papua New Guinea to see firsthand so many Europeans in administration, commerce, and missionary work who spoke of the native people as being "like children," and merely involved in "superstitious nonsense." They had come thousands of miles from their own land and yet would laugh and even sneer when the *locals* did not grasp the logic behind *white* customs. They felt no embarrassment, though, none, that *they* did not understand the logic behind the ways of their Melanesian *hosts*.

Sometimes they imagined that they *did* understand the local people — e.g., "children." Once they had the tribal peoples neatly labeled and classified it seemed impossible for them to see the rich and complex vitality and diversity, the extraordinary maturity and sophistication of the local cultures. This is not to say that Melanesian cultures do not indeed have at least as much or more superstitious nonsense as our own. But this is not a sufficient descriptor of either their world or ours. One might rarely find an alert white European who had lived there for years and could describe the fascinating things yet to be learned as "bottomless" and yet various scientists and art lovers shared my own experience that the vast majority of white people were not only uninterested in the local cultures but were indeed blinded by labels like "superstitious," "primitive," "childlike," "natives," or even "wogs." These words seemed sufficient to them to explain all observations, and what was complex structure and sophistication to the anthropological eye was simple chaos or pagan nonsense to the badly conditioned eye. I could learn *almost nothing* of substance from most whites, of the native people that they lived among and worked with daily. They had learned little of substance from the "natives" because they had given them low-status labels.

They even had a snickering term for any white who had gotten too close and sympathetic to the locals, "gone tropo."

Plato's Rationality and Truth

An articulate national of Papua New Guinea (P.N.G.) told me, "White people see things only one way. Only one way or another. I am not a good person who is sometimes bad, or a bad person who is sometimes good. I am both good and bad. But it is difficult for white people to see things that way." What was frustrating him is based on the fact that Westerners are idealistic in the Platonic sense.

I had already understood in an abstract way this effect of philosophy on our Western perception, but his independent observation and his frustration in dealing with it as a member of another culture gave the phenomenon a certain freshness and a very human dimension that made our conversation stick in my mind. Western language and thinking have been influenced so long and so deeply by the ancient Greeks that we automatically think in terms of presumed essences (Recall that Plato argued that reality was not in the material world and its diversity but in "essences," in its forms, ideals. Christianity incorporated Plato, especially with Augustine.): "the Truth," "true love," "our real selves," "true beauty," "the true purpose of . . . ," "*the* purpose of life," and so on.

Words like *deep-sea trenches* suggest that these exist, and they do. Words like *essentially* and *Truth* suggest that essences and singular abstract Truth also exist. But there is only evidence for mirage. In physics, for example, Newton and others were once convinced that all of physics is essentially mechanics: could be explained in terms of the singular Truth of mechanics. But it has become clear that there is no hope of this (Einstein and Infeld 1966). Westerners have a very difficult time with relativism and pluralism, and these have even become bad words for many in our culture. In the absence of instruction or reflection, we tend to see these as amounting to or leading to chaos or meaninglessness, and we lean toward absolutism in our language and thinking, even though in areas of modern democracy we have won some important intellectual appreciation of diversity.

Hayakawa (1950) argues that many people, even academics, believe that rational discussion is impossible unless we agree on certain essential definitions and first principles. There can be no "real" communication, no "true" morality, unless one *agrees* to *their* assumptions. Thus, by believing that there is no basis for communication except on their terms they in effect refuse to communicate. He calls these the neo-Scholastics. Though they do not identify themselves as any group, I certainly see a lot of such people in academia and this can be an element in our "balkanization." (Surely we are better off when we can *grasp* our own and each other's basic and hidden assumptions. But rational discussion can be had without *agreement* on these.)

I helped organize an interdisciplinary conference on the question of why many religious and secular humanists alike have become demoralized with, and have abandoned confidence in the possibility of, rationality. This is a complex matter about which I will have more to say (e.g., chapter 11). But for one thing here, many of them have seen reason as a single sort of ideal algorithmic quality or process, in opposition to uncontrolled emotionalism and chaos. Order/Reason versus disorder/chaos. Apparently, ideal Platonic Truth—singular, monolithic, exclusive, findable, and verifiable only through geometry-like logical Reason—is the only sort that these persons have had any interest in. It even seems disturbing to some contemporary scholars that the supposedly absolute Truths of Euclidean geometry were realized to be *relative*, and limited to a flat surface, in the 1800s.

Science has long since demonstrated that logic alone is inadequate to find or to identify truths (see also Piaget 1971), and to the irritation, dismay, and indeed to the disorientation of many, the best science (as critical thought) does not support many of the other common and basic assumptions of Greek thinking that many academic, religious, and laypeople still treat not as mirages but as fundamental to rationality and social stability. It seems that much of the academic community had not discovered this fact about science until very recently and they were not prepared for it. (It even dawns only slowly on some scientists that scientific truths are not Truth.)

The mental discipline characteristic of the great breakthroughs of science is not merely in the nature of simple reasoning or logical technique as one would at first assume, particularly in our age of technique. Disciplined, careful but inventive observation of the material world combined with analysis and calculation, ingenious experiments testing the intuitions of the highly prepared mind, all have led to abstract conceptions of the composition and workings of nature that have proved to have enormous and even terrifying power. But neither the conclusions nor the procedures of modern science can be verified by logical Reason (or logical "methods") alone, as Plato argued was necessary to know Truth. Thus, in this sense our science-based civilization does not have the "reliable" knowledge so long sought by Western philosophers—even though science explains very, very well what causes rainbows and why eels are fish and not snakes.

What I as a scientist had in mind at the conference by the word *rationality* (forms of critical mental discipline, including observational discipline, that have been verified empirically) was very different from what many academic humanists and philosophers had in mind (logical Reason that will reveal the Truth about life and the world).

To other scholars there is the feeling that the world "should" work rationally. There should be either harmonious stability or idealistic progress. Many of them still expect essentially a world of Aristotelian teleological design. And, as much of society does not seem these days to be involved in progress or harmony, these thinkers were shaken and were questioning whether life can be viewed "rationally." Neither this group of neo-Aristotelian scholars nor the neo-Scholastics above (and many may belong to both groups) seemed to be aware of the origins of their attitudes and definitions, or that their definitions of rationality were quite relative and narrow in the broad scheme of the history and sociology of ideas.

When Karl Popper and other philosophers conclude that science is an "irrational" process, they are correct in their technical Western *philosophical* sense: it has not followed any sort of Platonic model. But it is *not* irrational in the sense that it is whimsical, impulsive, emotionally motivated, chaotic, meaningless, and so on—as is implied if one thinks in idealistic and dualistic terms. An irrational number in mathematics is not an emotional or meaningless number—the same word may have different meanings to different groups of people.

Does Language Diminish Personal Experience?

Westerners lean toward dualism in language conventions and in thinking. We will tend to want to hear *the* other side or *both* sides in an argument, by such language constructing a mind set that there will be only two significant positions.

The interests of the individual and of society are separable for analysis, and while they often coincide, often they do not and may actually be in conflict. A society may be viewed as a sort of superorganism or regulated, cybernetic machine, with children as replacement cells or parts that slip into social roles that keep the whole thing going. Language and beliefs function a bit like the chemical messages that regulate the termite or ant colony (see also Douglas 1986). Dualistic thinking seems to be adequate for most functions of a cybernetic machine. In Darwinian fashion, those belief systems have survived and prevail that have been the most convincing and that can capture widespread allegiance. No more than modest truth and intellectual freedom may be necessary for popular belief systems, since these systems generally need only economic and political *adequacy*, and modest attention to reality, in order to function and survive. (This is very close so far to an economic determinist view, but see chapter 9.)

Most popular ideologies and belief systems give lip service to individuality, but covertly indoctrinate that what is best for the whole is also best for the individual, that behaviors and beliefs that promote the survival, stability, and well-being of the system are what we should insist on and strive for. If it is seen that forms of selfishness are good for the system then *these* forms will be encouraged by the powers that be, much as Mandeville in *The Fable of the Bees* argued they should be. My own position differs from this.

We are social creatures and it is indeed quite reasonable to be concerned about, and work for, the quality of the societies in which we function and from which we derive so much of our perception and identity and so many of our satisfactions. But some expect us to behave as if the System is all that matters in the end. I write here, though, to the individual. Each person has individual interests as well as societal ones and some will find it desirable and necessary to go beyond giving lip service to individuality and beyond spending our powers in consumption, escapism, and in competing for money, status, and attention. They may wish to work to develop their potentials for insight and creativity. It is no news that one *will* have to work to find harmony between individual and social interests, much as one tries to balance long-term and short-term interests. This fact may, though, bear repeated emphasis.

I have the luxury here of writing for an unusual audience, I am writing on the nature of mind, and for those who are interested for *personal reasons* in critical thought, good judgment, and wisdom. I make no pretense here of trying to solve society's considerable problems.

This book does touch on controversial and emotional public issues. One cannot discuss the functioning of the mind on the basis of brain-biology alone and without a consideration of values and perceptions—including the social, metaphysical, and ideological factors that help shape perception. This is not, though, by any means organized to be a call or a plan for social reform or to build a better world.

What I wish to discuss now is one example (dualism/idealism) of how certain habits of seeing and thinking may be adequate for many daily work-tasks and maintaining society and the status quo, yet restrict individual consciousness.

Our adversarial legal and political systems help to perpetuate dualistic mind sets. For a competitive and legalistic people it becomes especially attractive to divide issues and groups of people in simplistic ways. Society at large moves in a direction where the important thing is to make simple decisions, right or wrong, to be for or against something, to decide to try to win or not, to purchase or not.

Many important things do seem to be roughly divided into two categories—the sides of the body, the sexes, night and day. Two makes the smallest number of differentiating categories possible. Duality is an easy way for the child to begin dividing the world. The minimal instructions needed to control behavior in the child or adult are yes or no, stop or go, good or bad, safe or dangerous. There are friends or foes. Then one can begin to subdivide in a series of binary steps on this basis. Friends can be divided into superiors or others. These others can be divided into close kin or others. These remaining others can be divided . . . , and so on. The simplest, but not necessarily the most instructive, taxonomic identification keys[4] are constructed this way, and computers use a similar simple logic path in the sequential arrangement of states of on or off. But this in no way means that nature itself is actually arranged in systems of opposites.

In practice one tends to skip much of the subcategorization and to classify things one way or *the* other. The perceptual consequence is that we may not look carefully at things once labels are on them. We all know even adult people who see things, particularly values and morals, in black-or-white terms and who have trouble with shades of gray, to use a figure of speech. They might be of the opinion, for example, that some few things are good to eat and most others are not, and so gourmets must be strange people who are not "normal." Flavors or people may be largely either familiar or strange, OK or not.

There may even be a deep sense of power, control, and security in the ability to trivialize, alter, destroy, and bury away vast amounts of reality with words, and many of us are seduced by this game early in life. Even a child can pulverize whole philosophies, races, and cultures as "weird" or "foreign" or "stupid" so that they cannot command her or his attention or respect. It is one way of establishing control over the environment. Children's games may be preparation for adult social and economic life. They prepare the child to build a narrowly adaptive (not necessarily widely *adaptable*) conceptual framework that conforms parochially to the particular beliefs, values, and power relationships of the roles and spheres that she or he will be confined to in society.[5] But the price paid in loss of vision is dear. And it can become dearer if we aspire to be more open, relaxed, contemplative, or original persons. Triumph at adjustment to the social neighborhood can shut out emotional and intellectual appreciation of much of the world.

One of the great problems with political and religious ideologies is that they will dogmatize and institutionalize dualisms and stereotypes so that unsophisticated people, and people of action, can enter into and follow the movement. One sees the sinful and the saved, us and them, the faithful and heathens, workers and capitalists, rightists and leftists, good guys and bad guys, decent people and bums, whites and wogs, right and wrong, thesis and antithesis, revolutionary and reactionary, patriotic and subversive, liberal and conservative, Republican and Democrat. (Middle-of-the-road and extremist can even form a tricky dualism.) Such simple views may offer emotional security and/or paths to quick action but they also foster narrow and muddy habits of thought. Potentially capable minds may be blocked from developing their full adult abilities for thoughtfulness, perception, and humane relationships. Thought systems with simple rhetoric will obviously have wide appeal as they compete against more demanding thought systems. The individual interested in clear thinking is well advised to be wary of popular opinions and proselytizing groups of the right, left, or middle.[6]

Thus popular thought systems in the West promote dualistic thinking. It should be no surprise then that even teachers help to promote it. Educational psychologists too often tell us that that young people can only understand values, ethical issues, and such in simple either/or terms and so it is wasted effort to try to introduce subtle concepts. Let them "grow out of it" first, they advise, and meanwhile just deal with the students at their own level. But this is passing the buck (if we take the position that the goal of *education* is to promote insight and understanding, rather than *instruction* to train young people for roles in the System) and does not face responsibly the need for experience and mentorship in order to "grow out of it." Growth is not merely an inevitable physiological maturation. There are many intelligent people in our society who continue into old age to see things in black-and-white terms.

In a study of Harvard undergraduates Perry (1970) details how life confrontations and intellectual encounters are involved in causing most (some never leave the security of dualistic and absolutist thinking) students at his elite institution to become relatively liberated from dualistic thinking. There is no automatic process of physiological maturation. It is a process that good educators must find ways to stimulate, encourage, and support.

It is interesting and important that, while too many teachers cite Perry's study to support a hands-off attitude and retreat from education into instruction, Perry himself reaches very different conclusions. For example,

> [T]he study makes salient the courage required of the student at each step of his development. This demand upon courage implies a reciprocal obligation for the educational community. . . . This is a creative obligation: to find ways to encourage. . . . At this advanced moment of maturity he would seem to require not less support but more [by confirming the membership he achieves as he assumes the risks of each forward movement].

Along these lines, Hayakawa (1950) noted that, "The habit of trusting one's . . . verbal associations . . . is one of the most stubborn remnants of primitivism that remain to affect us. . . . and education that fails to emphasize this fact is more likely to leave students imprisoned and victimized by their linguistic conditioning, rather than enlightened and liberated by it." He continues:

> The establishment of definitions is therefore not what Plato imagined it to
> be, an insight into Eternal and Transcendental Verities and the basis of all
> knowledge; it is as the Confucians realized, a principle of social control. By
> elevating definitions, which are simply rules of language to the status of a
> Doctrine of Essences, Plato hit upon a formula that in application would
> have made social control absolute and airtight.

Scientists and scholars have made great strides in providing particular insights, but language binds the vast majority of people to overly simple and outdated modes of perception and thinking and prevents one from taking advantage of the progress contained in libraries. It is like a strong young person trying to climb a steep hill of loose rubble and making little headway because of slipping—the substrate will not support the potential progress of the able climber. Except with the substrate of language the individual may have no perception of the situation at all, since words not only are tools, but help shape one's reality.

Much of language does not produce communication in the deepest sense but rather it triggers internal images and associations. It is as though words usually do not *paint* new mental pictures in our heads as intended, rather in effect they push buttons that *display* pictures that were previously acquired. Thus, novel situations can be easily misinterpreted if one does not break the habit of depending uncritically on words and symbols as handles to classify and confirm limited sets of previous expectations.

It is impressive how extraordinarily observant hunting and gathering peoples such as the Seri or the Australian aborigines can be. Their astuteness challenges any professional field zoologist or botanist.

A layperson might say that the hunters and gatherers have a sixth sense, but this is not satisfactory. They do seem extremely alert and observant to their surroundings. Of course their *home* is a nature that is constantly changing, where no two years are ever exactly the same. Nature may have fluctuations something like the moods of a close friend to them, and they are intimately attentive to and familiar with it. Recent findings in paleoastronomy, for example, suggest that many "prehistoric" peoples such as the Anasazi paid astonishingly close attention to the sky, to the yearly patterns of shadows cast, and presumably to the relationship of all this to seasonal patterns of weather and plant and animal life.

Scientists have gone into *new* areas along with hunter-gatherers and here again their awareness stands out. Most examples would have little impact on a reader who has not spent years studying nature, but one anecdote may give the flavor of the sort of thing that scientists shake their heads over. A zoologist colleague took an outback aborigine to Sydney. The fellow had never been in a city before and Sydney was my colleague's home, as it was for others in the car. They got lost while driving about. When this was announced the aborigine volunteered that he thought he knew how to get back. It turned out he was right and the route did not involve retracing their path.

I would not postulate a sixth sense in such cases. There are many cues in nature, at least, that an alert person could use (Gatty 1979), and we need postulate only that the aborigine was exceptionally alert, even in the city, though he could not explain his methods. (This last is not unexpected. Several of my field biologist friends thought of themselves only as "having a good sense of direction" until independently they traveled to the southern hemisphere and became disoriented. Only then it became clear to them that they had the habit of using the sun for orientation. Others of us, incidentally, do not use the sun in this way.)

As a scientist I know well that anecdotes are of limited value. On the other hand these issues may be important to an understanding of how the human mind works and are worth mentioning, pondering, and further study and discussion.

In Walt Whitman's epic poem, *Song of Myself,* he would have us participate in a relentless alertness of the senses and mind that is breathtaking. Empathy, the senses, the probing mind, the gentler passions, are wide open and unshielded, roaming and gathering energy where they will. Now I write as a scientist and ask, did Whitman's mind actually function in such an invigorated way or is the poem only carefully crafted to make a point? As with the question of the alertness of "the primitive," the skeptic in me again emerges at the same time that my scientific curiosity will not let me leave the matter alone. If most of us are usually dulled as adults, is this merely an inevitable product of aging, does it reflect some elusive "remoteness from nature," or do our languages and the thinking and ideas they support keep blinders on us?

The context should make it clear that I do not mean that Westerners are mentally retarded or cannot find great joy from sunsets or sex, or both together, and so on. People in various groups simply use their eyes and brains to slice different servings of physical reality. There is a rather wonderful scene in *The White Rose* by B. Traven. An oil company is trying hard to buy a ranch, the White Rose, from a group of Indians in Mexico. The company representative gets more and more frustrated because the Indian leader can see no reason to sell. The lawyer (the *licenciado*) cannot grasp the Indian's world and sees him only as backward and ignorant. At one point he tells the Indian that with the money offered he could buy an automobile and get to Tuxpan in a half hour. But the Indian replies:

I've never wanted to go to Tuxpan in a half hour. I like to speak to people along the way and see how the corn looks and what the little ones are doing—I know them all—and I want to look at the branches of the blue flowers up close. I want to know whether the big turtles in the lagoon have laid eggs in the sand. And there's that heavy mahogany tree that snapped off four years ago and still won't rot away. . . . When I go to Tuxpan . . . I . . . rise early, around three thirty, and I get to Tuxpan by nine. That's time enough and I've seen everything along the way. I've talked to Raphael who has been putting a new roof on his house because the old one leaked. Moreover, I still arrive in Tuxpan plenty early. I don't need an automobile. I really don't, Licenciado.

Jamake Highwater (1981) a writer and artist of Amerindian background, takes the position that Western modes of perception and thinking are so idealistic, hierarchical, dualistic, and linear that we have become in a sense myopic. Then, it is that Westerners as adults are unusually dulled by routines of thought and behavior, not that the mythical primitive is unusually alert and sensitive.

Rather than consenting to the possibility that truth is nonexistent or pluralistic, [Westerners] have consented to the very complex and suicidal notion that within the oneness of the truth is an implacable dualism. Consciousness is therefore regarded as a delicate tightrope walk through a lifetime of polarized experiences.

The issue here is, in what ways do particular values and beliefs, imbedded in our language, possibly restrict our viewing of and our participation in our environments?

After becoming blinded, people begin to pay better attention to their remaining senses, suggesting that they had previously not used them as effectively as possible. You can experience some of this for yourself simply by carefully and thoughtfully going around your home blindfolded for a few hours.

After dealing with both Westerners and "primitive" peoples I am comfortable with the hypothesis that the senses and minds of the latter may be more open than ours typically are, in some contexts at least. They are not necessarily more open to abstractions of sorts that have little or no basis in their own experience and ways of expressing things. Similarly, we may not find *their* abstract concepts easy to understand. It is necessary to underscore this point, which is clear enough as a statement, because in practice we may take our abstractions for granted, and we may see people as biologically inferior, even in our own culture, when they cannot grasp what is obvious to us. But, associated with their value systems, they may simply not have the particular ideas and assumptions, perceptual and cognitive habits to begin thinking as we do.

Can one increase perceptual and intellectual alertness by giving up certain habits of thought? Can one move in some way beyond language and "expand consciousness" and sensitivity? Many writers from the ancients of the East to some modern scientists have claimed this.

I doubt that one could give up *all* language and also survive in a working and social environment. As a scientist I am not ready to give up Western thought modes on any permanent basis. They are convenient for construction and analysis of robust cause-and-effect models and I am not sure which language would be better for disciplined thinking. Every language has *some* perceptual and cognitive filters and blinders built in. But could one gain enough mastery over language and Western habits of thought and perception to take them off or put them on, when desired, like a hat?

At its keenest, the scientific tradition has appreciated that language is imperfect symbolism, communication, and behavior control, and is a danger to clear thinking when it is not understood. Thinking and judgment have decidedly improved where problems with language have been discovered. As a result, Western scholars have tried to fix definitions in order to cope with the problem. This helps, but we deceive ourselves if we think that we have brought language under control this way. One must begin by appreciating the problem, working to increase awareness of it on a case-by-case basis, and trusting good sense to help one rise a bit above it if one is diligent.

Dealing sensitively with a wide range of people may provide considerable liberating perspective, for then one learns how many ways people have of using words and one may see the variety of world views that can hide behind the same words. When I write scientific papers I like to send them out to a number of reviewers before sending them to a professional journal. I learn ever more of the surprising variety of ways to interpret simple words and statements. Indeed, it may be impossible to write for everyone and one can only try to write for as wide an interested group as possible.

So talking with and really trying to understand a variety of people may help to keep a perspective on language and concepts. Learning several languages should help if one is thoughtful about it, but there is a lot to be learned even by being active and alert within a single pluralistic society. (Without leaving the house, one can open up to the whole world, as Lao Tsu put it.)

A sense of humor may help too. People in families who pun a lot and can joke about even their cherished concepts often seem less apt to take words and ideas too uncritically. Perhaps word games can help one to gain a feeling for the fragile nature of the powerful world of language. Humor, it has been suggested, may be an important element in allowing us to maintain some underlying sanity as we bow to the need of taking seriously the particular social reality in which we are integrated and must find our identities and survival.

Words and language are our teachers of ideas and perceptions. In a sense they are our gurus. In Jodorowski's surrealistic film *El Topo*, the cowboy hero is directed by a beautiful woman to prove his manhood by killing gurus. In some schools of Eastern thought particularly there is the notion that to reach true enlightenment one must kill one's guru. This is not meant literally, but the idea is to cease to depend on and focus on the teacher, or anything else.

The film is a not-so-subtle allegory dealing with both Eastern and Western paths to wisdom. *El Topo*, the mole, digs out of the darkness, seeking the light, and in the end he is blinded, much as the men in the cave that Socrates described in the seventh book of Plato's *Republic* would be blinded by the light of Truth.

The macho hero on his path is able by force and trickery to kill one exotic guru after another. But we see his victories as senseless and ridiculous; they are literal acts only, with no meaning. The beautiful woman rejects him as a failure and shoots him. Years later he awakes from his coma in a cave full of cripples who beg him to lead them out, into a world of naked dog-eat-dog brutality. Still the macho hero at heart, he now cloaks himself in Christlike humility and sets out on another path of foolishness and disillusionment. The Western cowboy has never learned to deal with his ego and his needs for social rewards.

So how can one kill a guru? Should one discredit, destroy, desecrate, ridicule, wave faults about, shun, or otherwise signal to oneself that one has cut the cord and killed the guru? That does not work.

Dropping Ashes on the Buddha by Stephen Mitchell takes its title from a misunderstanding of Zen wisdom. The student must learn that the Buddha is nothing, to move beyond reverence, to cut *all* attachments. But the American student may feel a need to demonstrate progress to him or herself and others by, for example, blowing smoke and dropping cigarette ashes on a statue of Buddha. Of course, the act shows that one has learned nothing. The student is still attached to many things—to the need to demonstrate, and indeed, the student is attached to *the need to free oneself* from the reverence.

One only kills one's guru when one's understanding makes it no longer necessary to kill the guru. When one is finally aware. That is when and that is how. For the genuinely enlightened person, the practical issue essentially evaporates.

Words and language are our teachers and cannot simply be swept aside by mere will or device or technique. We can only try to understand them, the ideas behind them, and ourselves, well enough so that language naturally becomes secondary in understanding and communication. Then one would have the vocabulary to converse with Chuang Tzu.

NOTES

1. I am not endorsing *philosophical* nominalism, which can be as extreme and unrealistic as the semantic realism of the medieval and contemporary Scholastics (see for example, Armstrong 1978a, 1978b).

2. More deeply, an attack on "masculine science" is worth exploring as an example of the unlikely, if tacit, coalitions that can underlie broad attacks upon the authority of rationality and scientific thinking. There are moments of alliance between Marxists, fundamentalist Creationists, bourgeois feminists, deep environmentalists, pacifists, and philosophical nihilists. In short, a range of social critics have an interest in the dismantlement of what they see to be the present power structure—they feel that they may gain a better hand if the deck is radically shuffled. So they attempt to dismantle the reigning ideologies intellectually, as a nonviolent way of breaking open the present order. In doing this, one may sometimes seek odd political alliances among dissatisfied groups.

The groups may share some sense that contemporary Western thought is basically "rational" and "scientific" and that this somehow is at the core of the ruling ideology that they wish to defeat. In this, even leftist radicals and rightist fundamentalists may find some shaky common ground. Yet obviously the different groups have ultimate conflicting interests that make viable coalitions difficult. Thus, there is the potential for any one group to cooperate with others to meet immediate political objectives, but it is hard to realize this potential.

Many scholars, in trying to define our times, have claimed that a repudiation of the "authority of science" and its supposedly central position in maintaining the status quo is actually a *basic* feature of the culture that they call our "postmodern" condition. Foster (1985) and the authors in his volume make a useful distinction between the different factions who reject the supposed *modernist agenda* (a form of Enlightenment optimism including faith in an "official culture" that promotes faith in inevitable progress based on rationality). They distinguish a *postmodernism of reaction*, and a *postmodernism of resistance*. The reactors would include the neoconservatives who wish society to return to supposedly lost traditions. The resisters are a smaller movement who strive to critically deconstruct both modernism and other traditions to understand them and to avoid being swallowed up by the status quo. This is surely too simple a scheme for explaining all the opposition to, and criticism of, science or rationality, but it is a helpful point of departure (see also chapter 11).

Returning to criticisms of masculine science, there is a microcosm of this quandry within radical feminism. On the surface it seems that there are many social problems linked to capitalism, patriarchy, the idea of a single truth, bombs, pollution, masculine aggressiveness and dominance, racism, and dogmatic heterosexuality and that the apparent clustering is not merely historical coincidence, but that there is some *essential common denominator* to all of these. Thus it is assumed that if some common denominator could be found, a great stable political coalition could be forged between bourgeois feminists (looking for more money and power within the system), radical feminists (who see a need for more systemic change of the system), black and white women, lesbians, female pacifists, and deep environmentalists. It would not be sufficient to say, for example, that industrial capitalism was developed by white bourgeois men who just *happened* to have Judeo-Christian patriarchal values, and Calvinistic values (where economic success indicates a state of grace).

This political goal has generated several attempts, which have not proved viable, to develop a monistic theory of oppression and destruction. For example, there was the theory that capitalism, conquest, science, and the oppression of women all flow from a male desire for immortality since males cannot bear children.

A problem common to such theories has been that there are, for example, socialist societies that are patriarchal, militaristic, and scientific. The correlation between socialism and intellectual pluralism or pacifism is not good. Still, Marxist *social analysis* has proved to be a productive tool in understanding the socioeconomic status of women, and it seems to many feminists and Marxists both that stronger political coalition could be valuable. So the search continues for a *theoretical basis* to link feminism and Marxism, and this includes a close study of Western rationality and scientific thinking for both capitalistic and patriarchal elements.

This general project can get a bit complicated, since ethical aspirations call for *pluralistic* thinking, yet political goals would be advanced by a *monistic* theory that explains all woes. At the same time one wishes to avoid biological determinism. So ideally one wants a *socially* constructed (nonbiological) "masculine" force that necessarily produces patriarchy, capitalism, runaway technology, scientism, militarism, racism, dogmatic heterosexuality, and philosophical monism. This is a tall order. Lydia Sargent has collected a series of papers to restate these goals and review progress in *Women and Revolution: A Discussion of the Unhappy Marriage of Marxism and Feminism* (1981).

3. Dr. Suzuki makes a noteworthy comment in *The Training of the Zen Buddhist Monk*.

I remember being very puzzled by the way one says in English "a dog *has* four legs," "a cat *has* a tail." In Japanese the verb to have is not used in this way. If you said "I *have* two hands" it would sound as though you were holding two extra hands in your own. Sometime afterwards I developed the idea that this stress in western thought on possession means a stress on power, dualism, rivalry which is lacking in eastern thought.

4. For example, Dr. Robert Johns (1978) of the Forestry College at Bulolo, Papua New Guinea, has invented an identification key for trees consisting of concentric circles divided into appropriate choice-sectors. One starts in the center and moves through rings of choices to an identification at the periphery. One can grasp the organization of the classification virtually in minutes using this key. But with the traditional Western dichotomous keys it may take hours or days of use to grasp the structure of a classification system.

Johns told me that he developed the circular key because his Melanesian students found it easier to use than our conventional dichotomous keys. It is tempting to agree that they were not so used as we are to thinking in terms of a linear progression through a series of dualisms and found that convention even more artifical and difficult than Western students do.

5. Later in the youngster's life other layers of complexity are added in our culture.

Like the adult, the adolescent seeks self-understanding only up to a point. There are some things he would rather not know about, because they conflict with either the standards of the group or his own professed ideals. In addition, to acknowledge some impulses is tantamount to approving them, and approving them to acting on them. In short, a portion of the adolescent's (and adult's) self-knowledge is cast in the form of the "defense mechanisms" mentioned earlier—denial, reaction formation, rationalization, projection, displacement, and so forth. The defense mechanisms work in either of two ways. One can refuse to give a name to a wish, impulse, attitude, or idea, thus denying its existence, or else can give it a verbal formulation which masks its true nature and so neutralize it. We can see an early defense mechanism in the legalistic quibble of the school-age child: "You said not to hit him. I only gave him a little shove." However, the use of defense mechanisms in the construction of a stable verbal self-image, including elaborations by which one integrates partial self-insights, does not seem to occur before the adolescent years. (Church 1961)

6. The argument against this warning is that if people think too much about subtleties they will not be able to take decisive stands and we will never change the world. A plea for self-

critical and independent thinking that goes to basic assumptions would be attacked by activists from all parts of the spectrum. Leftists may see it as counterrevolutionary and reactionary, and rightists may see it as subversive or immoral, middle-of-the-roadists may see it as radical, cults may see it as closed-minded. Of course it is all of these to the many groups that base their common perceptions largely on slogans and simple commonsense or ideological beliefs, or on intuitive doctrines. To these, "clear thinking" means seeing the holes in the *other fellow's* "propaganda," and to be "open-minded" means asking others to refrain from criticizing *one's own* particular beliefs.

Even if representatives of ideologies can be made to agree that their positions have been "watered down" or "modified for mass-consumption," then some will argue that some simplistic slogans are better than others and we must make realistic choices and not simply sit around and reflect on subtleties. ("Spending time on fine points is wishy-washy. One can't *do* anything if one is always thinking about details."—I have heard it said in this context.) But I wonder if there is ever a danger that people will think too carefully. What sort of a better world we can expect from people and doctrines that eschew "subtlety"? It is surely posssible to think carefully and yet take active stands on issues with which one does not agree 100 percent.

CHAPTER 8

Diverse Searches for Wisdom

All who are content in what the eyes can see,
into blind darkness they descend.
They who content in higher knowledge
dwell in darkness greater still.
Other, they say, than the perceptible!
Other, they say, than the revealed!
—Thus we have learned from the wise
Who have explained it.
Shape and essence—
He who this pair can cojointly know,
With such eyes may navigate perishable nature;
With such wisdom may find the immortal.

Isa Upanishad, eighth or seventh century B.C.

All Organ Systems Have Limitations

The lungs are organs that regulate the body's exchange of gases with the environment. The kidneys are organs that regulate the balance of fluids in the body. But the lungs are not by themselves adequate at very high altitudes or under water. The kidneys, too, have their limitations, and vertebrates cannot long survive on a saltwater intake. Those organisms that have developed ways to *compensate* for the limited capacities of their organs have been able to survive in harsh conditions. Seabirds, for example, may have special glands in their heads that remove excess salt from the body.

The brain is an organ that balances behavior to the complex environment.

It is an organ that also has limitations. We expect ever more of it as societies become more complex and as demands increase and personal and societal aspirations rise, but still the brain cannot simply see truth. Nature is exceedingly diverse and complex. The imaginative brain automatically attempts to distinguish, select, and organize events from this complexity according to perceptual habits and thought systems that are largely social constructions and that work adequately for day-to-day survival. When the systems fail an alert person may not completely understand what has gone wrong but may begin to appreciate that truth is deceptively hard to come by. We can see others err conspicuously while remaining quite blind to their own shortsightedness, and we sometimes can even see our own shortsightedness in hindsight. We are inalterably prone to illusion, plagued by degrees of self-deception, vulnerable to suggestion and peer influences, easily misled by beliefs and language, provincialized by ego, and blinded by the false security of species of common sense that are only normally satisfactory.

In this chapter I will argue that much of the diversity of human thought systems can be understood as alternative efforts to exceed the peculiar limitations of the brain. Societies all over the earth, their wise persons in each, at least, have long realized that our ability is faulty to see the world as it is. (More robust or more sound constructions are not usually at once obvious, we might say with today's hindsight.) There *must*, it has seemed, be truths that evade our simple senses. Humans everywhere have sought intellectual tools to improve understanding and judgment, in large part to improve personal and social tranquility. It is a matter of *emotional self-interest*.

This chapter will look at the world's major thought systems from this simple but powerful perspective. How have different spiritual and intellectual traditions developed to try to cope with the challenge of finding reality beyond the imperfect senses? One may think of thought systems as attempts to build intellectual analogs of respiration devices, like scuba equipment, that allow our kind to explore ocean depths, high mountains, and the skies—perhaps to move beyond psychic pain and social strife.

The remainder of this book will examine the development of the world's major classes of religious and philosophic systems from this point of view. It will further study the nature of science. It will begin to explore the nature of modern Western cultural institutions devoted to the pursuit of knowledge and wisdom and the position of a thinking individual in our own time. This chapter begins to look at the major paths in the odyssey of the imperfect senses and reason in their pursuit of an improved ability to make good judgments, and to find wisdom for living.

The next section emphasizes that there has been no simple progression in Western culture. We should not expect it in others. There has been no uniformity of perspective at any one point. There are scenarios of disagreement (and struggle) lurking at all times, even when generalities can be made.

The Death and Rebirth of Empiricism in Europe

It appears that the world is flat, and as children many of us were taught that it was not until the time of Columbus that people figured out it was spherical. Perhaps common people did not accept the idea until then, but educated Greeks had early combined facts with logic *to reason* empirically that the earth *must* be spherical, and they used good methods to calculate its size. Hipparchus in the second century B.C. even used reasonable methods to try to calculate the distance from the center of the earth to the center of the moon. Columbus believed wrongly that he had reached Asia because scholars of his time had been mistaken about the unit of measurement that the ancient Greeks had used in their writings. So he underestimated the distance from Spain to Asia.

Educated Greeks were very serious about mistrusting their senses. They were acutely aware of the difficulty in knowing reality, and we have seen how this issue was explored by Sophocles in *Oedipus Rex*, as well as by Plato in *The Republic*. They saw a need to use careful study and reason to build a new view of the world. They began by realizing that they were frail and fallible beings and their genius was that they went on to develop careful intellectual and observational tools to attack their blindness and embrace the world. Their thinkers were dedicated and creative and explored several diverse lines of thought and enterprise in their search for truth and wisdom.

There came the "fall" of Rome—the destruction of pagan religious, commercial, and social systems and the construction of the Age of Faith. Emphasis shifted to efforts to develop and define "Christian" modes of community, commerce, and spirituality. Fundamental was an emphasis on an *afterlife*, for which salvation from supposed eternal damnation was a priority. This situation was not congenial to empiricism. For a thousand years there was profound mistrust not only of the senses and the intellect, but of conceptual constructions based on the study of the material world rather than on contemplation of the Divine.

By the Renaissance new community and commercial structures were being constructed, encouraging the exploration of new thinking. A result was a revival of the heroic spirit, a confidence in the imperfect senses and mind in a disciplined manner to break free of authoritarianism and timidity. Leonardo da Vinci wrote, for example:

> I know that many will call this useless work; and they will be those of

whom Demetrius declared that he took no more account of the wind that came out their mouth in words, than of that they expelled from their lower parts: men who desire nothing but material riches and are absolutely devoid of that of wisdom, which is the food and the only true riches of the mind. . . . And often, when I see one of these men take this work in his hand, I wonder that he does not put it to his nose, like a monkey, or ask me if it is something good to eat. . . .

Though I may not, like them, be able to quote other authors, I shall rely on that which is much greater and more worthy:—on experience, the mistress of their Masters. They go about puffed up and pompous, dressed and decorated with [the fruits], not of their own labours, but of those of others. And they will not allow me my own.

These rules are sufficient to enable you to know the true from the false — and this aids men to look only for things that are possible and with due moderation—and not to wrap yourself in ignorance, a thing which can have no good result, so that in despair you would give yourself up to melancholy. (*Notebooks* 10, 11, 12)

We cannot depend on the *undisciplined* senses or mind to reveal truth. Leonardo's notebooks and art reveal a great pioneer's efforts to invent observational and rational *self-discipline*, a painstaking analysis of his senses and of the smallest details of his world. To understand the meaning of his genius deeply one must be aware that Europe, whose official culture had rejected the senses for hundreds of years without finding absolute Truth, was returning to empiricism as a controversial vehicle to Truth. In that effort, out of those unique times came the birth of a uniquely powerful way of thinking: modern science.

To better understand the relative rejection of empiricism historically, let us back up and go through this story again in more critical detail. Recall that what seems obvious to the undisciplined mind can be untrue and misleading. Modern studies show that in large part this is because our brains must necessarily constitute sensory input into the realities in which we each live. These vivid constructions and classification schemes may be tied relatively securely or insecurely to the material world, but there is no way to tell, simply from the texture and intensity of individual experience, which *necessarily seems authentic* for each of us. If we each know only one world, then it is difficult even to grasp the *idea* that this is not "the one" reality that our language, heavily influenced by ancient monistic beliefs, tells us should exist.

Nevertheless, in perhaps every human society wise people have sensed and reasoned that there must be truths or Truth beyond everyday experience and common sense. For example, a creature may die. But then another may be born that resembles it, or it may reappear in a dream. Is the similarity an illusion, or was the death an illusion?

Greeks early staked out the two views on change. Heraclitus reasoned that the nature of things *is change* and that *the unchanging is an illusion*. Some cultures with languages dominated by verbs apparently see the world somewhat this way. Parmenides argued instead that *change is the illusion*. (Consider how far one can extend such a frame of mind as, "Oh, that's not *really* new; *basically* . . . " Or, "The more things change, the more they remain the same.") A similar view was to be advocated by Pythagoras and Plato and is the one that for better or worse would dominate Western thought even into our time. They reasoned that the meaningful reality is not in matter (materialism), but is in a world of eternal and unchanging numbers and ideal forms (idealism) and that the material form and event that we experience is a sort of illusion, a corrupt manifestation of the perfect and eternal.

The powerful formal arguments of Plato were to prove very useful to those Christian factions who came to control Europe. They were interested in eternal spirit, absolute moral laws, and afterlife; and following certain older Eastern schools of thought, they saw the flesh as weak, corrupt, and temporary. The senses were to be mistrusted, along with life's sensual pleasures. Jesus had preached the Kingdom of Heaven. "The world is only a bridge; cross it but build not thy house upon it." This religious emphasis on otherworldliness merged conveniently with the philosophical otherworldliness of Plato.

The *great* philosophical synthesis of Jesus and Plato on which modern Christianity is based was by Augustine in the fourth and fifth Centuries. But there had been earlier related efforts in the Judeo-Christian tradition. Even before Jesus, the Hebrew book of Job shows a strong interest in explaining how God might be both Creator and Ideal Good, much in the sense of Plato and earlier monistic beliefs in a fundamental Universal Spirit of Goodness, even if life seems corrupt and unjust much of the time. The Jewish philosopher Philo used a varient of Greek naturalistic ethics. For example, "Sexual intercourse is only natural when it leads to reproduction" (Cosby 1984). So if a man marries a barren woman he is "an enemy of God, . . . an enemy of nature." Likewise with homosexuality. Likewise if a couple has sex when the wife is not fertile. Such thinking would also be reflected within Christianity.

There were polytheistic Christians in the early centuries, but many Christians conformed to what had become the Hebrew tradition of being staunchly monotheisic and of resisting religious compromise. These refused to worship pagan gods, including emperors, along with their own, and so they were considered atheists and disloyal misanthropes and were persecuted for this. Yet despite persecutions, the faith had appeal in its promises of salvation of an eternal soul from eternal damnation, and in its promises of sure forgiveness from sin, and it grew rapidly in those uncertain years after the fall of the Roman Republic and during the Empire (MacMullen 1984).

The Christian *political conquest* of the Roman Empire dates from 312, when Constantine cast his lot with the Christians (previously largely in non-combatant positions) in the army and believed that this had helped him to defeat Maxentius. Christian revelation and ascetics soon became socially enforced (not at first legally *imposed*) dogma. (For example, under Constantine, the Christian Eusebius complained of the "unspeakable hypocrisy of those accepting the church and adopting the facade" for "fear of the emperor's menaces" and to win his favor.)

With the early union of Plato with Christ, Westerners as a whole were pulled vigorously away from concern with the here and now. Those Christian factions that came to dominate had long opposed and attacked the Greco-Roman culture (excluding views such as Plato's), with its pluralism and metaphysical tolerance, with its celebration of sensuality, an interest in the perfection of the human body, and even phallic worship—"the flesh" that the Christians eschewed had a distinct place in pagan religions. As the power of Christians grew, they went beyond merely ending gladiatorial spectacles and destroyed ancient art and literature that were deemed "materialistic," and as Gibbon, Spengler, Boswell, and others have detailed, the European image of antiquity was reshaped and adulterated, history rewritten, and the truth about antiquity was forgotten and replaced by myths for a thousand years, through the Age of Faith, and into our own time among the many uninformed.

In about 400 A.D., only 78 years after Constantine had made himself ruler, the schoolmaster Palladas wrote this short epigram, "A Pagan in Alexandria Contemplates Life Under Christian Mobs Who Are Destroying Antiquity."

> Is it not true that we Greeks are actually dead
> and only seem alive—in our fallen state
> do we fancy that a dream is life?
> Or are we actually alive and is life dead?

These were not simply conflicts of humanistic reform. For example, Christian churches and clerics held slaves and were even *forbidden* to free them (for they were seen as church property) at this time (Finley 1980). The fall of Rome, in the broad sense, was the rejection and overthrow of the spiritual and aesthetic world view of a culture in the midst of its economic, ecological, managerial, and political difficulties. By no means was it a simple military defeat or the simple incapacitation of leadership by lead poisoning or sexual excess, as children are sometimes told (Falco 1964, Grant 1976, Kaegi 1968, Katz 1955, Mazzarino 1966, Pirenne 1958, E. White 1927, L. White 1966). After all, political and economic structures may even collapse without there also being vast cultural changes. This happened repeatedly in Chinese history. Greek civilization survived military and economic conquest by the Romans. And so on with other civilizations. The fall of Rome was as much a social, political, and religious revolution and reconstruction as a series of military defeats. It was the replacement of philosophical and spiritual pluralism by a monistic absolutism, by an antimaterialistic, antiempirical, antisensualistic form of spiritualism, by an emphasis on life as preparation for a presumed eternal afterlife. Some scholars prefer to call it a cultural transformation rather than a fall. (Though then one should not refer to the "fall" of the Inca or Aztec empires either.)

The Germanic barbarians had mostly been Christianized, but not Romanized, long *before* they were *allowed* to migrate into the Christianized Roman Empire. Even the sacking of Rome in 455, 143 years after Constantine's victory, was only one phase of an attempt by have-not, largely Christianized barbarians and slaves to topple an ancient, bloated, if highly cultured, somewhat differently Christianized, and certainly more pluralistic ruling class and social system.

It was especially the sensual, materialistic, Dionysian leg of pluralistic antiquity that was smashed. In *The City of God*, Augustine shaped the course of European history and the character of Western values into our own time. He argued that the parts of paganism that could aid the doctrine of Christ should be preserved and that only the materialistic parts were bad. The large materialistic and empirical aspect of Aristotle's philosophy was "pagan" in this sense, while his logic, and the otherworldliness of Plato, were useful to the sorts of Christians who took control. These were under pressure to respond logically to pagan philosophers who pointed out that miracles could not prove the claim of those monotheistic Christian factions who insisted that there is only one god. *Any* god can perform miracles, the pagan philosopher Celsus argued. Miracles by no means prove that there is only one God. The Christian monotheism party adopted Platonic arguments in preparing their rebuttal and thereby became committed to the exaltation of this particular Greek philosopher (Wilken 1984).

Politics and class conflicts were as important as religious beliefs in motivating the Germanic immigrants and the leaderless mobs of the lower classes in the destruction and gradual decay of books and art, and eventually in the repression by Christian emperors of pagan civilization that led to what the Renaissance dubbed the Dark Ages. The so-called Dark Ages were ushered in by very well-meaning if very shortsighted people. They probably found it inconceivable that good *intentions* and good *motives* could in part contribute to what even their own descendents would one day judge to be unnecessary fears and new abuses. *Dark Ages* is an overstatement, though. Those centuries have been both damned and romanticized excessively.

From the point of view of its own leadership, the new era was not "dark" at all, but was enlightened and progressive. There *was* some splendid artistic, architectural, and social development during that period.

Religion also played a critical indirect role in political and economic deterioration. As strife between Christians and pagans mounted, it became difficult to discuss issues of common civil concern outside of the context of religious conflict. In the early Christian Empire, for example, pagans and largely polytheistic Christian factions would charge that a certain disaster or circumstance was divine punishment because they were not allowed to sacrifice at their temples to the traditional gods. But monotheistic Christians might insist that the event was punishment for "depravity" among pagans and still-partly-Pagan Christians, or that the time of final judgment had begun. (The myth that Rome collapsed because its upper classes were sexually decadent is a very old story. *Political* corruption may have been an important contributing factor, though.) Thus diverted by myths, it was difficult for the society to see the real cause-and-effect relationships at work and to take corrective actions. The result was serious economic, ecological, political, and military weakening.

The cultural revolution continued steadily. In 529 A.D. the Christian emperor Justinian closed the last of the major Greek universities, at Athens, finally to eliminate the pluralism of paganism and what was to his followers the undesired materialism of the civilization of antiquity. The worldliness of Aristotle withdrew along with Greek scholars to Islamic centers of learning.

Intellectual interest in the material world subsided relative to the spiritual, in part because of the asceticism that Jesus preached. In part also it was because they were always expecting the end to come. "The sun shall be darkened, and the moon shall not give her light, and the stars shall fall from heaven," as Jesus predicted in Matthew. "This generation shall not pass, till all these things be fulfilled." Since the end did not come in the generation of the disciples, each subsequent generation thought perhaps he meant theirs.

Issues of where reality resides remained in Judeo-Christian sects in one way or another, and they reached extremes particularly in some older sects. Even in some modern sects, such as Christian Science, material reality is seen as largely an illusion and as a corruption of the Ideal Good. In other Christian sects, the material world is not necessarily an illusion but may be treated as being not the most essential reality. Early in Christianity, denominations such as the Gnostics held *extreme* views on this issue and they created great problems for Christian political/theological unity.

> The present material world the Gnostics regarded as utterly alien to the supreme God and to goodness, and as therefore the creation of inferior powers, either incompetent or malevolent. The natural order of things

reflected nothing at all of the divine glory and of the matchless heavenly
beauty and toward it the Gnostic initiate was taught to acknowledge no
responsibility. . . . The world was in the iron control of evil powers. . . .
(Chadwick 1967)

The so-called Holy War, or *jihad*, of Islam originally referred to the obligation of maintaining the spiritual balances and harmonies in the material
world of politics, economics, and social life. Salvation of the social order was
at that time placed ahead of saving more and more souls. In this sense there
was a relatively more "Aristotelian" world view in flourishing Islam than in
the Latin world at the same time. Much of Aristotle remained to be reintro-
duced from Islam to Europe in the twelfth and thirteenth centuries, just after philosophy came largely to an end in the Near East under the influence
of al-Ghazālī (*Tahāfut* or *Collapse of the Philosophers*).

It was primarily Thomas Aquinas who began to rehabilitate the rediscovered Aristotle with the spiritual aspirations of Christianity. An eventual re-
acceptance of empiricism would be timely because a thousand years of faith
and formal logic alone had not led to One Truth, or to peace. Kings in castles
with armored troops, and not the meek, had inherited the earth. Inquisitions, crusades, and bloody sectarian wars tarnished whatever peace individuals had found in various forms of faith. Contending religious sects and
philosophies had proliferated even prior to the Reformation. In the reign of
faith and logic, the message of love and peace had too widely evolved into
one of repression, intolerance, authoritarianism, and hurt. A sense grew that
there must be better paths to charity, peace, and brotherhood. But who
could know the solution? Given their assumptions, perhaps Europeans were
doing the best they could.

In *Landscapes of Fear*, Yi-Fu Tuan writes that "Europeans in the Middle
Ages were insecure to a degree that it is hard for us now to envisage," for
material reasons, but also because the leadership gave a low ontological status to the physical world and high ontological status to a world of both good
and evil spirits. Europeans were ready to tremble at demons and dragons on
the basis of hearsay, and even late in the Middle Ages would carve gargoyles
and chimeras into cathedrals along with empirically real plants and animals.

Economic "recovery," travel, trade, and technological advances combined
gradually to expand European consciousness, particularly following the
Crusades and by the time of the Renaissance. They saw the great civilizations of other regions, and eventually Europe discovered its own pagan past
as well. Europe had renewed its contact with the East and had discovered
how limited its own economic and technical development had been
(though see L. White 1962).

Empirical science was reinvigorated and flourished when it was widely appreciated that the deceptive senses could be trusted to a qualified degree, if observation was careful and systematic. They learned that logic was valuable to a qualified degree, but only when combined with careful and systematic observation; for example, as craftspeople and inventors gained economic and political power relative to the coalition of clergy and aristocrats. Heretic burning and sectarian wars certainly did *not* end; that even got worse in many respects, but science held out new humanistic hopes as well as economic ones.

In Leonardo's notebooks we see the notes and sketches of not only the artist and inventor of machines, but of an inventor of *a fresh type of critical mind.*

All of the above history of the decline and reinvigoration of empiricism is of critical importance in understanding contemporary Western institutions, and attitudes, the development of scientific thought, and contemporary metaphysical and political disputes.

The complex interplay of well-meaning faith with shortsightedness, and the understandable but complex sociopolitical discontents that led to the so-called Dark Ages, has made such issues very difficult to discuss even for the adult public, and so schools usually avoid teaching them in any depth. Lossky (in White 1966), for example, observes that in our schools and press we "conveniently" lose sight of the basis for the major epochs of European history, this being "too embarrassing for the state of education in our age."

But for present purposes it is essential to keep history in mind. The mainstream of modern scientific Western thought is very much an ongoing effort to examine the material world by using the senses carefully once again. Even elegant abstract theory must in the end be supported empirically, in terms of the material world, or it is only acceptable as food for thought. With the rise of modern science, the empirical effort became tempered by an *extreme* carefulness that is best understood in terms of the unique history of European metaphysics, and as a reaction to a thousand years of skepticism that knowledge might be gained by the senses and from the material world.

Scholars have long wondered why modern science developed only once, in Europe, and why it had not developed also in in the great civilizations of India, China, or Mesoamerica for all their creative genuis and economic aggressiveness. I suggest that a *large* part of the answer may be, *not* to exclude local socioeconomic factors, that skepticism of empiricism had been so very profound in European scholarship during the early Christian era that when empiricism was reborn its practitioners had to be extraordinarily meticulous, double-checking themselves at every possible turn in order to erect models that were convincing to themselves and others. For example, priests knew well of illusion. Why look through Galileo's telescope? Why trust what one's "corrupt" senses think they see through some odd tube? In the end, whether or not Galileo really conducted his experiments exactly as he claimed (Segre 1980), he convinced people to critically observe the material world, measure and devise critical and ingenious experiments.

Next we can explore the point that in one way or another most societies have faced and coped with — the universal problem of deceptive senses and reason. The odyssey of the search for good judgment and wisdom has not been confined to the West, but should be viewed in a much broader world context.

Direct Ventures beyond the Senses

No one knows precisely how the worldwide diversity of cultural attempts at solutions to the problem of the illusion-generating brain and category/ hierarchy-imposing mind evolved, and just what forces shaped each attempt. Still, it is instructive to begin to develop a framework for thinking about the diversity. Diverse customs and institutions seem to have common elements and to be variations on a common theme — the central problem of how to be thoughtful humans despite our illusion-generating brains. This paradigm can provide starting points for detailed and critical scholarship. In this spirit I will next sketch the outline of an overview. It is based on facts but I only present and share it as a scheme that has been helpful to me in trying to organize thoughts.

Let us begin by designating a traditional, or *shamanistic mode* of systematic searching for realities beyond the senses. What sense could the early members of our species hope to make of what we now call dreams, illusions, and hallucinations?

Mammals dream and all have curiosity. When human language evolved to the degree that abstractions could be communicated, dreams must have been a most peculiar phenomenon to discuss and try to comprehend. It would be evident that individuals differ dramatically in their experiences, and dreams would be a most dramatic example. How did our ancestors cope with it when respected and trustworthy observers told of fantastic experiences? What did they make of it when the hearing of one person's fantastic experiences might be followed by their own dreaming of a similar thing? Did they argue over the exact nature of these other worlds and beings that seemed to come and go of their own power and yet be nightly with them? Surely some did, just as many people do today.

Dreams do often relate to waking reality, as we know, and sometimes we can even "see" or appreciate things more clearly in dreams than when awake. Important insights and discoveries have been reported to come in dreams. Did our ancestors long ago confront such things and take for granted that there must be abstract dimensions to reality? Searching deliberately for realities beyond waking experience may be as old as the ability to converse about abstractions and could have been based on trying to know what to make of dreams and waking illusions. How could one begin to form meaningful opinions about life in general outside of a framework that attempts to explain dreams, a significant proportion of our total experience? The problem is simplified for those who have come to explain dreams as "imaginary" and forget them quickly, but we should keep in mind that *this contemporary position is a very bold one* in anthropological and historical context.

Some cultures, such as the Australian aborigines, while remaining quite practical, have the opinion that the dream world is more real than the waking. An uncle may die in the waking, but live on in the dream world. Obviously, the waking world lies, some cultures have concluded quite "logically" (e.g., Kalweit 1988).

Cultural inventions to determine reality in this shamanistic mode involve primarily explorations of dreams, trances, and visions. These experiences are taken seriously and discussed extensively and critically. The group will search for meanings and will refine beliefs that they use in turn to interpret "waking reality." Beyond this they negotiate with spirits encountered in dreams and visions, and discuss strategies for differentiating deceptions by spirits as well as for obtaining favors from spirits.

Societies widely took the added step of actually cultivating their visions. There have been various types of *vision quests* among peoples of the various continents. It was found that sleeplessness, exhaustion, isolation, fasting, prolonged pain, hypersexuality or sexual abstinence, drumming and chanting, the darkness of caves or lodges, high expectaton, all could promote trances and increase the visionary experience.

Fasting to increase the possibility of having visions was also a custom in the Biblical world. For example in Matthew 4, it is written that Jesus followed his baptism by going into the wilderness and fasting for forty days and nights to bring on the devil and confront his temptations.

In most parts of the world people also use drugs from plants to further promote trancelike states. Usually these have not been used thoughtlessly for recreation or escape from workday troubles as too many "civilized" people now do, but they have been used carefully and according to custom at special ceremonies to explore the supposed alternative or primary realities. Also, various cultures have experimented with *dream incubation* — techniques to increase the chance of having desired encounters during one's dreams.

Exploration of these ephemeral worlds was very direct. One might visit and confront presumed spirits in a dream or trance and fight them to win an emotional or spiritual victory, or ask for guidance or for specific signs and information. A modern psychobiologist might see this as being, *within the context of an appropriate cultural interpretive system*, one way of tapping the powerful subconscious and intuitive capacities of the brain. The same experiences in an unsystematic context might properly be seen as only hallucinations or maddness.

Tribal people everywhere also found that some children are more sensitive than others. These children were encouraged, for the good of their people, to go through years of rigorous training to cultivate their sensitivity. Such *shamans* can enter trances easily. They are usually trained to be able in effect to depersonalize, to put aside the ego and workday perceptions, and commonly to visualize themselves and others as skeletons and as collections of organs. They are also trained to be able to empathize to an extraordinary degree with others and even with animals.

The mind can contemplate a bit more objectively by stepping outside of one's normal personality and role. Individual personalities include perceptual and relational habits, responsibilities, and abundant jealousies, ambitions, and stubbornness that are part of a perspective and identity that help one to function and maintain position within a group and to survive in the workday world. But these may also cloud one's perception.

Survival and objectivity are not necessarily partners. People will search and even fight for the reality that seems to "work" for them, not one that may in fact be more true but that might threaten to leave them emotionally uncomfortable or less enthusiastic about life. In the short term, at least, an ego-serving attitude toward reality may have high individual and group survival value, providing stability and security. But there are other times when egoless relative objectivity may provide advantage and promote survival.

Shamans were once regarded by Westerners as pure frauds. The colonialists saw them as political priest-bosses who instigated suspicion and ill will toward foreigners; they saw shamans as political rivals. Their missionaries saw shamans as agents of evil. Marxists saw them as exploiters of their people's ignorance. Much as they saw the European clergy as a parasitic privileged class in collaboration with kings and captitalists, they saw shamans as a backwoods privileged and selfish class.

Now anthropologists consider these views to be highly simplistic. There were sometimes grains of truth to some of these views, yet for the most part the shamans were and still are selected as children, trained very arduously indeed, and they do develop talents that help their people. A trick that is fakery to outside eyes may be a move by the shaman to help the people see his or her inner vision. One does not have to believe in their spirits to see practical skills and benefits flowing from their vocabulary and grammar of perception. Often their judgment concerning domestic quarrels, illness, politics, and hunting strategies is far above average. Apparently their training in depersonalization and empathy assist their judgment. They develop intuitive ways to provide psychological comfort.

In many societies dreams are taken very seriously and explored systematically not simply by shamans (Garfield, 1974; Tedlock, 1987).[1] Various sorts of dream explorations have been widespread among both non-Western and some recent Western peoples. Clearly, people will much too easily get carried away with dream interpretation, often to absurd extremes. But there is no doubt that sometimes insights will come in dreams and that dreaming can be probed and cultivated in quite valuable and enriching ways. Robert Louis Stevenson worked out ways, for example, to repeatedly return to his dreams in the process of developing stories.

All such explorations, from the shaman and vision quests to dream interpretation and dream incubation, seem to be well organized and to produce moderately practical effects — institutions, going beyond the senses and improving the quality of life. These vision quests and dream interpretations are in this respect the ancestors of theology, philosophy, and science.[2]

"Scientific" Divination

The faith that numbers, calculation, and geometry can give certain knowledge of the nature of abstract reality greatly precedes modern science or even ancient Greek science. One traces the origins of the *numerological/ astrological mode* to occult but disciplined studies of the nature of numbers and geometrical forms and the movements of the heavenly bodies. This mode developed into very different systems for divination from the examination of intestines, oracle bones, or conversations with ancestors and spirits in dreams or in altered states of consciousness. It is different from the explorations of human psychology and experience by Buddha, Socrates, or Confucius.

Many scientific and orthodox religious Westerners today would just as soon ignore astrology and numerology. These are skeletons in their family closets. So we must be reminded that these have been believed in by more people (even astrology alone) than any single faith in all of human history. Once they developed they quickly spread throughout the civilizations between Europe and Japan. They have been widespread among not only Christians, but Muslims, Hindus, Taoists, pagans, Confucianists, Buddhists, Zoroastrians, Jews, Shintoists, and Jains.

Preliterate people certainly knew that there were relationships between the cycles of movement of celestial bodies and weather and phenological patterns, as well as mood shifts. They would regard unusual events such as eclipses as portents. Ancient stone circles and temples were set out in relation to important regularities such as the position of the rising sun in its most northern or southern position, or on the day when light and dark are equal (solstices and equinox). It is not very difficult to reckon such things *without* numbers, mathematics, or geometry. Some scholars, though, infer beyond the above an impressive knowledge of mathematics, geometry, and astronomy for the ancient stone-circle builders. But these speculations are based on statistical analysis that can give false impressions of accuracy. So claims of advanced mathematical and astronomical knowledge for these very early people remain interesting but unsubstantiated (Chippindale 1983). Any explorations of numerology would be even more speculative for this very ancient era.

Counting systems have long been common all over the world. Mathematical *calculations* developed for commercial and engineering purposes apparently only as the large civilizations developed. There are problem sets and solutions going back to 1800–1600 B.C. in Babylonia. But with this practical interest was a parallel interest in the supposed *occult meaning* of numbers and their relationships to each other and to geometrical forms. This might have to do with spiritual qualities assigned to numbers and numbers assigned to spiritual qualities and gods, the combining of gods in rituals, the determination of geometrical proportions to honor gods or to harmonize spiritual forces, etc. The Pythagoreans were well known to be interested in such things and there are indications of older such interests in India and Mesopotamia. There are those who argue that mathematics, geometry, and astronomy developed *purely out of* numerology and astrology. The evidence for this simple scheme is not firm, though clearly there has long been a close interaction between the practical and the highly speculative interests in numbers, geometry, calculations, and celestial movements.

Personal astrology combines these last. The band of constellations across which the stars, sun, and moon move was divided into a zodiac. An early form of this reference system seems to be one Babylonian contribution to astrology. It was long thought that the Babylonians invented personal horoscopes and did not simply contribute certain elements to this. But more recent studies, Neugebauer (1969) argues, indicate that Greeks in Egypt invented *personal* horoscopes, as late as the fourth century B.C., where the positions of planets, sun, and moon in the zodiac were calculated for each person at the time of birth. Calculations supposedly reveal when one should take certain actions or not. Neugebauer's Hellenistic origin argument makes a strict distinction between *calculated* personal horoscopes, and astrological predictions based merely on birth dates (as Hittites, Egyptians and others had earlier done). Personal astrology spread across Europe and Asia and is still an intense part of community life in areas such as India.

For our purposes, the place and time of the origin of astrology is not so much of interest as is the appreciation that using the heavens for divination gained simply enormous prestige as precise astronomical knowledge improved. One might be told, in 243 days Mars and Jupiter will be in Leo and that is the best time for *you* to get married. What a comfort to watch Mars and Jupiter each night and *know* that despite any retrograde movements they would be in Leo at the appointed time. If there are to be problems in the marriage, perhaps they will not be as bad as they could have been, one might think.

Good records for the moon only date back to about 747 B.C. (Babylon), and good records and calculations for the planets come later. But advances in the determinations of even some crude period relations of the moon and planets must have had great psychological impact for credulous people living in uncertainty.

Astrology has been thought of as a religion by scholars. But they have also thought it a fossilized science. In some sense it is transitional between divination and science.

> The concept of predictable influence between these bodies is in principle not at all different from any modern mechanistic theory. And it stands in sharpest contrast to the ideas of either arbitrary rulership of deities or the possibility of influencing events by magical operations. Compared with the background of religion, magic, and mysticism, the fundamental doctrines of astrology are pure science. Of course, the boundaries between rational science and loose speculation were rapidly obliterated and astrological lore did not stem—but rather promoted—superstition and magical practices. The ease of such a transformation from science to humbug is not difficult to exemplify in our modern world. (Neugebauer 1957)

Various studies have been conducted to evaluate astrology. For example, the psychologist Michel Gauquelin placed an ad in a newspaper offering free horoscopes. To the 150 replies were sent all *the same* determination (incidentally, that of a mass murderer) and each person was asked how accurate their reading was. Ninety-four percent felt the description of them was accurate and for 90% their friends and families agreed. There is much self-deception in astrology.

Certain studies, for example testing for influences of Mars on sports champions, have suggested some effects, but then the effects would vanish in repeat tests (Abell and Singer 1981). As a typical Sagittarius, let me boldly draw the likely conclusion. It seems that the great attraction to astrology has been based primarily on its psychological appeal, and not on any verifiable predictive power. (The *Skeptical Inquirer* is the journal of the Committee for the Scientific Investigation of the Paranormal and frequently publishes critical analyses of current astrological and other paranormal claims.)

Divinations involving number and geometrical analysis, and those involving calculations of celestial movements, have often been pursued separately, but I have grouped them here for convenience and economy of discussion. (And alchemy, an extremely rich if complex subject, has been ignored here for convenience and economy.) They have been simply enormous movements with rather close cultural roots in the chain of civilizations that developed between Greece and China.

Today people still have the problem of separating claims *based* on science from scientific knowledge. For example, I explained earlier (chapter 4, note 2) that *constitutive/methodological reductionism* is a firm cornerstone of good science. But the *ideological reductionism* of many scientists today may have a similar relationship to methodological reductionism as astrology did to astronomy at the time of Copernicus (who calculated horoscopes to make ends meet).

Formal Philosophical Systems

With the development of philosophy within metaphysical concerns, we can discuss a degree of systematic, formal attention to thought itself, the *philosophical mode*, of seeking beyond the senses. In the shamanistic mode there is abundant ontological concern over what reality is and intense epistemological self-consciousness over how to find it, but this does not necessarily focus formally on consciousness processes themselves. All people have been concerned with reality and how to find it in connection with their ethics and politics. *No distinct line can be drawn* between shamanism, religion, scientific divination, and philosophy, but in the philosophical mode attention is focused not merely on schemes and techniques for interpretation of physical or nonphysical experiences. The philosophical approach to epistemology is drawn to the reliability of the mind; humankind steps back from the body of revealed beliefs, and begins to look *systematically* at the *nature of thought itself* and *to record the debates*. This sequence begins to appear in the Vedas, perhaps fifteen centuries B.C., and is more distinct in the concluding Vedas, the Upanishads, composed in about the eighth and seventh centuries B.C. Philosophy, at its most vigorous in both East and West, has involved a systematic questioning of accepted and popular beliefs, and the invention of various theories about thought that were *themselves questioned and refined*.

The more formal critical systems, for example in Jainism, Buddhism, and the earliest Greek philosophy, all began, interestingly enough, in the late sixth century B.C., with strong hints of intellectual influences between India and Greece. Confucius was not a speculative philosopher, and there is no question of Indo-European influence, but it is interesting that Confucius began to travel and attempt reform about the same time. The amazing flowering of philosophical thought among Indo-Europeans in about the sixth century B.C. and the systemizations of thought that developed in the fifth century were comprehensively diverse at the very outset. Raju (1985) argues that such diversity is (in India alone) evidence that philosophy was not a simple ideological response to given sociocultural (or economic) factors.

> When a history of Indian philosophy is written, it will not be right to follow the authors of western philosophy and to show how changes in historical and socio-cultural factors led to the founding of this or that school of thought. Whatever be such factors that obtained in India, the fact that all the schools started simultaneously shows that the factors had very little to do with the differences among the schools. The same factors could not be the causes of all the schools, if they were to be causes at all. The motive of all the schools—theistic, atheistic, and materialistic—was the search for the ideal of life. This search implies that the seekers were not satisfied with the life—material, ethical, and spiritual—they were living day to day. . . . We have . . . to attribute this search for a deeper meaning of life than could be found in day-to-day existence to a keen and critical sense of happiness somehow developed by the people. . . . What is man's life? What is its meaning and purpose? How is man to plan his life so that he can attain his ideal? If life is part of reality, how is he to know this reality? . . .

There are ample reasons to believe that the "wise men" of the sixth century B.C. were directly or indirectly familiar with the seductive notion that abstract things like geometrical forms and numbers and absolute relationships between them really exist as spiritual forces outside the mind. Thus, the notion may have been encouraged that one could more generally know spiritual things with *certainty*, as *objectively demonstrable* truths, through steps in reason rather than by appeal to personal subjective experience.[3]

With hindsight we know how incredibly appealing such ideas might have been. We also know how inadequate pure logic nevertheless is to attain the mirage of certainty. One must suppose that if at some point intense and "irrefutable" logical claims to certainty were made, this might at once have encouraged other widely differing points of view to develop into competing systems (chapter 11). Disciplined and intense debate do seem to have happened rather suddenly in both ancient Greece and India in the sixth century. Later, something similar began to happen in early Christianity—there developed vigorous competing claims to a supposed single absolute Truth—but much of the proliferation was stamped out. Such happened again about the time of the Reformation and perhaps only the rise of a modernist secular agenda checked the intensity of conflicts.

Despite the variety of formal philosophies established both East and West by the sixth century B.C., we can distinguish a different general East-to-West trend. In both areas the original goal was to find personal peace and virtue, but the priority of Western philosophy often became to determine where cosmological reality is located, in matter, in ideas or numbers, or in a spirit world—to develop a thought *system* to determine the nature of this, and to determine what one's political and moral position in reality should be. Eastern philosophies in the Vedic and Buddhist traditions mainly gave such (ontological and epistemological) issues secondary priority and explored most what it is "to be" and how to attain a state of enlightened "being" existentially. This may seem like religion, but it is not necessarily wedded to doctrines, beliefs, or devotions.

Early in this chapter and elsewhere in this book, logic and empiricism in Western philosophy and science are discussed. So in this section I shall focus next mostly on some general features of Eastern philosophies, especially in the Vedic and Buddhist traditions (though the diversity in each is a highly complex matter). India has long remained quite pluralistic and, though materialism has been opposed, no single faith has been institutionalized as official dogma, even eventually in those territories conquered by Islam.

Careful analysis of reason itself, somewhat in the manner of Socrates and Plato later, was underway by the time of Buddha in the sixth century B.C. Philosophy in this sense traces back to the Upanishads of the eighth or seventh centuries B.C. While Aristotle analyzed reason and founded *formal* Western logic in the fourth century B.C., setting a standard in Western logic that was to go unchallenged until Francis Bacon, and again in the present, the Buddhist detailed analysis of thought took a different course. It "rejected" reason, and did not return to development of formal logic except slowly in the sixth and seventh centuries A.D.

> At the time of Buddha India was seething with philosophic speculation and thirsty of Final Deliverance. Buddhism started with a very minute analysis of the human Personality into the elements of which it is composed. The leading idea of this analysis was a moral one. . . . The external world was also analyzed in its component elements. It was the dependent part of the personality, its sense data. [A millenium later, during India's golden age,] in accordance with the new age, the condemnation of all logic which characterized the preceding period, was forsaken, and Buddhists began to take a very keen interest in logical problems. (Stcherbatsky 1962)

So as Western thought before the Renaissance had "rejected" the senses for a millenium, Buddhist thought had "rejected" logic for a millenium.

The early mainstream Eastern mystics had begun to reason even prior to Buddha that it was necessary to move beyond logic and words to *intuitive* understanding. It is largely through experience, intuition, and meditation that truth is to be found. But—and this can be a delicate point to grasp—the main Eastern philosophies are *not therefore irrational* in any simple or obvious sense. It is *through* reason that one must *abandon* logic and reason. One must move *beyond*, and this cannot be done simply by ignoring reason and knowledge. Thus, as one sees in the *Isa Upanishad* alone, worship of intuition or the delight in knowledge can lead into blind darkness. Somewhat similarly, one accepts the authority of a guru virtually completely. But the goal is freedom of the spirit from all external authority. The process moves one beyond.

There is a *form of logic* to the instruction given, the discipline taught, by a guru or Zen master, even if its goal is to destroy logic and goals, to awaken one from the dream that these are seen to be (for example, Kim 1981). There is nothing intrinsically mysterious about this, it does not necessarily require an occult belief in spirits or the supernatural, it is simply an attitude and a way of being, a psychology, that is very nonfamiliar to Western attitudes and ways of being.

Our minds are indeed filled with selected thoughts. Ego accepts some, constructs others, closes the mind off to much more. The goal of the Eastern wise man is to be able to find moments of clarity and being and awareness that are free of ego's collection of thoughts that focus or distract the mind from its full potential of consciousness. One frees one's consciousness from the thoughts so that one is completely aware of one's nature and the world without the filter of ideas and longings.

An unreflective person will *not* necessarily become free of thoughts and vexations over the long run. If the waking experience of each of us is a form of a dream (a narrow slice of reality, constrained and distorted by the subconscious, in the larger sense arbitrary), then the task is to find ways to wake to full awareness, uncorrupted consciousness. Ignoring the problem will not help us. Then we only stay "asleep."

Simon Gray seems to explore some aspects of this "sleepwalking" in his play *Otherwise Engaged*. The main character has his mind on playing a new recording of *Parsifal*, but visitors keep stopping by, wanting someone to listen to their problems. The characters are, sadly, never fully involved with one another, the visitors have their minds on their own problems, unnoticing that their host secretly wants to get to his music.

One visitor says that he is a typical Englishman, never so happy as when he is in his automobile, caught in traffic in the rain, dreaming of what is past or of what is to come, recalling anticipations or anticipating recollections. We are all like that much of the time, dwelling on problems in our heads or with our minds on the past or future, not fully open to and aware of the here and now. Liberation from this mode of mind operation is a goal of "no-mind."

In representative Eastern philosophies, truth is not seen as a cosmological reality that resides in statements, words, ideas, or formulas, in models of physical reality. It is not an epistemological issue that can be proved mathematically or logically. It is in a state of mind and of being, a "psychology," a personal awareness of what it is "to be." Salvation of the spirit cannot come simply from *believing* something deeply, or even from *following* certain instructions. For those who are religious, rituals may be obligations, but (in the "higher" paths) they do not produce spiritual salvation. And one may or may not be categorized as an "idealist," a "materialist" or a "spiritualist." That too is usually secondary.

Truth is seen as a quality of *living human consciousness.* Truth is not seen as something that can be captured in phrases or printed on paper and put away in a library, separate from experience. Rather, with some direction from a guru or master one has experiences in active life (as in forms of Zen) or in sedate meditations (as in forms of Hinduism and Indian Buddhism as well as Zen). Awareness expands to its various possibilities and then wisdom, higher consciousness, enlightenment come more or less spontaneously and unconsciously. This usually involves temporary or extended abandonment of the ego and formal logic. In this egoless enlightenment is spiritual salvation.

Of course this is extremely difficult to do, especially for mainstream Westerners. Outside of certain monastic and mystical traditions, Westerners are used to thinking that truth is in ideas and beliefs grasped by the ego, rather than in egoless aware experience, and Westerners tend to believe that our reasoning minds are the essential us. (So that, for example, it makes sense for there to be a "me" that *has* two hands, a mind that *is* me, and a body that *it* controls.) One calls the Eastern paths to wisdom "disciplines," since this implies the finding of awareness in states of being and living rather than simply in ideas, words, and descriptions.

The states such as "no-mind" or the "silent witness" are also somewhat difficult for Easterners. The average person in the East is affected by philosophy, just as we are in the West, but is not necessarily willing or able to go deeply into it. Indeed in traditional Vedic cultures one devoted the early part of one's life to learning survival skills. They were expected to devote the middle part of their lives to meeting obligations to family and society.

After their children were grown they might explore contemplation more deeply in the forest, and eventually some would cut all ties to family and society and devote themselves completely to contemplation.

For the intent and determined persons, diverse schools of meditation were devised, ranging from the classical contemplation of the calm Buddha to the less well-known frenzied "release" of the whirling dervish (see, for example, discussions of Tantra by Bharati 1975, and the interesting collection of dynamic meditations assembled by Rajneesh 1976, a respected religious scholar who became quite controversial personally, and somewhat outrageous, in his late years).

Western culture tends to mistrust the unconscious and spontaneity. (Some feminists tend to see this mistrust merely as part of the domination by "masculine" values.) The ego is so much the core of Western identity and so essential to pursuing the goals that an acquisitive and competitive society sets for us—and thus meeting the conditions for earning some degree of decent treatment, respect, and power—that many can scarcely imagine the possibility of suspending it. Part of the interest of Eastern thought to some Westerners has been that it does attempt to deal with the cultivation of neglected intuitive and subconscious potentials and sensitivities.

In any event, through disciplined suspension of conscious reasoning and the ego one can experience realities beyond those that ordinary people do. In some schools these realities are not necessarily supernatural or visions, but are mostly in the nature of heightened awareness, relative objectivity, joy, and a sense of liberation.

But in some schools insights *are* experiences of some "innerspace"—which some would interpret as alternative worlds in a literal sense. I tend to see these last supernatural realities as objectively imaginary but subjectively real to the individual and perhaps genuinely interesting and informative in the right hands and context.

In correct usage, the term *mystical* refers to a direct and personal intuitive or insightful inner experiencing of God or of normally unavailable truths. It has become a very negative word in the culture of materialists, or among Christian sects that stress salvation through loyalty or devotion, but it does not necessarily have supernatural meaning.

The goals of the philosophies of the East have tended to be individual and practical, to find liberation from the vexations of *life*, rather than to avoid a presumed eternal punishment, to earn a presumed comfortable afterlife, or to draw abstract conclusions about Truth. For example, in China, humanistic Confucianism, more custom and law than philosophy or religion, held together the civic order while Taoism served the individual, served to liberate one from tensions and frustrations including those caused by civic values, perceptions, imperatives, demands, and restrictions.[4]

Such Eastern philosophical *goals* were similar to those of the Greek philosophers who were searching for the Good and the Right. A basic feature of most all philosophy, except of much recent academic philosophy in the West, has long been the question of how best to live. Only gradually and recently did Western philosophy grow independent of primary spiritual concerns with how we should conduct ourselves and feel about our lives. In the West, the philosophical study of the material world evolved into science, which we have now come to think of as apart from philosophy. Today, symbolic logic, along with semantics and such, may be the most active subjects in American academic philosophy. Both science and philosophy as they are presently divided in the West give less emphasis to ethical and moral sentiments, and have, at least largely, left these to religion, literature, economics, and politics. Specialization invades all of life. ("Good Lord, if you want to find out how actually to make ethical decisions then the last thing you want to do is take our class in ethics," a philosopher colleague advised students.)

Allen Watts, author of *The Book: On The Taboo Against Knowing Who You Are, The Way of Zen*, and *The Two Hands of God*, argues, in *Psychotherapy East and West*, that Eastern philosophies are more in the spirit of modern psychotherapy than they are like Western religion or philosophy. Avoiding a detailed response, I am somewhat in agreement. Western philosophies have certainly informed people's values, but Eastern philosophers tried to go right to the heart of the problem of how to live life free of vexation and to develop various practices rather than abstract theories mainly.[5]

It is of course only a starting point for discussion to refer to "East and West." Christianity, Islam, and Judaism are quite diverse, especially when one considers all the small monastic and mystical traditions. For example, some forms of Christianity seem to overlap contemporary Eastern solutions to the problem of vexations, with their *indifference* to, or *detachment* from, the flesh and to troubles. Other forms of Christianity seem to cope with the problem by making a virtue of the *endurance* of pain, suffering, and strife, as one strives to emulate Christ and prove that one is worthy. Thus, certain Hindus may walk on hot coals or pierce their skin with skewers to demonstrate their indifference to the possibility of pain—they do not aim to feel pain. But some Christians are directed *to feel* guilt and pain, perform acts of penance, and so on, and thus co-opt vexation rather than try to reduce it. So the similarity of the Christian penitent to the Hindu is mostly superficial. But in some sense both represent mystical extreme variants of their diverse traditions.

Modern Empirical Science

Another major intellectual effort was the *modern scientific mode*, of seeking beyond the senses (see also chapter 11). It was entered when great thinkers began to regain confidence in their own senses. People like da Vinci and Galileo found that with careful and critical attention, they could make observations that even moderate skeptics could repeat and share.

Both idealism and materialism have been historically critical for modern science. From the Pythagoreans and Plato, the seductive ideas of Truth, of universal abstract realities, a fascination with numbers, and a mistrust of the undisciplined senses were handed down. Their Truth was to be found through Reason alone, though modern science is irRational. From Aristotle and others, on the other hand, came not only formal logic, and theories of causation, but the *critical* precedent for careful observation of the material world and its objective and detailed study, *empiricism*. Though, modern science is in no way merely trust of the senses.

When should we *trust* our senses? When should we *mistrust* them? To accept the Copernican theory one must mistrust certain sensory information and trust other data, for example. There is "pure" theory in science today, and there is pure observation, but as a system of knowledge, science accepts abstractions, or observations, only by constantly testing them against the data of material reality and the interpretations of the emerging conceptual framework.

But science created a crisis for the Platonic/logical schools of philosophy that has never been resolved. Academic philosophers may sometimes call the incompatability of the two systems *a crisis in science rather than in philosophy*, since science cannot define its methods or provide absolute Certainty *in their desired terms, logically*. It is irRational.

Science was accepted by mainstream European thinking not because its findings or methods were philosophically appealing or even satisfactory. *They were distinctly neither.* Yet the power of the approach was beyond dispute. It transformed life. It redefined *rationality* for all but the most technical philosophers and religionists. I do not mean to say that science should be accepted because it can make microwave ovens, and Plato, Christianity, and logic cannot. Rather, with science individuals can understand for themselves why the sky is blue and that whales are mammals and not fish, or rabbits are not rodents. There is no need for reasonable people to fight and argue over such things anymore. Science was accepted because it became clear that over the long term, by combining careful thought and organized experience, one could expect intellectual results (chapters 9, 10, 11).

Science of a sort (but identifiably science) was present in the ancient world. But it was not quite so meticulously *cautious*, experimental, precise, and skeptical of dogma, authoritarianism, and popular beliefs to the degree that modern science was forced to be. Modern science aspires to be fiercely skeptical of its own research material. Scientists, as a subculture at least, regrettably tend not to turn their critical standards inward to their hypotheses of professional self-identity, or to study systematically their general self-consciousness and values or the institutional influences on their thinking. But commendably they will try hard to falsify their own *research* hypotheses to weed out the bad ones.

Indeed, research usually turns out to be much more interesting when we prove ourselves wrong than when we are right. When scientists are wrong, then they may be forced to move out of a flawed conceptual mode that has been limiting thinking on a particular issue or in a given field of research.

It is sometimes argued that Christianity and the book of Genesis were essential to the evolution of science, making the world material and unspiritual and placing humankind in charge. I doubt that this was necessary, since the Greeks began their vigorous analysis of nature with the distinct belief that they were entering a spiritual realm. Even modern scientists have had both atheistic and religious attitudes toward nature. One does not have to believe that the world is unspiritual to be a careful empiricist.

Science deals with truth as a body of *emerging abstract concepts about the material world that are based on rigorous observations of the material world*. Some scientists do treat the physical, mathematical "laws of nature" as though they actually cause things to happen, rather than that they are merely the most powerful language ever constructed to describe the ways material things behave. Yet few major scientists would say that a formula or idea that accurately describes physical phenomena is a true Platonic Form. Writing of mathematics, Kline (1985) concludes, "It is paradoxical that abstractions so remote from reality should achieve so much."

Even the most useful and splendid scientific truths are as a body emerging and in a sense tentative, for science is in principle largely open-mindedness and critical thought. As a practical matter there is, for example, no doubt whatsoever that the earth is a great ball that spins, and also moves around the sun. These motions of the earth are facts by all but the most technical philosophical criteria. Yet principled scientists *try* to remain open to the possibility that the solar system has some other structure, and will try to think of critical evidence that could overturn current dogma no matter how absurd a long shot that may seem. Biologists attempt to construct *natural* systems of classification, for example, schemes that attempt to reflect the actual genetic relationships among plants and animals. But they realize that in certain cases continual study and revision will be necessary for some time. Even the solid cases are consciously tested for robustness.

Science is an *evolving* set of questions, perceptions, methods, values, lessons, conclusions, and goals. It is *not* unbending Truth and rules as is too often taught and as many would like to see it. This is a confusion with the old Platonic idea that Truth must be found through a logic, like the rules of geometry. Science is as fragile as culture. Science finds truth in the emerging system of accurate, tested descriptors and predictors of nature. This is neither the Truth of Plato, nor even that of Aristotle, though it owes historically to both.

Science and scientific thinking were once closely linked to a quest for wisdom and gave the world more firm reason for hope that we can extend our abililty to see and know and become wise than did any other effort in history. The Age of Reason was a time of great faith in the potential of disciplined thought to help the progress of individuals and societies toward Greek and Judeo-Christian ideals of fairness and brotherhood. As the mainstream of European society redefined reason, then came the great reforms in justice, the end of heretic burning, and movements to end slavery and to educate women. Wisdom may have seemed within grasp.

Much of educated world society has now come to see colonialism as a bad outgrowth of the days of the Enlightenment. In fact, though, European expansionism and enforced hierarchy began much earlier. One might recall the forced establishment of Latin principalities throughout the Middle East by the First Crusade in about 1097. Yet certainly, while avarice and missionary zeal drove colonialism, disciplined reason and advanced technology did give the colonialists a powerful advantage over the conquered. Supposedly scientific, but in fact ethnocentric, notions of cultural superiority and the march of progress were part of the supposed moral justification for conquest. The new ways of organizing the mind did increase the ability of Europeans to gather wealth and power, to analyze and think ahead, and to dominate their own compatriots as well as people of other cultures. Science and scientific thinking can be used for antihumanistic as well as humanistic ends.

Let us not forget that if Wisdom and Compassion have evaded us, we also owe to those times many positive dimensions of our own being and values. Out of this Enlightenment optimism came the idea of modern democracy and the belief that European people of diverse views could find ways to govern themselves and live together in a humane manner even after centuries of inquisitions and neighborhood persecutions of heretics.

It worked very well for a very long time. In our economic, political and ecological dark days we have doubts about the world and the myths that misguided reason has helped build. Yet we should keep in mind that well-used reason, though not the pure Reason of the philosophers, has largely proved itself. But we have tended to forget the goals and spirit of the Enlightenment and its historical context. Indeed we have tended to forget the historical basis for democracy. Many know it today incorrectly only as a balloting system for electing officials, or as an economic system.

Some would like to see a more simple view of the history of science and modern thought than I have started here and will continue in chapter 11. They might claim that religion was merely superstition and modern thought was merely a breaking loose from the bonds of superstition. Or, science originated as a natural manifestation of a desire to control and dominate nature. Or, modern science was *merely* an invention to help garner more and more wealth and power. Others would like to believe that science and modern thought are merely new religions replacing faith in God with faith in reason, replacing a sacred priesthood with secular liberal professors. The degree of truth in such positions is so little that these popular notions are gross distortions of history and of human nature.

Much of the history of science did have to do with combatting superstition, and this often brought it into conflict with established beliefs that were often wedded to religion. Skepticism is critical to good scientific thinking, though various scientists may hold discredited beliefs, even with superstitious tenacity, about progressivism, naturalistic ethics, essentialism, methodological absolutism, explanatory reductionism, pure objectivity of the agenda of the science establishment, and some even hold a sort of neonumerology that Truth is in numbers and formulas.

But religionists, too, have often enough been of a skeptical mind and so skepticism certainly does not place science into *categorical* opposition with all religious thinkers. The *major* religious thinkers were skeptical of many of the popular religious beliefs of their times. Consider, for example, Kierkegaard's criticisms of modern Christianity from within, arguing that piety, while comfortable and popular, cannot be enough to allow people to be certain that they are not fooling themselves, that they are in fact sincere in their belief in Christ (see also Martin 1986). Such certainty, he argued, can only come from being willing to actually follow Christ's example—a radical and life-altering departure from the popular idea that salvation can come from belief, prayer, tithing and perhaps proselytizing.

Scientists often have had personalities driven by curiosity, to know a thing, to reduce mystery to predictability, and in this sense to gain control over and dominate it. But curiosity is universal, so if science is *merely* a natural manifestation of curiosity, then why did modern science evolve only once? Necessary but not sufficient.

It is true that the economic climate of Europe was favorable for the development of science and technology as Europe emerged from the Age of

Faith and became more wealthy and ambitious. It is true that science has often served social wants and needs. Particularly since Francis Bacon, it has been seen by the ruling economic, political, and military establishments as a source of power and wealth. As kings once "wedded" themselves to religious leaders, those with power in our own time have been negotiating the exact relationship that they will have with their new mistress. But the desire for differential power and wealth, while *not* universal, are widespread; other cultures have been wealthy and ambitious, and modern science only developed under peculiar historical circumstances.

Corporate Science of the Multiversity

In the *corporate scientific mode*, reality remains (ontologically) in the material world and in continually tested abstract models that form a metalanguage for understanding it (chapter 11). But there is conflict between traditional scientific ideas about how to find truth (epistemologies) (as in chapters 9, 10, 11), and newer ones. The newer ideas esteem advocacy, competition, specialization, management, and mastery of ritualistic methods.

The *findings* generated by contemporary ("postmodern"?) science are certainly of interest to all of us. But can contemporary science be used as a *model for clear thinking* by the aspiring *individual?* The answer is possibly not, or only with the greatest caution. The search for wisdom may be moving into a new phase. If so, it is not clear if it can be a genuine advance or will be yet another massive blunder.

The individual seeking improved powers of judgment is cautioned against using contemporary scientific thinking as a model without appreciating both the controversies on this subject (chapters 9, 10, and 11) and the fact that science is diverse and changing and that one can easily be part of the scientific career structure without having much in common with the founding intellects of science. We may stand on the shoulders of giants, as Newton said, but we do not therefore necessarily stand taller.

Changes have been taking place in science and scholarship. Today science is complex in character and in a state of flux. There is not (and never has been) one "scientific method." Any brief characterization must be as cautious as any brief sketch of diverse Christianity or Hinduism.

Institutions of higher learning are evermore collections of specialists who have few conceptual matters to discuss together—multiversities. Scholarship generally is becoming more and more a matter of specialization with goals of steady, high output. It is becoming evermore managed and corporate in nature. People enter it looking for secure, routine, status-granting careers.[6] It is plagued by trendiness. Society does not leave as much room as before to support the passions, adventures, and visions of creative individuals, except as they fit into program goals. It seeks to recruit, train, manage, and direct capable workers and fund-raisers. It has become a highly expensive, managed, and nearsighted career system.

Science was never completely value-neutral, and corporate science is quite value-biased in many ways (e.g., chapter 4, note 2; chapters 6, 9, 11). Economic, social, and ideological priorities may mask as intellectual priorities. Political policy may mask as an intellectually derived, "objective" organizing perspective. Society has paid huge amounts to study marine geology, for example, because of its economic potential. In the same years scientists who desired to study conceptually interesting aspects of microbial ecology or lizard anatomy could not find direct research support (and often had to fund their studies from grants for other purposes) because science managers little valued such knowledge.

Along these lines, it is awkward for the political system to face the ecological and socioeconomic complexities of problems. But if problems can be portrayed as demanding simple technical adjustments, then parties can appear to be dealing with them simply by funding reductionist research and engineering programs. Thus, ecology (as an example within biology) has only been funded, and a thoughtful and determined effort made to develop it, at a small fraction of the scale dedicated to reductionistic molecular biology, with its promises of "magic bullets" to solve social problems. Starting in the 1930s, the Rockefeller Foundation dedicated itself massively to the development of molecular biology and a simple approach to social/behavioral problems (Abir-Am 1987, Fuerst 1982, Kay 1988). The federal government took over the programs when its role in funding basic research became significant following the Second World War (Yoxen 1982).

Yet the intellectual (and social) issues at stake within biology were and are equally profound for conceptual ecology and for molecular biology. On purely *intellectual* grounds they have merited equal priority to attempt to develop them.

In terms of *social need*, prophylaxis would be as useful as prosthesis. But from a *political* point of view it is easier to discuss, promise, and champion technological fixes, no matter how remote, than prevention. In *theory*, for example, the greenhouse effect, groundwater contamination, deterioration of the ozone layer, disruptions of marine ecology, desertification, and widespread problems with deforestation, watershed, erosion, and even overpopulation were once all quite preventable. Politically, they would not have been easy to deal with, though.

Moreover, even prosthesis, if it is to succeed, should be based on a firm understanding of ecological and socioeconomic complexities. But the visibility of even suboptimal technical prostheses makes these politically useful.

> According to an old story, a lord of ancient China once asked his physician, a member of a family of healers, which of them was the most skilled in the art.
>
> The physician, whose reputation was such that his name became synonymous with medical science in China, replied, "My eldest brother sees the spirit of sickness and removes it before it takes shape, so his name does not get out of the house.
>
> "My elder brother cures sickness when it is still extremely minute, so his name does not get out of the neighborhood.
>
> "As for me, I puncture veins, prescribe potions, and massage skin, so from time to time my name gets out and is heard among the lords."
>
> (Cleary 1988)

There has been priority manipulation by politics and economics for centuries, but commentators agree that "Big Science" began to develop after World War I and that this face has asserted itself dramatically since World War II. We see little idealistic independence and fire anymore; indeed those may be regarded as unbusinesslike. There may be little crosstalk across disciplinary boundaries. Individuals may have only one or two areas of serious professional interest.

Nevertheless it has been possible for a few scientists to think deeply and broadly and even to speak out against the interests of the system that supports them—for example, to point out that the population explosion and industrialization resulting from the unbalanced promotion of science and technology are causing problems such as a greenhouse effect or ozone-layer deterioration that will upset the earth's climate. Sociologists of science who try to describe statistically normal behavior may miss seeing the really interesting dynamics of diversity and change still within our professional culture.[7]

Yet, I would not recommend to anyone that wisdom can come simply from specialization of one's potentials, mastery of techniques, or vigorous and chauvinistic advocacy of one's opinions. (Though some would argue that this is how forward-looking scientists should and do primarily use their minds today.) Science has had an enormous impact on human thinking. But perhaps one cannot use it, in what some see is its presently evolving form, as a model for the pursuit of truth or wisdom *by the individual*. This is not to say that scientific *discoveries* are not useful in replacing ignorance and superstition. This is certainly not to say that multiversity corporate science cannot make discoveries. *It does*. These are quite different matters. But I would recommend that the *personal* lessons to be learned are from science in its more individualistic and idealistic modes (chapters 9, 10, 11, 12).

This chapter has illustrated that as a species we have long wandered about on diverse paths over the intellectual and spiritual landscape. The search for improved ability to make judgments and to find wisdom has been intense because serious issues are involved. Religion, philosophy, political theory, and science affect our thinking, perception, and action in the most intimate ways. They may stimulate or suppress, channel and direct, our capacities for love and justice, sex and violence, joy and despair. They determine how peacefully and happily we can live together.

Human beings often enough have raised their heads from their social routines, beliefs, worries, labor, strife, the daily priorities of survival, from cozy fantasy, and have wanted much the same things: to be able to see enough reality to make their lives a little more certain and a little more free of unnecessary anxiety, to know who and what they are, to know the shape and size of the labyrinths that they seem ever to be building for themselves.

Diverse cultures have made some simply remarkable progress over the centuries. A variety of rich traditions have developed. But we have been lost sheep for the most part, whatever enrichment we have gained along the various paths. We have wandered into blind desert canyons of the intellect and have made mistakes on massive scales (chapter 9). Despite the fact that most mistakes have *in principle* been avoidable, and that *potentially* we can understand ourselves as never before, there is no sign that the masses of people, intellectuals, and the rich and powerful alike will stop making misjudgments soon, given how our brains are constructed and how our thinking is influenced by social dynamics.

Over 25 centuries ago the Upanishads pointed out that no-knowledge leads to absolute darkness. Satisfaction with the present state of knowledge can also blind us — much as Oedipus was blinded by his false belief of his own identity — and the false security of such satisfactions can plunge us into a darkness of an even more profound sort. One recalls the words of Wittgenstein: "Some philosophers (or whatever you like to call them) suffer from what may be called 'loss of problems.' Everything seems quite simple to them, no deep problems seem to exist any more, the world becomes broad and flat and loses all depth and what they write becomes immeasurably shallow and trivial."

The prideful sheep of nations too often divide into self-righteous yet often confused, hating, and fearful camps; the powerful of each faction too often act as though they have the answer, hesitating to talk deeply with one another about basic assumptions or to rethink their own basic beliefs. And multitudes are even ready to die for one seductive doctrine or another. If an individual can keep his or her own thinking straight and mind open, that is a lot.

NOTES

1. It was claimed that the much-discussed Senoi of Malaya would carefully review each other's dreams, looking for significant symbols and events, talk these over each day and then try to direct subsequent dreams to explore implications from the previous night. In essence, they were reported to have gone beyond what "primitive" people commonly do less effectively and to have developed dream interpretation, dream incubation, and lucid dreaming to a highly sophisticated degree. The Senoi were reported to be remarkably happy and well adjusted, free of psychoses and neuroses. They would teach children to confront subconscious fears and frustrations directly in dreams, in the theater of the subconscious, and to overcome them. They would teach them to exercise the creative imagination. "You were falling? How excellent. Do not fear. Next time spread your arms and try to fly. Fly to a wonderful place and meet fine new friends. Have them give you gifts, perhaps fine new beautiful songs, and then come and share them with us when you awake. A monster chased you? Fabulous. Next time know that you can conquer him and make him your friend or helper."

It was Kilton Stewart who published that the Senoi had developed the skills of dream interpretation, dream incubation, and lucid dreaming to an extraordinary degree and had used these quite systematically to create an exceptionally healthy society. Anthropologists who have studied the Senoi have not confirmed his reports. Stewart apparently developed theories of mental/social health from his own experiences as a psychoanalyst and hypnotist, and in healing ceremonies and tribal dances. But he incorrectly attributed his theories to the Senoi and painted an untrue utopian picture of their society.

Domhoff (1985) argued that many people have eagerly adopted Stewart's ideas not only because they seemed to make some psychological sense, but because they seemed to be based on an alternative authenticity, on some supposedly proven ancient wisdom.

Will Stewart's theories work? They are extremely difficult to test. Dumhoff concluded, "They may or may not prove to be very important in the long run, but we should remember that they have been useful for a few people even if the jury is still out on the degree to which they can be put to work by people in general."

2. In discussing this traditional mode, prior to the development of formal sorts of logic, I do not want to slight "paranoia." It is very important. But it is seldom institutionalized (in the present sense). People have surely long known that they cannot necessarily trust each other. Then the smart thing is to act as though the other person can be taken at face value, but to have a second opinion of the person and his or her motives in reserve—act on the senses, but invent a second abstract view of the individual. I will speculate that this has long been done informally among family and friends. The primary area in which (mild) paranoia or suspicion may have been institutionalized would be with regard to traditional or potential enemies. This could even have anticipated the shamanistic mode of the development of concepts of worlds beyond the senses. A good deal of discussion of abstract threats may have taken place out of fear of the intentions of neighbors. One party would invent an image of the other. Such attention to abstract possibilities may have improved judgment and survival from time to time. Whether or not such fears ever bred a professional class of paranoids, I do not know. In many ways war is the art of deception, and military men in our own societies, and war chiefs, must in a sense dwell in such a mental world, and if they did not we would be let down by them. This question would be interesting to pursue, but will be set aside in the present discussion since it is not now obvious to me that military/political paranoia has led to important, well-developed sys-

tematic thought systems or cultural institutions (with distinctive ontologies and epistemologies) that should be separated from others.

3. It will be very difficult to know just what happened in the sixth century B.C. An external economic or political cause that influenced cultures stretching from Italy to India is indeed hard to imagine.

We know that obsessions with the spiritual nature of numerology and musical tone were issues not only for Pythagoras and Plato, but are suggested in the earlier Vedas and in Egyptian and Middle Eastern cultures. Speculation over the metaphysical meaning of numbers and forms was extremely disciplined and mostly part of the secrets of schools of "holy men" and "philosophers." Travels among the spiritually minded of the area in ancient times is known, but its extent and impact is not known from historical sources. I suspect that occult interests in numbers, and esoteric interests in calculation and proof, had prepared the minds of metaphysical thinkers, in cultures across that broad land area, to more critically, if often nonmathematically, question reality and conventional beliefs and to speculate boldly on abstract entities that were not dream creatures. This would be a very difficult path to trace, since cloistered discussions, unrecorded travels, and subtle intellectual influences would have been involved.

Yet it would be too easy to overlook the possible influence of the habit of calculation, and indeed of the notion of mathematical proof, on the minds of determined metaphysical thinkers. It is today clear that relatively sophisticated mathematical thinking was not merely the product of Greek genius in the sixth century. During the 1930s Neugebauer published studies showing that Babylonian mathematics was quite advanced. They even knew the theorem of Pythagoras well over a thousand years before he visited Babylonia. Even the legend that Pythagoras sacrificed an ox to honor his discovery is probably a confusion of history. He was strongly opposed to animal sacrifice, and the legend may refer to the real discoverer who at some point became confused with Pythagoras.

Seidenberg (1962a, 1978, 1983) analyzed sacred formulas for constructing and reconstructing altars of various forms and magnitudes, depending on the ritual, in India. He concluded that the Indians too were familiar with the theorem of Pythagoras before the sixth century. It was important to have exact proportions and areas for their altars. Which shapes and sizes were important for particular purposes were subjects of serious theological decisions that involved serious contemplation of the metaphysical "realities" represented by geometry and number. Suppose there is a plague. We are told to sacrifice to Indra, the wind. Atmosphere is five. So to begin with, the altar should have a particular multiple of five layers of bricks. Or, "he should pile in the form of a triangle who has foes." These may be extremely simple examples of the motivation to rebuild altars. What is clear is that Vedic wise men were quite familiar with relatively sophisticated calculations and were also intensely interested in the presumed underlying reality of geometry and numbers, as Pythagoras and Plato would later be, though no one's thinking is completely and precisely known. In Plato's description of Atlantis, it is clear that a lot of occult numerology is involved but the precise meaning is obscure.

The earliest mathematical texts offer actual geometrical proofs. The Babylonian clay tablets, for example, show problems and answers: "4 the length, 3 the breadth. What is the diagonal? The magnitude is unknown." This at least shows the habit of thinking in terms of there being *correct* abstract relationships.

Van der Waerden (1983) argues with Seidenberg for a common origin between 4000 and 1500 B.C. for the mathematics of India, Babylonia, Greece, and China, based on the details of mathematical knowledge in each culture. They agree that the origins were in religious ritual. Van der Waerden infers that the origins were in astronomical/astrological observation and the ancient construction of European megalithic monuments. But there are problems with the stone-circle data (Chippindale 1983).

4. In India, in the main Hindu and Buddhist traditions there was generally an attempted liberation from the clinging to ego, so important to Westerners, and of the life of society and of the flesh. Even among the common people, most art has been directed to the gods and away from worldly life. Some people take paths that are quite removed from the social order. This was once even commonly expected of older men. Religion/philosophy is sober stuff, and Buddha seldom laughs.

In China, however, Taoism emphasized unity with nature and liberation through acceptance of this unity. Buddhism crossed the Himalayas, eventually mingled with Taoism, picked up its *joi de vie*, its humor, its love of paradox, and we encounter the "laughing Buddha" (Pu-tai) and Zen. The "enlightened" can even be a devoted warrior and we find the development of amazing martial arts abilities. In much of India menial chores would tend to be thought of as a distraction by the flesh from enlightenment, whereas in China and Japan menial chores might through Zen become a source of mediation, enlightenment, and enjoyment. There is the tea ceremony, the flower arrangement. Art is filled with scenes of nature and of secular life.

5. More recent Western psychotherapies for "normal" but troubled people have rejected dependence on drug therapy, and they also have attempted to "go beyond" intellectual insight. One may begin with intellectual insights into the childhood sources of repressed feelings of pain, guilt, shame, love deprivation, worthlessness, and so on. But this is mostly a step in helping the feelings to come to the surface. Thus one can then observe nonjudgmentally (virtually as the "silent witness" of Eastern thought) the contexts in which deep feelings are triggered and observe the sometimes irrationality (in adult context) of the associations between current events and feelings, and the ways in which such feelings influence adult attitudes, perceptions, and behaviors. This direct *experience* of self in turn forms the basis for positive self-growth and ongoing discovery when combined with the exploration of more healthy adult outlets and attitude and behavior pattern options *for the particular individual*. This contrasts with older Western forms of psychotherapy that stressed the value of intellectual insight largely alone, or mere "release" of pent-up feelings, and that attempted to hold up a limited number of models of healthy behavior for everyone.

6. Harding (1986), following others, argued that scientists shifted from being motivated to change society, to being motivated to find secure niches for themselves within it following the establishment of royal academies in 1662 and 1666. Then they began the myth that science is value-neutral, objective, and unmanaged. I do think that a great deal of diversity was retained through World War II, and that a marked reduction has taken place since then.

7. Clearly, not all sociologists of science have tried to identify an "essence" of science, for example, in the bald, competitive advocacy that we see so much of, and sell this as "the" quality that has always caused scientific discovery. Eiduson (1962), for example, studied the psychological dynamics of forty distinguished scientists of her era and reported how their lukewarm emotional childhoods had a variety of complex effects that contributed to their creativity and careers. They looked for teachers or other adults to compensate for emotional gaps with parents. Such childhoods gave them incentives to intellectual rebellion, channeling of rebellion, building of intellectual fences, development of self-sufficiency.

She was concerned, along with C. P. Snow, Ashby, and some other sociologists of science about "the growing schism between the intellectual and the scientist." She noted that "the whole trend of social values is away from the contemplative, away from concern with the complex, away from the sense of calling, dedication, and single-minded purpose" that characterize the older generation of scientists. Does it matter? She suspects that the diversity within science will provide some safeguards. I too suspect that science will survive. I have merely asked here if its emerging character is a good *model* for the *individual* who desires to develop critical thinking skills.

Gilbert and Mulkay (1984) analyze the structure of actual scientific discourse and its role in the construction of scientific culture. See also Latour (1987).

Price [1986] documented and described the growth of "big science," by the application of quantitative techniques in *Little Science, Big Science and Beyond* in 1963. Following this, Klaw (1968) took a more narrative approach to an analysis of the growth of big science, and the character of the scientific community, in *The New Brahmins*. Greenberg (1971) in *The Politics of Pure Science* probed the growing role of "pure science" as an instrument of government. (Or more recently, Dickson 1988). These books are interesting especially because they were written during an acceleration phase in the growth of big science and the multiversity when it still seemed that there were simple social choices available.

Ravetz (1971) saw the new science as inadequate for the pursuit of natural philosophy. Even those who enter not just for secure careers but "as a refuge from the intellectual and moral squalor of ordinary society find, in their advancing years, that they are involved in administering just another bureaucratic establishment. They are enmeshed in the demands of society and the State; they must accomplish the administrative and social tasks of getting high-quality craftsman's work out of a set of manpower-unit employees; and in participating in the leadership of their field, they must cope with the insoluble practical and moral problems which emerge when corruption sets in." As a scientist who loves his work, I believe it is not always so bad. Yet our folkways do direct us to look so much at the up side and to be so defensive of the down side that our careers and thinking are too easily manipulated by political leaders *within* science and academics and bankrollers from *without*.

Suppose Ravetz and others are correct that "the innocence of academic science cannot be regained." What of the immediate concern of this section of the book—can multiversity corporate science contribute more than facts to a liberal arts objective, to the development of personal critical thought skills? I doubt that it can be modified to *deliver* these to unprepared students and other citizens. But I am optimistic that an astute and self-confident person can *find* helpful and instructive models even in an era of multiversity corporate science (chapter 12).

CHAPTER 9

Is Relatively Good Individual Objectivity Possible?

Faith is to believe what you do not yet see:
the reward for this faith is to see what you believe.

St. Augustine, *Sermons*

Philosophy says human nature is interesting.
Well it is not. That is all there is to say about that.
It is so easy to be right if you do not believe what you say.
Please listen to that.

Gertrude Stein, *The Geographical History of America:
or the Relation of Human Nature to the Human Mind*

I want to continue the discussion of the nature and development of critical and ethical thought and the contributions of Western science to these. Despite widespread doubts of science and critical thinking in our time, we should not throw out the baby with the bath. They can contribute much to individual intellectual potential and ethical aspirations.

Against this optimism is the belief that individuals are ideological, biological, economic, or historical puppets that can never control their own destinies for the better. There is also the notion that science is merely impersonal method. We should ponder these matters next. This chapter and the next deal with subjects that are germane especially to these beliefs. Then we can return in chapter 11 to a discussion of the development of critical and scientific thought. I shall argue that the development of those was long value laden and, especially early, was linked to ethical aspirations, notwithstanding socioeconomic factors that contributed. After this and the next chapter it may seem more reasonable to critics of individual objectivity that individual aspirations for peace, justice, and happiness could be significant historical factors. It may seem more reasonable that what is superficially only memorized method (scientific thinking) might be integrated broadly into the life of a healthy individual.

Thinking Individuals or Social Puppets?

I once heard Margaret Mead define the middle class in America as, "that 90 percent of the population that believes that they are special."

The audience broke into laughter. It is clear to many if not most educated people that in our society we hold many pretensions about being individualistic; this is obvious enough that we can even have a healthy laugh about it.

Of course we may resent it when coworkers, bureaucrats, or clerks treat us impersonally, as just another warm body, case file, or number. But this has to do with notions of human dignity and respect for the individual's personal circumstances. Aside from that, we appreciate that we have much in common that can be known from our social and economic classes, age, sex, social roles, and backgrounds. Such factors enter into our daily decisions, for example, when we seek out others and pursue personal or business relationships—we think we will be able to get along with some *types* of people better than others. We think of ourselves as the *type* of person that certain others should like or not, respect or not.

Most of us have modest notions of our individual specialness, and there might seem to be little to think about here. In fact, though, individuality is an intricate and important subject. It is tied to issues in the world of ideas and policy, such as the so-called crises in values, meaning, truth, objectivity, and to versions of so-called great men theories of history. Different groups have different tacit or even open definitions of individuality, objectivity, truth, and such. Individuality is also a political issue in that people from across the spectrum have strong views on the linkage of ideas of individuality to political policy and theory. It is an economic issue, since one's beliefs in certain styles of individual expression encourage one to be eager and willing consumers and workers, while beliefs in other styles might not. Among some, both advocates and critics, the word *individuality* has even become nothing more than a synonym for egoistic self-identification or selfishness. (Abercrombie, Hill, and Turner 1986, for example, review discussions that explore this dimension of the issue; also MacPherson 1962.)

Some of my serious and thoughtful colleagues say that the dispute over individuality is *the* major intellectual and moral issue of our century.

Here I focus on a single critical element: the possibility of *individual objectivity*. Biological individuality is not in dispute. Sexual recombination of genetic material ensures that each higher organism is unique. Scientists, such as Roger Williams in *Biochemical Individuality*, have discussed this at length. We have not only different faces and fingerprints but substantial differences in our individual biochemistry.

The question is to what extent upbringing and culture (or instincts) tend to make cohorts in a society so alike in perceptions and values that they are incapable of objectivity or independent thought and action. Can we be only passive elements in creating seemingly thoughtful social and historical events? Do these events only unfold because of impersonal laws of economics and politics (or genetic determinism)?

We should note, too, those critics who take the position that we can only be individuals if we accept Christ as our personal savior.

Last, in extreme sociobiological views, genetically determined drives and instincts override any possibility for individual objectivity. If that is so, we cannot think for ourselves since our behavior is determined by genetics from behind a curtain of rationalizations.

Schwartz (1986) has seen the main problem coming from a *loose alliance* of seductive scientific theories. There is *economic man* driven by economics, *conditioned man* driven by the behaviorist reinforcement schedules of Pavlov and Skinner, and *sociobiological man* driven by instinct merely to survive, spread his genes, and promote his genetic inclusive fitness.

I shall not try to deal with each theory, since typically the problem is with vulgarizations of idea systems that are not themselves so ridiculous. But vulgarizations cannot be brushed aside, for history shows that they commonly have more influence than do original positions. I detect that in this case, as economic, behaviorist, and sociobiological theories seem to become more interesting (each obviously has at least *some* role), the specialists apply them too enthusiastically, and they and others will also sometimes use the theories to trivialize their competitors and critics. Such drumbeats from academic tribes that must compete for funds and prestige in effect may seem to play in harmony. The ear may hear the conclusion that individual objectivity is simply impossible.

Sociological critics of individuality look at the masses of people and see all varieties of value systems and ideologies. Followers of each system accept their own way as right and as *the* way to live and die. And if we learn our perceptions and values as we learn our language, from the accident of birth into one of many cultures, if we are puppets of our culture, then how can we claim to be individuals or to have any objectivity?[1] And if we are mere puppets, then how could any persons ever have helped to shape their own history? Critics say that we only fool ourselves.

Economic determinists, in particular many sociologists, anthropologists, and Marxists, have discovered that many values, perceptions, and character traits do not vary arbitrarily from one society to another, but are crafted in ways that support the particular economic system. One's perception of the world can be said to be a "false consciousness" that blankets one in cozy myths while hiding the true conditions of servitude to the economic system (Berger and Luckmann 1967; see also Heilbroner 1988).

All critics, economic-determinist researchers or not, of any society or group of individuals tend to develop some theory of false consciousness, no matter what they call it, to explain why other people do not see eye-to-eye with them. The oldest version may be the idea of the possession of minds by devils or spirits. A more recent term is *brainwashed*. Some like the more neutral term *programmed*. Indeed, how could an analytical black person, homosexual, woman, third-world national, or other member of a disadvantaged group, when they see others look down on them and see merely stereotypes, and use facts *selectively* to justify prejudices, not explore one theory or another of false-consciousness?

When such valid observations of self-serving perceptual constructions, by persons disadvantaged *or* empowered, are pushed in their implications to the extreme, they leave little possibility for individual objectivity. This last too common conclusion, itself a perpetuation of the stereotyping habit, is the extreme absolutist frame of mind that I wish to deal with here.

Let me state my own position at the outset. It is in part a considered reaffirmation of certain Enlightenment conclusions.

1. Selfishness, egoistic self-image, or a conviction of unique personal destiny (the "individualism" of some) is *not intrinsically linked* to the issue of the possibility of individual *objectivity*. Selfish, isolated people *can* be quite nonobjective, and objective people (even social critics) *can* see themselves as deeply socially connected beings.

2. Concepts such as *false consciousness* can have methodological utility and are not absurd except in the extreme. They are too often used in the extreme. I am clearly not an idealistic individualist who cannot "adequately comprehend how much of our nature is a reflection of our social selves" (Tucker 1980). But neither am I at the other extreme that always sees everyone as puppets of the socioeconomic system. I do not accept that there are essentially only two positions on this issue.

3. To some degree, individual objectivity is common, as when we will talk over a problem with a friend in the hope that she or he will be able to see our situation more clearly than we can. But at a level where we demand evidence of truly unprecedented objective insights, individual objectivity is rare.

4. An exploration and cultivation of individual potentials for objective and creative insight took shape in the Renaissance and Enlightenment and can and should continue to be encouraged for the sake of the individual's

effectiveness and welfare. This also serves society by generating valuable new insights and creations. There is a great deal that we can accomplish even if philosophically pure objectivity evades us.

5. We have learned much from science about how to think critically, if our goal is to be as objective as possible. We should gather evidence and data before becoming firm in our views. We should not form convictions quickly. As Augustine long ago observed in another context, conviction leads us to see what we believe. So it is much easier to think critically if we do not take our notions too seriously and we remain ready to change our minds. We should be alert to cause and effect. We should not be too quick to explain away disturbing information. We should not accept an idea merely because it sounds logical or because everyone believes it. We should not allow a belief in the inevitability of progress, or the spectacle of social success, to cause us to mistake intellectual fads, attractive novelty, or the aura of respectability for genuine advance. We should look carefully at the variety of conflicting views on a subject and then weigh the cohesiveness of the evidence for each.

We should pay particular attention to facts that would falsify our theories and beliefs. Ideas that conflict with our point of view should not be readily shunned, even if we have been successful with our methods and philosophy. We should concern ourselves with how well our theories and methods of observation mesh with the body of empirically established knowledge. We should try to see if one theory makes unusual predictions that the others do not, and then test those predictions. We should be careful and exact in making observations, so that we can be sure of what we see and so that others can repeat them. We should understand how easy it is to be quite enthusiastic about hunches and convictions that may prove to be false, as well as the true. If we make mistakes, let us try to learn from them, not call it bad luck or look for someone to blame. We should understand the ways in which our minds can play tricks on us. And so on. In these ways we can potentially make some degree of progress toward individual objectivity. Impersonal methods can only supplement; they do not replace the above.

6. Most people are focused on domestic affairs and will not be much interested in following this path (beyond the degree to which it has been a systemic part of their socialization). Perhaps in the thick of the days when one risked being mobbed and killed for witchcraft or heresy, or had to choose in sectarian wars, there was exceptional incentive to find ways so that people might be able to agree in their thinking. But in times of relative safety for the average person, many people prefer to follow a leader or a peer group on questions of what to think and what to believe is right and wrong. They want confirmation of their beliefs, not general enlightenment. Nevertheless, they should not be kept ignorant of the issues.

Moreover, we tend to have habits of self-deception, even by following example, and actually *cling* to self-deceptions to avoid facing moral dilemmas.
Indeed some would argue that *the* chief and most common moral weakness
is to pretend that we *do not engage* in self-deception (Martin 1986).

7. The problem of multiple frames of reference is not necessarily evidence that individual objectivity is impossible. When two parties disagree,
one is not *necessarily* wrong. If one sees a giraffe as a fifteen-foot tall animal,
and the other sees it as a warm-blooded mammal, both are being objective
and we can even benefit by taking individual points of view, or frames of
reference, into account. But if the other sees the giraffe as six-foot tall, or as
five-legged, then one party *may be wrong.* Many unnecessary disagree-
ments take place over issues that are not so clear-cut as giraffes, when people insist that only *their* frame of reference is valid, and so the other party is
not being objective.

8. Marx was primarily trying to shift emphasis from the common view of
his time. For Marx, there was no natural form of the individual. Each society
may shape the image and self-image of individuality differently. He actually
saw in capitalism the potential for a partial liberation of human capacities.
But many of his followers and other economic determinists have taken such
words as the following to an unintended extreme. "The mode of production
in material life determines the general character of the social, political, and
spiritual processes of life. It is not the consciousness of men that determines
their existence, but, on the contrary, their social existence determines their
consciousness." (His own complete position was actually more involved
than this.)

Much of the discussion in previous chapters might seem to be a defense of
an extreme economic determinism. My words may push seemingly familiar
buttons for some readers. Rather, the previous discussions only help one to
see why an extreme model of false consciousness *seems* to have a solid basis, and on the contrary will help one to understand below why it would be
wrong.

It is clear that the social system intensely tends to mold perceptions and
personalities in a society. Indeed, many do seem hardly capable of seeing
outside of their conditioning. If critics of individuality held only this idea,
there could scarcely be an informed rebuttal. Too often, though, people
from all parts of the political spectrum actually do talk and behave as though
this is a black-or-white issue. Consciousness is seen as something that is
taught, a software package inserted into an otherwise functionless computer. But clearly the child must organize much of its perception of the
physical world and of its own body, and culture can only interact intensely
with this learning process and with the process of the development of a subjective world.

9. There is always enough conformity and imitation around to point to as evidence that people are *incapable* of any objectivity. In many cases, though, I suspect that the situation is much as Euripides described it in about 425 B.C.

No man on earth is truly free,
All are slaves of money or necessity.
Public opinion or fear of prosecution
forces each one, against his conscience, to conform.

(*Hecuba*)

Many of us smother our individuality and do not let it grow. It is not that we have never known it to any degree, and *cannot* know it. Dante offered a similar perceptive thought in the *Purgatorio*, with regard to a sense of justice. "Many have justice in their hearts, but slowly it is let fly, for it comes not without council to the bow."

10. Western culture provides many excellent resources for individual development scattered about in a field of pitfalls. Eastern culture too has much to offer. Diverse cultures are rich in potential lessons. Certainly, though, no culture or ideology can guarantee us an illusion-free perspective. Scientific thinking and the institution of science are much too often abused and we must be concerned about that, but they have proven that they can be powerful allies in gaining some small elevation of individual objectivity. We should focus criticism of science much more carefully than is often done.

Next I propose three main lines of rebuttal to the critics of the possibility of individual objectivity or effectiveness. In the first two, individuality is largely deduced; in the third it is largely observed. To begin with, one should think very carefully about the brain and the issues raised in chapters 3 and 4.

1. I have detailed that the social forces molding perception and subjectivity face the considerable difficulty of trying, as it were, to synchronize inherently expansive, intrinsic, *imaginative* operations. A large part of the individual's subjective existence in a complex world cannot be standardized and *must* be unique. In this sense each person has unique *potential* to develop an identity, talents, creations, and to make social contributions. Individual potentials can be smothered, ignored, misdirected, or nourished. On theoretical grounds we expect internal realities to differ greatly among individuals, and when we watch children and adults closely, they do seem to confirm this expectation. It may be the norm that members of a given culture will disagree on at least some noteworthy percentage of perceptions. In these differences in perception are the seeds of skepticism and of coherent individual points of view that *effort* can develop. Indeed, the interesting fact

is that we can learn to communicate and interact much of the time as though we actually do think alike. The society's conventional views and idioms must be tacitly negotiated among its members in order to accommodate the most common perceptions. Our individual capacities for imagination can be socially inhibited or enhanced, but it is hard to see that they can be *completely* dominated by social indoctrination.

2. Not all conditioning in life is Pavlovian, in which the dog salivates when a bell is sounded. The environment does not *merely* act *upon* us. We usually select the pieces of the complex and varied environment that will act upon us, much as the Skinnerian rat chooses to press a lever or not for the reward that will condition a behavior. Thus, one child will pick a favorite uncle that may influence her greatly, she may have individual experiences with a frisky pet, she may have a rather restless disposition and prefer to watch birds while her playmates nap. Her sibling may favor an aunt, not care for pets, take long naps but stay up late at night and listen to the men in the family talk politics, and so on. It is quite improbable that two individuals, even if they had identical genetics and emotional dispositions, would have identical conditioning and experiences even in a village, let alone in a complex society. In such is the potential for people to have different points of view and to evaluate information differently even if they have been raised together. Even identical twins raised together and treated as a pair do not see eye-to-eye on *everything*.

Whatever the pressures to conform in behavior and thought, any society will retain some diversity among the minds of its members. There is always the potential for one to stand intellectually a bit back from the group, to have his or her own perceptions and opinions, and to some extent be a free thinker. Though we may not see this if we are not alert. Indeed, many Athenians saw Socrates as an old crank, not as one of the important minds in history. At the cocktail party we hear a dozen conversations, but when we begin to listen to one, the others become not exchanges of information but merely irritating noise. So our brains filter things out of higher consciousness once we have decided that they are unimportant. We can forget that the loud sounds of traffic outside our window are there. We may miss what is being said on the television if we are engaged in conversation. If an ignorant or prejudiced observer of other cultures can see all dark people as stupid savages (e.g., Drinnon 1980, Sheehan 1980), then one so persuaded can also trivialize individuality. But if one has lived among diverse sorts of people and has worked with them in their element, one can see the individual perspectives and insightful skepticism from which bits of objectivity are cultivated.

3. When we look, there do seem always to be individuals who stand somewhat outside their culture and who can question aspects of it. I am re-

minded of a Seri Indian youth who was telling us about the time his band had caught a leather-backed sea turtle. He very seriously explained how according to custom everyone closed their eyes tight while they ate it (because they would die otherwise). I wondered how he could have been sure of that if his own eyes were closed. I calmly had him asked if he had closed his eyes. He turned from the translator and, breaking into a charmingly sheepish grin, looked at me knowingly, as though I had caught him at something deliciously naughty. "I looked," he answered in Seri. "I didn't believe I would die," his grin broadened. He was quite illiterate and unChristianized and unindoctrinated in Marxism. His skepticism and willingness to risk death quite probably came from his personal capacity for insightfulness. I might give other examples, where the glance was less memorable, but the point is clear. It is certainly possible for a person to stand outside his culture, at least on one set of issues or another, and to develop his own point of view. At some level, children do this all the time. Buddha and Socrates and other important thinkers simply did this on a grand scale and organized their cases in convincing manners.

If individual insights are to grow, they will face obstacles. The social forces that shape our beliefs and perceptions can be powerful. Nonconformist views can be assisted if there is a) a personal social support network, b) societal values that encourage originality and dissent, or c) some individual system, such as judicious journal writing.

Otherwise it is hard to maintain an original system of thought or one that runs contrary to canonical beliefs. Readers of this book will find original insights flashing in their minds as they read and think, but surely will lose many of those before long without some social support in developing them, or systematic journal writing, for example. I am fascinated by the "heretics" throughout history who developed nonconformist convictions, cogent criticisms of orthodoxy, often alone, despite the threat of bone cracking and the iron-maiden. Some strength of personality or streak of social independence must also figure in.

Anthropologists tend to publish *norms* of belief and behavior, but my field-seasoned anthropological colleagues (and enough publications) say that there are autonomous skeptics in every culture that they know of. "Curiosity, inquisitiveness, questioning, seem to be characteristic of our species. There are simply different degrees of social support for skepticism in different societies. But one sees it to some degree in some individuals of apparently every culture," as Professor Eugene Ogan put it. Winston Smith of Orwell's *1984*, who could glimpse for a time past even highly sophisticated brainwashing, is not an improbable character. Mark Twain's Huck Finn is not so improbable—poorly educated and superstitious, but often able to spot humbug, hypocrisy, and foolishness. The issue is how broadly and deeply individuals can develop relatively objective viewpoints, not whether they can and do generate them.

This ability may develop with age and experience as one watches and ponders the repeated failures of conventional myths and enthusiasms. But there is no simple pattern; some independence of thought is evident even in childhood when a person is physically most dependent. Babies must learn for themselves perceptual organization, details of body control, and thought, that we do not even know how to teach. The mammalian brain is an active, inquisitive organ; it is not a blank slate on which society may write anything at all that it wishes. Of course we *profoundly* influence the development of babies in many conscious and unconscious ways. Nevertheless we can see plainly that children learn and integrate *for themselves* many things about the physics of their own bodies and of objects, about their social environment, about time and space. Even after they start to learn language, we can watch children make inventive verbal and cognitive associations, say "cute things" that they have not been taught, and poke about in places where they are not allowed. Magician/author James Randi tells me that children are harder to fool than are adults because they are often more open-minded about what they are seeing. Mathematicians often claim that major works are more likely to come from younger mathematicians, who are relatively more open-minded. (On the other hand the issue is delicate. Younger people are preferred as soldiers and sought by cults because they are usually easier to indoctrinate.) The brain, with millions of years of unbroken survival behind it, reaches out to interpret the environment and the self. Culture is an ever-present part of that environment. But it is not the only factor in the equation, and people can be part of their culture and yet have independent opinions on a variety of issues.

The arguments that I raise here will not quiet the core critics of individuality. Their positions are tied to larger world views, interpretations of history, and strong social attitudes. So in the remainder of this chapter we will explore the natures of economic determinism and historicism to learn more about the main theoretical critics of individuality. But first we must say a bit more about the many definitions of individuality.

Respect and support for individuality is probably a basic trait for humans. It is simply the case that individuality expresses itself differently in various cultures and receives different types and levels of support or discouragement. For example, among the ancient Greeks there was a wide degree of individual *freedom of abstract belief* so long as one followed social obligations. (This is commonly true for small tribal societies.) In later Christian theology, all sorts of individual behavior might be forgiven so long as one generally *conformed in certain abstract beliefs* and had certain *acceptable motives* (as we might distinguish murder from manslaughter, or allow for repentence; the door is even opened for the "Sunday Christian"). This is in contrast to cultures in which punishment was or is more an eye for an eye and a tooth for a tooth. If you kill someone even by mistake in New Guinea, tribal law says you must pay in full at once, usually with your life, even if it was a traffic accident.

When I was in Papua New Guinea, tribal values were still strong, and one of the worst things that one Melanesian could say about another was that the person was interested in advancing him-or herself as an individual. The "good" for them is to work for the security or advancement of the extended family and tribe. A man or woman may delay marriage for many years and work far away from their loved ones in order to help their people. In those Melanesian cultures where one finds the important person, or *big man*, he is respected for his political skills, wealth, strength, etc., *as a member of his line*, not as an isolated being. This is almost opposite from the Anglo-American view, in which very high respect is given to children and adults who promote themselves as individuals. We give lip service to meekness and brotherhood, but the fact of our actions can be to esteem the bold, independent, and ambitious. We even hear ever less in America of how the individual's long-range interests converge on those of community and humanity, particularly since the doctrine of unelaborate self-interest that started spreading as popular wisdom in Washington, D.C., during the early 1970s, and with the repopularization then of versions of mercantilism and laissez-faire.

The Melanesian directs his or her individuality in very *different* ways than Westerners do, but that does not mean that they are not individuals or do not have original insights. Different personalities, characters, and creative and leadership strengths are seen and fully appreciated by the people, and they fully appreciate and encourage intelligence, insight, dependability, originality, a sense of humor, courage, tact, caring, leadership ability, and such. These are openly discussed in picking leaders, negotiating bride-prices, or deciding which children are worth the expense of sending to a missionary school. They make extremely careful decisions because survival may depend on the qualities and judgment of individual tribal leaders, and family welfare may depend on good marriage decisions.

By placing a price on ceremonial rights and such and by negotiating these prices for each individual, they ensure that individual traits will be fully examined and discussed. Individuals do make a *great* difference in these societies, even more so on average than in ours, since they are organized into smaller effective survival units and people are less replaceable and expendable. It is obvious to them and to me that, while there is much sharing of perceptions, each person has unique perceptions and contributions to make to the group and to the loose perceptual consensus. Heated arguments within a group are common because often some see an important issue differently than the others. But on the surface it may seem that since they are different from most Westerners and since we assume ourselves to be at some acme of individuality, then they are necessarily less individualistic than us. This is an easy trap to fall into.

Witcutt, for example, in *The Rise and Fall of the Individual*, focuses on attacking Communism because it "aims at the complete destruction of the individual." So he must explain what individuality means, and he takes the position that it began with Christianity. "The individual is above the race in Christianity because the Christian does not exist for this world alone; he aims at immortality." Salvation is a personal matter and hence Christianity directs our attention to our individual relationships to God. Then individuality is produced.

This sounds "logical" enough but there are many problems with this sort of position, not least being the near disappearance of what most of us would call freethinking during the Middle Ages, when Christian faith reigned supreme. A pagan, Jew, or Christian heretic might not agree that the individual was placed above the race. And then too we must ask, was Socrates, long before Christ, not individualistic? And so on. But I want to focus on where such positions can lead. Witcutt misreads the tribal situation. "So when the missionaries came they found that the primitives to whom they preached could not understand the idea of individual salvation. They could only think of the tribe as a whole saved or damned. To them the tribe was one thing and its members, on their own, non-existent." This curious reading of the situation leads to unfortunate conclusions such as, " . . . in the primitive tribe, man had not yet become a true individual. He remained, as it were, half animal."

One grain of truth in this misinformed nonsense is that the villager sees the riches in life to be in social bonds and rituals even more than we do. The rewards in their lives may be heavily in the daily world and in daily experience, rather than in individual possessions or in the afterlife. The tribal people I have met and seen are happy at the smell of a flower, the touch of a friend, the knowing of stories, sunrise, a good joke, love and security, and so on. There is a solid basis in this attention to nature and in the closeness to one's people for respect of human dignity and individual differences. This is one of the pleasures in dealing with tribal people.

Missionaries have described to me, sometimes impatiently, Melanesians as being like children, not because they are intellectually slow or lack keen wits, but because they do not share fully our culture's adult priorities. For most of the day they are smiling and seem to be happy with innocent rewards: the beauty of nature, music, art, food, and the company of family and friends. ("My face actually hurts from having to smile back at them so much," an Australian visitor commented.) But *childlike* paints a distorted picture, since there may be endless political conversation, attention to ritual and duty, concern over "voting power" in the form of pigs and other properties, and even some adultery, theft, rape, and war. They cannot be pigeonholed into our humanistically primitive system of classification. But in any event, the priorities of the group are terribly important to the individual even in what we would see as personal matters such as marriage, childrearing, and property.

A second grain of truth is that the tribal individual's information about the world strongly reflects that of the group, since his or her information sources are limited—while among literate people one may read from a potentially vast literature and may develop a much different preception and sense of identity than do others in the group. If McLuhan is right, this may be changing, though, with mass communication, where through the mass media we nearly all participate in rather similar editorial views of the world. This could lead to a return to a tribal sort of consciousness and the advent of what he calls the "Global Village." In a sense then, Western middle-class people would not nowadays be much different in individual consciousness from tribal folks. True or not, the idea gives perspective to the present issue. Literacy and illiteracy, information or its lack, can alter individual consciousness profoundly.

The sort of misinformed perception that Witcutt illustrates leads easily to the view that individuality is only possible through Christ—a convoluted idea. Though, I am sure that were it not for their faith then at least *some* Christians (as with other world views) would not have had the emotional support to resist peer pressures and amplify their own ideas on some issues. Worse yet, though, views such as Witcutt's can lead to the view that a belief in salvation through Christ *necessarily* confers individuality—an improbable and obviously untrue idea. There seem to be as many thought conformists among Christians as among any other group of secular or religious believers.

Individual Objectivity or Emotional Self-Reliance?

Americans, and others in similar societies to be sure, treasure dearly the image of *self-reliance*—persons who can take care of themselves physically, emotionally, intellectually, and materially. They have financial and social goals for themselves and compete largely as individuals (or as mates) and are not held back by an obligation to share financial or social gains with siblings, uncles, aunts, and cousins. They do not necessarily negotiate with relatives on career, marriage, or child-rearing decisions as do Melanesians, for example. They *may or may not* have well-thought-out opinions and values that differ markedly from those of family or friends. In fact, many Americans tend to be more comfortable with persons who, beyond endearing quirks of personality or a slightly oblique sense of humor, are not markedly different from themselves in basic opinions and values. What we usually value as individuality is in fact a particular sort of *emotional* self-reliance.

Americans tend to see ourselves as materially and socially independent even though we rely on police, firefighters, bankers, doctors, neighbors, relatives, automobile mechanics, teachers, librarians, postal carriers, lawyers, friends, travel agents, grocery and shopkeepers, plumbers, ministers, journalists, television commentators, employers, clients, and so on. A Mexican farmer may be much closer to self-sufficiency than the Anglo-American today, yet the latter will usually see him- or herself as more independent. There is some mythology to our self-image of self-sufficient individuality. Many works, such as the important study on individuality by Hsu (1981) comparing Americans to Chinese, treat this subject in detail. Our society is quite complex and we must in fact very much depend on each other.

A sympathetic or unsympathetic critic of our society would say that our self-reliance has become just a nice way of saying that we have become very tolerant of loneliness and alienation. Immigration, frontier expansion, and industrialization made the dislocated and fragmented family more common, and uprooted one from community cultures with old and involved roots. Then Americans adapted, they argue, by making a virtue of the angst of alienation, somewhat as centuries ago some Christians made a virtue of enduring physical pain and suffering. Cowpokes don't cry and their horses are their best friends.

There is some truth to this analysis. Our ideals of individuality are relatively recent in history; their start can be seen in the Renaissance and the Enlightenment. It had to do with the idea that the individual could train one's mind and *grow* through one's own experiences. Thus, one could form opinions and values that were potentially superior to those dictated by peers and the authority of church and state. This intellectual individuality is clearly not quite the same as emotional self-reliance, though there is some linkage. This first mode of development had many sources, including increased contact of Europe with other civilizations, improved printing, a reaction against the control and conformity of thought in the previous centuries, capitalism, and the rise of a literate and economically strong middle class. *Economic* individualism was at the same time promoted in Europe.

Then, especially in the nineteenth century in the United States, came the reshaping of an agrarian and frontier society by the Industrial Revolution, population increases, and urbanization. People evermore would have to specialize and hire out to survive in an increasingly complex market. The ideal of individuality was thereafter redefined in times when each year there was more pressure toward economic adaptation in life and thought than toward personal and idealistic contemplation. Jefferson's ideal of a nation of contemplative farmers was undermined by the very technological change that he had helped to promote, as my colleague historian Ed Layton argues. Great numbers of upwardly mobile people then had to balance any aspirations for themselves and their children for intellectual growth through education and reflection against preparation to compete for jobs and social roles in commercial and governmental corporate structures.

One raises children with an eye toward their economic survival and advancement in adult life. Roles for increased numbers of middle-level managers and professionals developed and these managers in turn had to be managed. In the broader sense one does this by hiring people who are "types," with particular sets of skills and personalities that employers and nonfamilial coworkers can depend on. Humanistic ideals of individuality and individual respect became even more than in previous times complexly mixed with socioeconomic pressures to specialize and conform and to see other people more as means to ends than as ends in and of themselves.

Some vulgar Marxists see in today's "individualism" merely selfishness and false consciousness, which stand in the way of the presumed progress of the collective and support supposedly dehumanizing capitalism. They see the rise of the middle class entirely in negative terms—groups exploiting those below while comforting themselves with myths about the sources of wealth and privilege. This traces back to Marx's recognition of self-congratulatory yet myopic "bourgeois egoism." (Recall also Rousseau.) Bourgeois values, morality, and perceptions may be seen by vulgar Marxists, though, *entirely* with disdain.

Westerners clearly *do* confuse and mythologize middle-class individuality and other aspects of self-identity. Indeed, family and community systems generally also speak and think well of themselves even when they are insular, shame based, or otherwise unhealthy. It is not always clear when emotional independence and social mobility have become warping forces and when liberating forces. Indeed, it is generally unclear when a community is anywhere near as nurturing as its members may believe. Middle-class people are not so independent, special, objective, moral, or superior as they were raised to believe. Too many will ignore or rationalize social injustices and societal mechanics.

But this must not obscure the fact that the ideals of individualism, and even the myths, also have given many the personal support and encouragement to nourish insights that vary considerably from the common perception. Indeed, they may provide the support that allows one to rearrange peer alliances from time to time and to pursue new individual experiences that can serve as the basis for personal intellectual growth. People of the Enlightenment fought to make available opportunities to develop individual thinking—public education, public libraries, freedom of the press, freedom of speech, freedom of assembly. I do not doubt that these have been used largely to pursue economic goals, improve social status, and for diversion. But this is not to say that they have not also served hoped-for purposes and helped some fine thinkers to develop.

Economic Determinism

"God only knows what man is; I only know his price" (Bertolt Brecht, *Supply and Demand*).

I first learned economic determinism in college from economic *conservatives*, and from Marxists shortly thereafter. It is certainly not merely a leftist perspective, as some suppose. Glasgow's Professor of Moral Philosophy Adam Smith (1723–90), the philosophical hero of many economic ultraconservatives, began to set out how economics could shape human values and perceptions (but see also Mandeville, chapter 6, note 6). He also mixed metaphysics with economic theory on a grand scale.[2] The "laws" of social conduct are economic, is the claim, and these laws will drive society toward historical "progress."

First, Smith showed that people pursue their individual self-interest, but the market determines their behavior and values.

> It is a curious paradox that thus ensues: the market, which is the acme of individual economic freedom, is the strictest taskmaster of all. . . . Economic freedom is thus more illusory than at first appears. One can do as one pleases in the market. But if one pleases to do what the market disapproves, the price of individual freedom is economic ruination.
> (Heilbroner 1980)

Economic determinism has proved to be a powerful and valuable analytical tool in the social sciences. We would certainly be much more naive without it. What concerns me here is the tendency to see in it the single truth about life and to draw extreme conclusions from it.

Economic determinists can be in bitter conflict with biological determinists. Can *the* taskmaster that directs human behavior be reduced to the economic system or to a genetically programmed "human nature"? This is the needless competition of two reductionisms and my criticisms of both follow earlier comments on reductionism (e.g., chapter 4, note 2). *Methodological reductionism* can be a valuable analytical tool, and a cautious analysis of underlying economic or sociobiological contributions to human or animal behavior can certainly be interesting and enlightening. *Explanatory reductionism* though, can degenerate into discredited essentialism.

Again, one criticism is of the "nothing-but-ness," the exclusive claim to truth, the claim that others are "nonscientific" or mystical. When economic or biological determinists become extreme or even ideological, they claim explicitly or tacitly that social values and behaviors are nothing but economically serving, or else are nothing but devices to gain a competitive advantage in spreading genes. Anyone who argues that the situation may be more complex may be dismissed as being brainwashed by the opposing ideology, of being a romantic, of having mystical illusions, etc., but certainly of being "nonscientific," or "nonobjective." The middle ground is attacked from both sides. Polarization results. The pursuit of truth becomes a matter of defending one economic system or another (for economic determinism), or of defending individual competition and selfishness, male dominance, etc. (for sociobiology).

For me, it is often a sign that an attempt is being made to parlay success in methodological reductionism into explanatory or ideological reductionism when the practitioners make no serious attempt to integrate their findings into higher schemes of complexity and to construct pluralistic models. If they *merely give lip service to this, or advocate the power of their point of view* against others, then this may suggest that they have lost the ability to critically evaluate the ideological seductiveness of their methodology.

Market forces do not simply lead to *self*-deception and *self*-determined myopia. Cynical ideological capitalists are apt to support popular beliefs and merely technical education, essentially following the advice of Machiavelli (chapter 5) *because they see certain beliefs and lack of insight to be good for business.* The influential Mandeville long ago argued, "To make the Society Happy . . . , it is requisite that great numbers should be Ignorant as well as Poor." (How ignorant, how poor, is adequate to ensure manageable and cheap yet effective labor in a given economy?) Those in power may or may not convince themselves that the beliefs are true or that merely technical educations are humanistically adequate. They may think something along the lines of, "the people need to believe such and such or, my God! the system may collapse. We need to encourage those beliefs. Get my checkbook."

A celebrity in my part of the country was James Hill, the railroad king. Though a Protestant, he gave large sums to the Catholic church in Minnesota in the belief that this was necessary to pacify his workers, whatever he might have believed this would mean for the salvation of their souls. Matthew Josephson, in *The Robber Barons*, quotes Hill's official biographer:

such payment "is as much a matter of business as is the improvement of farm stock or the construction of a faultless railroad bed." And he quotes Hill, "Look at the millions of foreigners pouring into this country to whom the Roman Catholic Church represents the only authority that they either fear or respect. What will be their social view, their political action, their moral status if that single controlling force should be removed?"

The common sense of diverse cultural chauvinists tells them that their various commonsense beliefs are true and right, and they become the natural allies of the ideological capitalist.

In economic determinism (Emerging as it did from an Aristotelian design teleology mind set, natural theology) the presumed wisdom or plan of nature became the presumed wisdom of the economic system. Purpose and social progress became economic progress, "is-ness" was "ought-ness," all parts serve the greater good (as in Mandeville). Leave the market alone, be cheerful about "necessary evil" and *in time* (as in Smith) the "Invisible Hand" will guide everyone to Valhalla.[3]

Smith wrote in relatively good times. But in a few years European economy got worse and Valhalla seemed not likely to result from a free market. The stockbroker David Ricardo argued that goodness for all would not eventually result automatically from selfishness and competition.[4] Tensions would increase and left to its own devices the system would tear itself apart. The Reverend Thomas Malthus rationalized misery as a necessary evil, yet he also argued prophetically that unchecked population growth would eventually void the promised utopia. The message was that humankind could not afford to sit back, pursue self-interest only, perhaps institute some welfare programs, and assume that the Invisible Hand would in time turn selfishness into the golden age for *all*. Smith, Ricardo, Malthus, and other economists were followed by Karl Marx.

Like Smith, Marx saw individuals' behavior and values shaped by economic self-interest and necessity. However, mixing versions of the design thinking of Smith and Hegel with versions of the pessimism and interventionism implied by Ricardo and Malthus, he saw a very different inevitable historical destiny than had Smith. Things would not slowly get better for the common good as each person behaved selfishly. Instead, opposed class interests would resolve themselves through revolutionary clash in a dialectic process. The supposed logic of nature would assure that the workers would win. As Heilbroner (1980) put it:

> There was no longer a contest in which one side or the other ought to win for moral or sentimental reasons or because it thought the existing order

was outrageous. Instead there was a cold analysis of which side *had* to win. . . . In the end they could not lose.

Unlike the Utopians who also wanted to reorganize society closer to their desires, the Communists . . . offered men a chance to hitch their destinies to a star and to watch that star move inexorably across the historical zodiac.

Marx's motives were utimately humanitarian, as were Smith's. But both left their followers with certain ideas that had a life and trajectory of their own. If history moves inevitably in a given direction, then it has seemed to many to follow that individual will and reason cannot ever have been very influential in human affairs. If the motives of naked survival and the local factors that appear to offer security and status determine our behavior and perspective, then individual will and reason cannot be important.

Similarly, if consciousness and values are merely the products of the economic system and social class myths, then no perception can be free of values and assumptions—so, they argue, as a lawyer might dismiss legality based on the wording of a written document, objectivity is a logical impossibility.

Many believe there is no point in trying to develop individual capacities for objectivity, because that would be simply aspiring to self-delusion. Similarly, objectivity, truth, and so on can only be approached by the progress of the impersonal system, driven as each of us gives oneself over to greed, competition, or one ideological creed or another.

Colleagues who *study* Marxism assure me that there is a genuine philosophical contradiction in "orthodox Marxism" on individuality (Why should one be compassionate about mere puppets?), as there is on the question of revolutionary struggle. (If historical destiny will eliminate capitalism, then why should Marxists have to work so hard consciously to bring about a revolution?) Other "academic Marxists" tell me that it is mostly the "vulgar Marxists," emotionally driven, with only class conflict and the itch for revolution on their minds, who wear blinders and who dehumanize and deindividualize liberally.[5] They claim that the movement is as diverse as Christianity, with its fanatical elements on the one hand and its level-headed elements on the other, both factions claiming the same basic texts. Such issues are far beyond the scope of this discussion, since it is not our purpose to clarify the endlesss tangles of philosophical and ideological disputes. The issue is that in public and academic life one commonly encounters arguments that the possibility of much individual objectivity is a myth, and the origins of these arguments need to be understood. Extremes of economic determinism are vigorous sources.

Historicism

A powerful ally to the deindividualization that too often springs from economic determinism of both right and left is what philosopher Karl Popper
calls "historicism" in critical books such as *The Poverty of Historicism* and
The Open Society and Its Enemies. Popper traces much of its modern form
to reshapings of Plato and Aristotle by Hegel (1770–1831).

Historicism is more or less the deep belief in *historical destiny* that
comes from the idea that the course of history is controlled by a simple
logic in nature. The "great man" in this view is the one who supposedly sees,
and in whom lives, the spirit of his times—somewhat as one rides the stock
market—and who pushes "destiny" along. (More traditional great men theories hold that an individual actually changed the course of history, or
helped it to move in some direction where it might not otherwise have pen-
etrated.) Then people *anywhere* on the political spectrum can get fanatical
when they believe that history/destiny is working through them, as was the
grim case with Hitler and Stalin, and with various industrialists and reform-
ers as well.

Intellectual individuality and objectivity shrivel. The world becomes one
of competing ideas and forces that are merely *voiced* through persons who
speak for their times. The course of history is determined by the conflict of
ideas and flesh and a relentless logic of nature and human affairs. The individual is helpless to do anything but cash in on the trends of the times.

Modern democracy worked as well as it did when people accepted that
although they might not agree with one another, they could live together in
peace if they had mutual respect for one another's basic rights and potentials and for the inevitability of differences in point of view and opinion. It
was accepted that if people were willing, then many differences over practical issues could be kept in perspective by reason. Democracy today instead is too often seen only as a technique of governing—voting for representatives and such; but where it worked well it was a certain spirit, or
attitude about how individualistic people could live together. Even Huck
Finn grasped these values in their elementary form: "It would have been a
miserable business to have any unfriendliness on the raft; for what you want,
above all things, on a raft, is for everybody to be satisfied, and feel right and
kind towards the others."

Now both ends of our political spectrum, with plenty of lawyers leading,
seem tacitly to have bought the Hegelist idea that, "the stronger cause will
defeat the weaker and that the progress of humanity is furthered by physical
and moral conflict" (Thilly and Wood 1957). Conflict becomes seen as
inevitable—even essential. And respect for individuality and individual po-

tential seems at risk of fading evermore as all political extremes evermore see persons as pawns in ideological/economic struggles. It is hoped that some numbers retain something of the humanistic Enlightenment spirit of confidence that individuality is more, and can be potentially much more, than a slogan or myth. If not, we are all at risk of becoming part of a self-fulfilling prophecy of the various Hegelists.

Popper (1950), argues at length that while the extreme left wing of Marxism and the fascist extreme right more or less consciously base their philosophies on Hegel, the middle has been unconsciously much influenced. One also sees this for oneself in academia and among decision makers. It is clearly not only Marxist academicians who buy into the pseudorigor of Hegel. Popper attributes much of the popularity of Hegel to his obscure style of writing, which comes across at first as being supremely highbrow, and hence supremely appealing to competitive intellectuals. One might add that Hegel's intellectual breadth can also be appealing. We have seen reasons for the appeal and attractiveness of design-teleological schemes in general, and beneath historicist rhetoric the average person finds a sort of familiar and comforting (if misleading) design-teleological common sense in ideas of historical logic and destiny.

Popper attributes much of Hegel's initial visibility to his deliberate promotion by the Prussian state. Hegel was installed as the official state philosopher. He and his followers were commissioned by King Fredrick William III during a period of political reaction to the American and French revolutions.

The successful spread of commissioned ideas beyond a sphere of direct political control, as with Hegelism, is not such an unlikely thing to happen. We do not have an absolute ruler to commission ideas in the United States, but we do have a number of political/philosophical think tanks where writers are commissioned to elaborate ideology and to propagate ideas that will influence academic and public thinking. Many of these are bankrolled by extremely rich and influential people who evidently feel that they get their money's worth from these writers and researchers (see for example, Rothmyer 1981).

I make no pretense of a formal philosophical analysis of Hegel himself. There is plenty of that in the library for the reader who requires it. Much of my concern is with current ideas and attitudes of antiindividualism that can be traced back to Hegel. And on the broader issues, I shall try not to duplicate the extensive analysis of Popper; for surely the concerned reader would wish to examine it in full. Popper's critique is not trivial and it is lively reading—"And the whole story of Hegel would indeed not be worth relating, were it not for its more sinister consequences, which show how easily a clown may be a 'maker of history'."

Though this is gentle compared to what Schopenhauer (1788–1860) said of him. "Hegel, installed from above, by the powers that be, ... was a flatheaded, insipid, nauseating, illiterate charlatan, who reached the pinnacle of audacity in scribbling together and dishing up the craziest mystifying nonsense. This nonsense has been noisely proclaimed as immortal wisdom by mercenary followers and readily accepted as such by all fools. ... The extensive field of spiritual influence with which Hegel was furnished by those in power has enabled him to achieve the intellectual corruption of a whole generation."

In any event, I am necessarily suspicious of doctrines that liberally attract fanatical devotees. Their widespread appeal is emotional, offering a sense of power in the form of a logical weapon that can be disarming and even destroying against the less sophisticated adversary.

In my experience there are types of writers and researchers who have strong impressions of where they think history must be going and will thus advocate their cause with extreme vigor—competition and conflict drive "progress." They believe that others are either doing the same or else are too stupid to do so. They see enough of both dishonesty and stupidity to reinforce their views and attitudes. To their ends they may even appropriate the ideas of others and present them as new, sometimes not even crediting the originator. (If individuality is an illusion, then originality must also be.) But originality may not be their *forte* even though it may appear at first and to the nonspecialist that they have generated new insights.

Such attitudes and values have become the natural allies of careerism in academics and elsewhere, providing supposed moral and intellectual sanction for ruthless competitiveness.[6] Little but ambition is needed to to drive such personalities, but if pressed they often fall back on ideas that can be traced back to Hegel. It is obvious why such ideas sell well and become common sense among people set on making careers in systems that are being pushed by managers to grow ever more competitive. If historicism were not available, then careerists would find some other convenient ideas to justify their behavior. (But, of course, the particular rationalization that one adopts has particular philosophical implications that will tend subsequently to mold perceptions and values in nonarbitrary ways.) This is how false consciousness develops.

The rise of such personalities on faculties has been personified in fiction such as the cutting but humorous novel, *The History Man*, by Malcolm Bradbury. This is a cult classic in some academic circles, about a shallow but quick and glib, indeed popular and charismatic, careerist social scientist of the above sort who always wins in the growing academic rat race.

Historicists keep the idea afloat that it is inevitable that someone would have had particular insights as *the logic of history unfolded*. Often this is obviously true; examples are not hard to find. But often their position is based on faith. They take to the extreme the obvious point that new ideas have a cultural and economic context, and argue that creative people were not really creative but were necessarily merely the first to articulate the logic emerging in their times. For example, it is easy to suggest that Mozart exhausted the possibilities of classical music and so music was ready for a revolution. Beethoven merely played out the inevitable role of revolutionary. If Beethoven had not been born then some others soon would have had all of his musical ideas. Such arguments will usually seem at first to make sense, since *in hindsight* we see vividly the logic of events leading up to the *status quo.* It will necessarily require a good capacity for critical imagination to grasp alternatives, or they will have the character of abstract and fuzzy "what ifs."

As a student, I too found it easy to pick out examples of ideas and discoveries emerging from the thrust of their times, and a *sort* of historicism seemed reasonable to me. I was busy learning to see sources of cause and effect about me. As my personal experience grew and my ability to handle subtlety improved, I was better able to evaluate such ideas critically. This sort of thinking seemed to explain less and less, or I should say I continued to find it useful but its limits became conspicuous, and I had to reject the priority of much of it.

There is cause and effect in history and it is good to appreciate how our own present was formed. Arguing the inevitability of a Beethoven or a Napoleon is a stimulating mental game that helps college students in dormitory sessions to keep events and tendencies to hero-worship in perspective as they expand their thinking. But to seriously buy implications of inevitability whole cloth involves an incredible leap of faith. Many artistic, intellectual, and economic traditions around the world have gone into a long stable phase *without* dramatic change or progress. And if they were to change, it may not be obvious in what *direction* they would change. The inevitability view also ignores the fact that great artists and creative thinkers were often quite unappreciated in their day and were not voicing the spirit of their times. Starving artists such as van Gogh are a cliche. Gregor Mendel, the "father of genetics," could not get recognition for his important work and he soon gave up research. Many would say that such persons were ahead of their time. We can say that they were imbedded in their time, but certainly many were not products of it in a simple sense.

A materialistic alternative to historicism is that there are many possible alternative histories and no one clear destiny is prescribed by any mythical logic in nature. There are many possible stable states or trajectories for a society. Any of the alternative paths or states would look logical in hindsight, since history is not exempt from material cause and effect and the paths were indeed constrained by the available material resources, geography, perspectives, and so on. The victory of one idea over another does not prove that the winning idea was correct. A people may set goals for themselves, but nature itself does not work toward particular goals. History follows one path or another also due to small contingencies ranging from storms that wipe out armadas to early deaths of leaders, or determined human intervention may select one path over another. Not grand design in the logic of nature, but not chaos either.

We have no need to imagine a violation of the constraints of economics and geography. In practice these are in any event usually not precisely quantifiable or mapable by the historicists or economic determinists. If pressed, uncritical technicians of the words of Hegel, Marx, or others fall back to broad rhetoric and have claimed to me that Marx, for example, only insisted that, "men make their own history, but they do not make it as they please; they do not make it under circumstances chosen by themselves, but under circumstances directly encountered, given and transmitted from the past." It would be hard for anyone who does not believe in magic to disagree with that. But does the "dialectical rigor" that is sometimes so chauvinistically brandished mean in the end simply that there is no (non-Hegelian) magic at work in history? Dialectical materialism and historicism are more than that to their more immodest sloganeers, of course.

Many of those who argue the inevitability of history seem to be using selective hindsight, a form of Monday morning quarterbacking where hindsight is 20/20 vision. They fall into a psychological trap. A "logic" can always be found in a sequence of events in retrospect, and then it can be claimed that the outcome was rational and hence inevitable. But advocates do not commonly weigh all of the possible alternative rational outcomes. This would take critical imagination and that is not a talent that all adults have. So alternative scenarios may be dismissed as mere speculation. Thus, the historicist's thought system may attempt to insulate itself against any possible refutation. We may have another example of a self-sealing premise, as discussed in chapter 5.

As long ago as 1840, in *Democracy in America*, de Tocqueville argued that history is shaped by *both* individual and impersonal causes, and he detailed psychological traps into which historians can fall. "They take a nation arrived at a certain stage of its history, and they affirm that it could not but follow the track which brought it thither. It is easier to make such an assertion, than to show by what means the nation might have adopted a better course." An exaggerated emphasis on "inevitable" causes by the historian "can always furnish a few mighty reasons to extricate them from the most difficult part of their work, and it indulges the indolence or incapacity of their minds, whilst it confers upon them the honors of deep thinking."

One approach to estimating the role of the aware individual, or one who diverts the course of history, is the "thought experiment." We can ask, what if Galileo had never been born or seen a telescope? And so on. Such thought experiments are important for gaining perspective, but they are not above criticism by opponents. The time was ripe for Galileo's discoveries, they will try to detail, and someone else would have made them. Once I bought such arguments, but in observing events closely since my early college days my views have had to change. Now I think that such thought experiments are valuable and should not be dismissed so easily.

Modern Western Europe owes its nature to the Marshall Plan of economic and political reconstruction following World War II. At that time the ruined continent had many possible destinies. That the United States should, in an unprecedented move, help it toward becoming a major economic competitor in the future, did not seem at all reasonable to many at the time. The victors might have made Europe into a sort of agricultural buffer zone, for example, instead of reindustrializing it (cheese and beer republics instead of banana republics). It might have all been cut up into ethnically odd packages as the colonial powers sometimes did to keep territories dependent, instead of largely along viable old national boundaries.

Some felt it was logical that the United States and United Kingdom should not end the war at all, but should move on and attack the USSR. A state for the Jews might have been provided in some other region than explosive Palestine, or nowhere at all. France might have given Vietnam independence at the end of World War II, thus preventing later continued Western involvement. The United States might have decided not to go to war there, thus avoiding the enormous dislocations that rocked the economy and democratic idealism and that caused the abandonment of the gold standard. The United States might have tried to keep the Philippines at all costs. At the time there were good arguments being made for and against each course of action. Would Greek culture have spread so far without the dreams of Alexander? Suppose Cortez had been killed by Montezuma? Roman power was expected to crumble in the last days of the republic; suppose the Caesars had not been able to keep it strong for additional centuries? What would have been the history of science in Germany if von Humboldt had not constantly had the ear of the king? Suppose Jefferson had decided not to make the Louisiana Purchase? Suppose Augustine had died early and had not written *The City of God*?

Philip K. Dick's renowned science fiction novel, *The Man in the High Castle* (1962), takes place in a California where the Anglo-Americans must try to understand the minds of their Japanese masters (Japan conquered the western United States in World War II and Germany conquered the eastern United States). There is a neutral zone in the center, in which a mysterious writer is said to live in a high castle. He has written what the public agrees is an absurd alternative-history science fiction novel (*The Grasshopper Lies Heavy*), in which Japan and Germany were defeated. *Grasshopper* involved so many improbabilities that most dismissed it. (Imagine Rommel being defeated in North Africa and the Russians turning back the German army!) It required that if President Roosevelt had not been assassinated in Miami after a year in office, he could have become one of the strongest presidents in history. He could have then pulled the United States out of the Great Depression so that it could have become strong enough to fight the Axis years later. It required that Roosevelt could have been reelected in 1936 and so was still in office when Hitler attacked France and England and Poland. Only thus could the isolationist United States have had a president popular and strong enough to put it fully into the war against the Axis powers. In short, *Grasshopper* set out a series of events counter to the historical inevitability of Axis victory, and it certainly could not be taken seriously.

The improbable scenario illustrates the problem with trying to figure out if there really were great personalities who changed the course of history, by thought experiments in which we ask, "What if so and so had died?" Then people must make absurd speculations in order to defend potential histories that do not follow "natural laws" of economics and politics. And accordingly, *Castle* argues convincingly that we should ignore *Grasshopper*, which maintains with so many "what ifs" the irrational speculation that the Axis might have been defeated by the United States, United Kingdom, and France. Sound advice.

Dick's playful thought experiment runs even deeper than exploring the question of historical inevitability. His heroine argues that the idea that Germany and Japan might have lost the war if Roosevelt had lived *is completely rational*. Dick, a student of philosophy, thus takes this whole issue one step further. If several histories are rationally possible (and if we have the false consciousness that some claim), then how do we know for certain that the reality we experience is the True reality? Our senses may deceive us. They only see the shadows on the wall of the cave. And, we cannot use the philosophical criterion that The Rational is The True, because we cannot know The Truth through reason if several paths are equally reasonable.

Dick's heroine makes her way to the man in the high castle and tries to find out which philosophically possible reality is True. But even the author is not sure. So they appeal to the *I Ching* and it tells them that their intuitions are correct. But neither is this divinatory means of determining Truth satisfactory to the man in the high castle.

Convoluted philosophical dilemmas are, of course, what we come to if we insist on trying to find Truth through Reason alone, as Plato insisted we must. Hegel argued that nature and reason are one and that what is real is rational and what is rational is real (Rationality becomes more or less the laws of economics to many economic determinists.); history follows a logical inevitability. But we are actually confronted in life by different apparently rational choices. And when we can recognize this, we do usually have to choose by intuition, which is not philosophically satisfactory—life deeply complicates the Platonic or Hegelian idealisms. Dick's characters are often convincing, coherent, individualistic people with quirks, humor, and intelligence who putter along satisfactorily and sometimes beat the odds in worlds where reality is never quite certain. Real life is a lot like that.

Any Guides from Science?

What can science teach the individual about critical thought, about the possibility for relatively good individual objectivity? What can its personal message be to the liberal arts student, for example? I thought the answer would be simple when I started writing this book, but I quickly realized that this is a complex issue.

Science is the project of many different sorts of people today. And science today is not what it was, especially since the escalations in society's attempts to manage universities that occurred after World War I and especially following World War II, and again following Sputnik, and again with the Thatcher/Reagan agenda for institutions of higher education in the midst of economic confusion (also chapters 8, 12).

It would be reassuring to believe that things are the same as they always have been. This would take pressure off of everyone involved to listen to complaints and to take action. Thus, it has been tempting to dig up any possible evidence to debunk the character of the great figures of the past, in attempts to show that any suspicious emerging values and standards are not in fact unusual and hence should not be worrisome. This psychological need creates a market for a sort of highbrow historical muckraking that we have been increasingly seeing. It is a market in which cynical ideas will sell and support careers, though very, very little dirt has been dug up that the experts can take at all seriously, or that illuminates the creative process; yet the attitudes and myths are becoming popular far beyond what the evidence will support.

"We taxpayers expect professors to be like the *Microbe Hunters*. But from what we read and see it's more like *Blind Ambition* [by John Dean of Watergate fame]," my neighbor observed.

A petty side of science has long been evident. Sinclair Lewis wrote his prize-winning *Arrowsmith* as long ago as 1924, about the grimy conflicts between idealism and politics in medical research. Even the romantic but still-inspiring and instructive *Microbe Hunters*, by bacteriologist Paul de Kruif (who worked closely with Lewis on *Arrowsmith*), in 1926 opened with a quote by Blakeney: "The gods are frankly human, sharing in the weakness of mankind, yet not untouched with a halo of divine Romance."

The sociologist Max Weber as long ago as 1919 wrote of science that was becoming bureaucratized, with status and power not always based on quality ("If the young scholar asks for my advice . . . one must ask. . . . Do you in all conscience believe that you can stand seeing mediocrity after mediocrity, year after year, climb beyond you, without becoming embittered and without coming to grief?" (*Science as a Vocation*).

But of late there has emerged the desire to paint petty competition as *the stuff* that has always generated scientific insights.

At the core of *Betrayers of Truth: Fraud and Deceit in the Halls of Science* (Broad and Wade 1982) are a troubling number of recent scientific frauds that had come to light, especially in medical research. The authors, well-known science journalists, saw these and related serious problems as resulting from an increase in competitiveness and careerism, especially since the 1960s and 1970s. Many of us in or close to science and academics had also seen this happening (Kohn 1986).

But Broad and Wade also try to put this modern phenomenon into some broader context for their readers and they seem to try to accommodate a particular school of "science watchers" (who include sociologists, philosophers, and historians of science).

They compiled every historical accusation of fraud or indiscretion that they could find, going back to the ancient Greeks (and asked the readers to send any that they missed to the publishers). They took the position not, for example, that the management philosophy of science and society is in error, but that cheating has always been endemic to science.

Very few of their older examples, though, are convincing and accepted by experts, and for these handful the shame is not at all clear since standards for research and publishing have changed significantly over time. For example, taxonomists once ignored or even discarded specimens that did not conform to the type specimen, since variants were considered to be "exceptions," trash of a sort, spurious deviates in terms of essentialism and idealism, and hence were seen to be worthless and even misleading. Today the informed biologist has learned to appreciate that individual variability is extremely important and interesting. For similar reasons, Newton and Mendel, working before the development of statistics, may well have "cleaned up" their data, honestly seeing variability as distortions of the ideal truths that they felt they had in their numbers. This matter, serious by our standards, was in their age hardly the same as fraud.

The book has already been widely criticized in these respects (for example, Joravsky 1983). I mention it here to illustrate a common uninformed attitude that is floating around in academia and elsewhere. The authors accept the invalid examples and the poor examples both, and seem to ask themselves, "If dishonesty is endemic to science then how can science work at all?" Here the answer is seemingly provided by the present-day science watchers of the careerists, and the apparently well-meaning authors accepted their position. (I attended a national conference on this subject, including one of the authors and a number of sociologists of science, and my scenario is not entirely deduction.)

Their model is that there has never been any degree of personal objectivity in science or anywhere else. Some people essentially guess right and the competitive system sorts out the good ideas from the bad. This is a sort of cartoonish Darwinism (true Darwinism would be much, much, more subtle), where each scientist advocates his or her own ideas and, presumably like mutations, there are plenty of these around. Then competition causes everybody to get stepped on by everybody else (the "Invisible Boot" of Broad and Wade, to recall the Invisible Hand of Adam Smith), like the forces of mortality in nature. Those ideas that work survive, as in natural selection. This notion is logical enough, but it is quite inadequate to explain real life.

There is a practical danger in this sort of view. We constantly must fight to deal with psychological bias and to keep up standards of intellectual honesty and objectivity in academia. My graduate adviser used to tell us, "Science is only a set of guidelines to keep us from lying to ourselves and others." The enterprise is a constant struggle against self-deception, faddishness, and subtle and overt venality. It is easy enough to *imitate* the methods of others, and much of science is that—the application of technique. But creativity and precious episodes of mental discipline and humility do not come without effort. They are hardly the natural states of humankind or the general climate of science and academics. We *do* have serious problems that we should face up to. But establishing and maintaining high standards is delicate and difficult enough without aggressive, blundering cynicism. It hurts the *foundations* of the effort, *not merely the image*, when young researchers are persuaded that efforts to maintain high standards are merely empty ritual to perpetuate a self-serving myth of objectivity and that these values and ideals have never actually been important to the progress of the enterprise. It hurts scholars and the public if they believe that attempts at objectivity will be ineffective and that science is an entirely impersonal process that dwarfs the individual and one that they cannot learn and improve from.

Many of us spend a great deal of time with graduate students, trying to share our experiences and work with them (and with ourselves) to cultivate their intellectual honesty (and our own), to develop checks on self-deception and egotistical stubbornness, and to guide them to research subjects where their talents for creative intuition can be cultivated. I see graduate students and established colleagues repeatedly come to *their own* conclusion, without the Boot, that they are wrong about something and then shift their research. Many of us seek out criticism of our ideas and will travel to meetings and even to other countries to try to get the best criticism we can. This self-scrutiny is hardly the bald advocacy that the critics paint, and yet it is still happening all the time in some fields. It is hard to say if this is the statistical norm, but it is at least an important part of the culture. I even have a few colleagues who repeatedly solicit severe criticism, even though we often end up hardly speaking after they get it. But I know they will not stay mad for long, because they are more interested in the truth than in their bruised egos, and they respect honest criticism.

For some good scholars, ego does drive the creative process, but not as the critics have painted it. There are some sorts of very good yet very egotistical scientists who are driven not merely "to promote their own ideas" but to play the difficult game of finding truth, and to win at it. If they have to learn to control their own egos at times to beat the odds, then they will make even this great personal sacrifice. But their egos may be so very strong that beating a human opponent would be boring compared to the personal satisfaction of actually prying loose difficult secrets from nature.

There are all kinds of science and all kinds of scientists. Some scientists even today function best as lonely individuals with only a loose network of colleagues who can offer hard but fair criticism. Others function best in various "corporate" systems of research teams or in tightly structured poltical networks, where it is hard to say where ideas come from, what the ethical checks and balances are, or how authenticity is verified internally.

It is careless scholarship, it can revert to essentialism, to study scientists in a big modern lab or who are part of a well-organized "old-boy network" and assume that one can use these results to repaint Galileo or Newton simply because they too had colleagues. Too few of the science watchers seem to be sensitive to the diversity of the modern enterprise. We certainly need good studies, but many of those that I have seen have been quite inadequate. *Laboratory Life*, by Latour and Woolgar, describes the efforts of two science watchers who actually went to work in the laboratories of the Salk Institute. Despite shortcomings, at least it is a step in the right direction.

Good ideas are by no means distributed randomly throughout the population of researchers as some science watchers assume. A scientist or scholar who has one *very good* idea is apt to have more than a fair share of additional good ideas. This demonstrates that some people do have generally better, relatively more objective, insight into a set of issues than do others. Galileo and Newton surely had human failings, they may have been chicken thieves for all it matters, but they each had many seminal ideas. Darwin had so many that we have scarcely been able to explore all of them. The question that needs to be examined in a book on the human mind is: Why have important thinkers had so many great insights while other bright people have not? I will detail in the next chapter that this has involved the mobilization of personal qualities, the particular humanistic incentives for this, as well as the cultural, economic, and technological support available to such efforts.

Many science watchers argue that progress could not have come from a cultivation of the human potential for objectivity, a struggle with character and ego, and high regard for truth, because objectivity is a myth, great scientists were very ordinary people (*ordinary* means they were like middle-class Westerners that the science watchers know, and has nothing to do with ordinary Eskimos, shamans, or ordinary great scientists) who faked liberally and were seeking primarily to advance their personal status. They imply that it is nonproductive even to think about guidelines for seeking truth, since everyone is and always has been out only to promote themselves, can only voice some political doctrine, and no one has ever *really* cared about truth nor been clear eyed enough to be effective. They portray the relatively independently thinking individual as a myth that makes one feel good. They try to explain intellectual *extremes* in terms of convenient social *norms*.

Why are anti-individualistic views so popular? This is clearly a complex matter, which might even involve such difficult considerations as mass communications, where one is typically exposed to important ideas in a very superficial and often disjointed manner. The illusion may be created that even substantial individuals do not have much intellectual depth. Equal-time media stagings may imply that one person's opinion is not much better than another's. The wonder and workings of original insight are seldom presented with sufficient richness. Implicit media messages could help to make simplistic philosophies of nonindividuality seem plausible. But it is hard to know precisely.[7]

In large part it may be that simple models especially attract certain types of personalities that must see the world neatly arranged, even at the price of forcing square pegs into round holes. As though exasperated with supposedly fuzzy concepts that they cannot get a firm grip on, such people turn to a rather simple perceptual system that works most of the time. Some subset of the sorts of science watchers who concern me may simply be influenced by a desire to make the world out as more tidy than it actually is.[8]

Civilization is not *simply* programmed and exploited puppets. Our art, music, cuisines, social arrangements, values, customs, myths, dreams, and ideals give our lives joy, support, and daily meaning. Humans are not simply abstract ideas; we are flesh and blood with human needs. It is an enlightening experience to live in a native village for awhile and see what things bring smiles and laughter and gentle pride, despite dirt floors and cold and bugs, or even in a very poor village with hunger and sickness or the lack of hope that life can be anything else. There is the reality of having one's own identity and social value. There is love and there is beauty in nature and in art. There is using all of one's wits and one's skills to make beauty in food, music, and art for family and friends, and in this can be caress, and this is when art and love can become one.

One may strip these meanings away, and ignore the realities of the connectedness, satisfaction, and passion that they bring to life, and reduce them to being "nothing but" adversary and conspiratorial forces; but this is not necessarily great scholarship.

Great civilizations might each be thought of as nothing but alternative modes of economic enslavement, but this insight seems somehow to be needlessly incomplete or even hostile when one considers that they also represent modes of adaptation of human temperament, aspirations, passions, energy, and intelligence—of the human spirit if one will—to the realities of survival. Given that we must adapt to *some* economic system, our species has found ways over the centuries to add beauty and opportunities for individual expression and happiness and the exploration of individual potential to the crowded road and the daily grind. That is no small accomplishment, and while great creations do not justify the misadventures of societies, neither are extreme materialist or spiritualistic analyses fair nor candid that trivialize individual insight, creative acts and dreams.[9]

Despite the considerable skepticism of reason and objectivity that one encounters, and the cogency of *much* of it, one should remember that many accomplishments have been made and we do now have some good ways to deal with differences in perception if we work at it. There is the familiar story of the three blind men when they were presented with a completely unfamiliar animal to feel, an elephant. One argued that the animal was a trunk, the second argued that it was an ear, and the third argued that it was a tail. Today a growing number of people, even professors, will tell this story, even with a smile, and say that it proves that objectivity is *impossible*.

Many who have responsibility or power and encounter social or management problems that they cannot solve may find it tempting to believe that differences in belief or opinion are necessarily irreconcilable. But this stance is absurd and can become a crippling self-fulfilling prophecy. All the elephant story illustrates is that objectivity is an uphill struggle, since what one thinks is much influenced by one's initial point of view.

If the three blind men were familiar with *good* critical or scientific thinking, they would describe their experiences to one another and discuss the differences. Each would consider that he might be wrong or only have part of the answer. He would swallow his pride and seek a resolution. The fellow at the tail would try to repeat the experience of the fellow at the ear and eventually would move over and feel it himself. Soon they would realize that the animal had a complex shape and they would come up with a pretty good model of what it actually looked like. (Try to imagine how far they would get if each blind person simply advocated his beliefs as strongly as he could.)

A critical thinker would *not* conclude that the world is made up of separate, equally valid individual realities and that objectivity is therefore impossible. Obviously the consensus elephant of the blind scientists would be a "social construction," but not in the relativistic sense as critics sometimes use this term.

Many people have heard differently and will call my scenario a hopelessly noble or idealistic belief. They focus on the competition in academics, on the cheating scandals and careerists that we have all seen. They may see the enterprise through the filter of popular ideologies of human nature—that people are basically and incorrigibly selfish and competitive, stubborn, fast-talking but small-minded. But they miss seeing how much of the above scenario of comparing observations on the elephant actually still goes on in science and may be critical to its progress even today. And not just in science. Humanity certainly has learned how to get past the problem of the blind men groping at the elephant. It is not the easiest path for the unenlightened, lazy, impatient, or one hungry for status, but it works.

For the sake of discussion let us surely agree that we do not have perceptions that are free of assumptions or values. But in any event experience shows plainly that some assumptions that enter into our perceptions are pretty good and others are worthless, and with work we can sort out the bad ones and build better ones. One can show that Redi's view that maggots are the young of flies is more objective than the view that maggots are a separate species of creature from flies, a species spontaneously generated by rotting meat (chapter 4). The first view proved to be in keeping with the reality of the external world, and the second was wrong no matter what its advocates saw and believed.

There may be no philosophically pure objectivity, but science is not a matter of logic or philosophy. Redi was right. Period. His was a true and rational perception by all but the most stubbornly technical philosophical criteria. Whatever one has heard of the supposed words of Kuhn, Feyerabend, Kant, Locke, Einstein, Freud, Marx, or Hegel, this will not change the fact that we now have available more objective views of maggots, moons, and humankind than were possible before modern scientific and scholarly thinking. Critical thinking will take some work for individuals to understand and apply well. We need to better understand the ideological forces in science and academia, and the higher educational system lacks coherence and is usually not too helpful. But these are separate, if massively important, issues.

Human thinking, the *mind*, can not be understood completely with knowledge of the biological *brain* and base emotions, without considering the different sorts of *habits* and *ideas* that are important in perception, in the processing of information, the triggering of emotions, and in the making of judgments. They lead one to open or to close the mind to information. They cause one to pigeonhole events and people into a limited set of mental boxes. They lead one to close off imagination or open it. They lead one to see complex things or to be blind to them. They lead one to face feelings or to stuff them. They lead one to develop individual courage and potential or to give up before one has had a chance to start. Ideas and habits can help one along, or they can hinder one as easily as a brain tumor or chemical imbalance can.

Pyramid Films' *A Private Universe* begins with scenes at Harvard University graduation. Graduates are asked why it gets hotter in the summer and they answer incorrectly that it is because the earth gets closer to the sun. The educational film proceeds to demonstrate how people, even an educated elite, can fail to grasp something as simple as a physical model of the solar system because they interpret straightforward information in terms of incorrect basic assumptions.

Whether or not one's parents, peers, and educators believe that responsible individuality and individual objectivity are posssible as more than slogans can affect the direction and strength of one's personal aspirations and confidence in oneself. Individuality is a sacred word in our culture, so we always give lip service to it, and there is no shortage of egoistic mythology. But unfortunately, much of what we hear, expect of each other, and are taught actually discourages the cultivation of independent thinking and acting.

A research colleague in the child development department at my university told me, "Yes, I can't even talk about it with most of my colleagues. Three-quarters of the faculty who should be thinking about the constructive cultivation of individuality *refuse* to touch it. The rest don't have time to get into it very deeply. It is no way to make a career. No grant or consulting money in it. Too controversial. They see the issue as too difficult and too fuzzy. It is too hard to get out publications that people will care about. Faculty attitudes don't seem to change even once they are protected by tenure; they still can only think that the administration wants to see lots of papers and grants. Look, the university offers several classes in the *socialization* of children, but *no* classes in *individuation.*"

I asked if that was not a bit frightening to parents and others. But he doubted that parents really want their children's individuality cultivated. "They seem to want youngsters to be well under control and to acquire their own family and community values, to grow up to get secure jobs. They may even fear individuality as chaos. Most people, even in our society that threatens to blow up the world to defend individuality, would see your concerns with *real* intellectual individuality as foolishly altruistic." He suggested, that, in his field, "our Marxists don't believe there *is* such a thing, so they are certainly no help. The capitalists want a docile, socially well-defined work force. The neutral faculty can't lead because they are competing for federal grants. And nobody in multinational corporations, labor unions, or in Washington *really* wants to promote individuation for the average citizen. I suppose they think it might make society even more difficult to manage. There is *no constituency* demanding it."

This story underscores a major point of this book—that one who wants badly to cultivate his or her own individuality in a constructive and more than superficial way will have to take personal responsibility for the effort, because these days institutions, philosophies, ideologies, and religions are not set up to help one do that.

NOTES

1. Let us note the flavor of some of this debate. For example, from Tucker's (1980) effort "to try to persuade Marxists that there is something to be learned from liberal writers," a passage where he argues that so-called holists (i.e., collectivists, for our purposes here related to structuralists and economic determinists) have set up a "straw-man" in attacking individualists.

> Individualists do not suppose that people are not changed by their values, but only that sociologists should not dispose of the individual entirely when providing social explanations. But this is what holists must do, for if they are to be convincing, they must show that the acting individual is merely a puppet whose goals and aspirations are provided by the social system. Once the rival perspectives are stated in this way, it is not at all clear that the holist has the more plausible assumptions.

Marx's own position on individuality was not so rigid, or simple, as has been the position of so many of his followers who seem to see this as a black-or-white issue. Marx believed that people are capable of transcending the laws of economics and history. He believed that after having their eyes opened individuals would be quite able to become "self-active, creative subjects, that is, highly unsatisfactory objects for exact scientific observation and prediction" (Wood 1972). No one seriously doubts the personal compassion of Marx himself. Indeed, a great part of the initial appeal of Marxism to some idealistic intellectuals has been that one side of Marxism shows remarkable sensitivity to conditions of dehumanization and to the human potential for individuality.

The problem, here, is that his compassion has gone so very easily astray in the hands of others who mold ideas to suit their own common sense and agendas. (As a militant recently told me, "OK, there are all sorts of contradictions. But when you see a condition that you hate and just have to fight and change, then you will grab at anything that seems to help make *some kind of sense* out of things and that will get you moving. You don't just give up because there are logical contradictions." As a reputed Marxist theoretician commented, "*History* will decide if I am right or wrong, if these philosophical problems are important.") Confusion and contradiction seems to have happened in part because much of Marx's compassion was for the salvation of future humanity. He seemed to place the future freedom of society ahead of the individuals living around him. That might have seemed to Stalinists to legitimate killings and repressions, for example.

Wood goes on to claim that Marx viewed the role of social sciences as setting the stage for a fostering of creative individuality, but that his "followers" in the social sciences, and in American social sciences particularly, have fallen back to a pre-Marxian deindividualized determinism. Why? "Any other conception of man would tend to complicate matters for the social scientist with his faith in precise prediction, quantification, and measurement."

Let me stress again that I am only concerned now that invalid but highly influential arguments against intellectual individualism in our era seem so often to be associated with economic determinism, Marxism, historicism, behaviorism, and biological determinism. Another author might dwell on whether or not this represents a misinterpretation of texts, or whether there is a contradiction between the young Marx and the old Marx, and so on. I will merely point out that there is fertile material here for academic research.

For additional reading, the papers and literature cited in the *Hastings Center Studies*, vol. 2, no. 3 (September 1974), may be of interest on the subject of individuality. Because of its diversity, Marxism is hard to generalize about in a brief discussion; the reader may wish to be aware of *Marxism After Marx*, by David McLellan (1979) as an introduction to some of this.

2. The popular version of a philosophy may have little to do with the words of the philosopher, and we may recall how the teachings of Jesus, Buddha, Marx, and others have been twisted by various factions. This discussion is not a minute analysis of Smith, Hegel, Jesus, or Marx but an examination of the power that their ideas have in certain contexts today. I refer in this text to the versions of Smith that are used today. In fact, though, Smith himself had rather humanitarian views.

The Wealth of Nations was liberally quoted to oppose the first humanitarian legislation. Thus, by a strange injustice, the man who warned that the grasping eighteenth-century industrialists "generally have an interest to deceive and even oppress the public" came to be regarded as their economic patron saint. Even today, in generous disregard of the details of his philosophy, Smith is usually oversimplified as a congenial *conservative* economist, whereas in fact he was more avowedly hostile to the motives of business than most New Deal economists (Heilbroner 1980).

Nevertheless, Smith did mix economics with natural theology. He did believe in a form of design teleology and historical destiny and that good would *eventually* result as a byproduct of selfishness. In this way he sought to reconcile the "self-evident" emerging ideology of the middle class, that devotion to economic gain and acquisitiveness were virtuous, with traditional Christian ascetic values and with the plain observation of analytical Christians that the emerging economic system was not producing harmony or more moral people.

3. Are competition and economics the simple logic that supposedly drives and directs nature and "determines destiny?" For human affairs, competition and economics are major components but not the only ones (as in the text). But the same may be said for communities of plants and animals as well. Biologists see some simple competition in nature, but also cooperation (e.g., Axelrod 1984).

Competition is not universal. Populations are often kept in check by predation, disease, and weather. Much evolution and behavior may function to *reduce* competition. Competition can result in extinction and not necessarily in progress. Extinction is the rule and is the fate of most species over geological times, though not only because of competition.

Scientists now question the "balance-of-nature" theories of the past. They have come to us not so much from critical observation and tested fact as from the old design and teleological tradition of natural theology that long ago informed natural history studies. There does not seem to be nearly as much balance in nature as was once thought. Nature is perhaps most often in flux—due to variations in climate, the movements of predators, pathogens, and competitors in and out of areas, and perhaps also due to genetic changes. Only a few would argue anymore that competition produces overall balance or progress. All three terms are extraordinarily vague and poor in the present context.

The entire concept of progress in evolution is far from simple and straightforward. The longest surviving species are not necessarily the best competitors in any conventional sense. For example, the Virginia opossum is a clumsy beast but perhaps the most ancient mammal and the top survivor, yet not conspicuously "superior" to cat or ape.

Biologists are testing economic theory as it applies to organisms, where energy expenses and profits replace cash in mathematical models. While such economic theory seems to have limited heuristic value for studies of nature, we are finding that nature has other priorities as well.

These subjects are in no way as simple as familiar language and common sense might have seemed to indicate. Biologists have found that their common sense of the past had been misleading.

4. I will put on my biologist's hat now and say a few words about selfishness and altruism, though this is difficult since there is the urge to write volumes. Organisms, in pedestrian terms, are basically "selfish"—the bee pollinates not from altruism to the flower, but for nectar—and humans are mostly no exception. But there are a seemingly infinite number of strategies for being selfish. It may be selfish to play and to explore the environment. One can see that it is in some sense selfish for social animals to groom and preen and even care for one another, and to share their food, space, and other resources. Apes are social and it may bring ape-peace and stability to the individual to have members of its group content, smiling, and happy. This may be in the long-range self-interest of most individuals. But these are very different sorts of selfishness than in semisocial species that may fight to gain and defend exclusive rights to a territory.

Organisms have diverse and conflicting long-and short-range self-interests that need to be balanced for survival. When people, scientists included, argue simply that immediate profits must come first *because humans are basically selfish*, one may be quite certain that they have not thought the issue through very far.

Altruism as such may not exist in nature, but apparent altruism may instead conceal long-range self-interests of the individual (e.g., Axelrod 1984). Those who wave the flag of self-interest (the many biologists among them included) may too often make a fetish of looking at short-range interests, at conspicuous anthropomorphic competition, and immediate gain. Thus, some complain that welfare legislation is an attempt to impose unnatural altruistic behavior on taxpayers. Yet one can argue that the apparent altruism conceals some tacit understanding that welfare is in the long-term self-interest of even the rich, since it would be in the short-term interest of destitute and disenfranchised people to steal and revolt if society did not help them. Such issues are often reduced to an irrelevant dichotomy between selfishness versus altruism, but they should instead be debated seriously as issues of what constitutes the wisest strategy of self-interest. To say that "we are all selfish" is about as enlightening as to say that we are all chemicals. It is virtually a nonstatement.

5. Polanyi (1962) argued at length that the contradictions in orthodox Marxism in fact create an emotional *appeal*. For example, they allow one to be moralistic and at the same time to

have contempt for myopia or hypocrisy in bourgeois sentimentality and morality. It gives license for one to be both moral and amoral. "By covering them with a scientific disguise it protects moral sentiments against being deprecated as mere emotionalism and gives them at the same time a sense of scientific certainty; while on the other hand it impregnates material ends with the fervour of moral passions."

Arthur Koestler, in *The Call Girls* (a 1973 novel in which academic careerists of all sorts try to solve humankind's problems at an exclusive meeting), amusingly sketches a conspicuously glib, trendy, intellectual quasi-Marxist. (The baboon comment below refers to a discussion of humans as naked apes.)

> "How do you reconcile mysticism with your Marxist dialectics?"
> "But perfectly. It is the synthesis of the opposites. When you partake of the magic mushroom or the sacred cactus sauce in the sacramental dialectic mood, it is a feast of spiritual gastronomy and you understand the secret of the universe which can be expressed in a simple motto: 'Love not Logic.' " "Love eh?" grunted Blood. "That's why your baboons carry bicycle chains."
>
> Petitjacques smiled with Mephistophelean benevolence: "The medium is not always the message. The Apocalypse must precede the Kingdom. Chopping heads is more effective than chopping logic."

Historicisms of one form or another must go to awkward rhetorical lengths to balance philosophical contradictions and inconsistencies. And in the end, expediency repeatedly and comfortably wins.

6. I should make it clear that I speak here of an extreme type. In the broader sense I mean by careerist only what Grant and Riesman (1978) have described.

> Especially in the great expansion of the fifties and sixties, people had come into academic life with very little sense of vocation, but often with the expectation of a quite obvious step upward in the economic system, as well as the sociocultural system—with no calling for either teaching or scholarship, or any great interest in doing more than keeping their jobs.

Most such people seem to want to do a good job and can become reputable and honest researchers in their specialty. They do tend to have loyalties more to administrators, their specialties, their consultantship ties, and granting panels (or rarely to alliances with an ideological or political clique) than to history and society, and this leads to profound complexities when a faculty professes to be governing itself for the good of the ideals of the larger institution.

7. Some argue that the very complexity of modern life and the press of vast masses of human flesh require us to depersonalize and dehumanize each other and that dehumanizing ideas and language are convenient and appropriate to the world we live in.

Even if there is this apparent necessity, or path of least resistance, then at the individual level those of us who want to cultivate intellectual autonomy need not let convenient ideas and language talk us out of these goals, on the mere commonplace that they are *unattainable* by individuals willing to work hard for them. Moreover, I am not convinced that crowding and complexity *must necessarily* depersonalize. This position seems not to allow for imaginative solutions, and to me the question is how we might adapt to a complex world. Could we use our creative potential to find ways to permit ourselves to think of one another humanely in a complex society, or will we allow the system to follow a dehumanizing path and turn us into docile, if frustrated, impersonal elements in it? One may recall the Ik, in Turnbull's *The Mountain People*. Starvation drove them to become evermore competitive, selfish, mean, nasty, and unloving to nearly any degree. But not all poor or starving peoples have let this happen to their humanity. Obviously here I can only scratch the surface of this profound and complex issue.

8. In *Forms of Intellectual and Ethical Development in the College Years*, Perry reports that in their simple form, the sciences seem to offer refuge to personalities that have problems

fully maturing along a path where in the later stages one deals well intellectually and emotionally with the shades of gray in life. When we are younger the tendency to see things in black and white is great, and with challenging experiences some grow out of this, while others do not. It takes both intellectual talents and emotional resources to confront squarely a world of gray, shifting categories and events. Oddly, the sciences and simple religion, in which the realities seem to be in tidy clean *ideas*, may attract similar personality types in this regard. The naive and strident call for "scientific rigor" (meaning some simple formula for research) in the social sciences and humanities by some types of scholars may have as much to do with personality as with the inherent merit or applicability of particular methods and doctrines.

The ability to withhold judgment when necessary, to be able to live with years of uncertainty, was once regarded as essential to good scholarship. Now, pressure to produce papers quickly may have actively selected, or at least opened the door wider to, a type of can-do personality who is not only not shy about looking at the world through a small window, but who actually sees no troubling loose ends.

This interpretation could explain why so many bright and vigorous academics paradoxically become so seemingly disoriented when it comes to complex but important issues. They may be mentally quick and have enormous energy but seem to find ways to avoid getting deeply involved in some of the more important but for them vague and difficult issues.

9. From a guide for Lutheran missionaries in New Guinea (Frerichs 1957): "It is absurd to say that, when the missionaries launched out into the deeps of New Guinea's heathenism they found a south sea island paradise. . . . Before Christianity came to New Guinea the Papuan had no answer for the three basic questions of life: where he came from; why he is here; where he is going. Without the right answers to these questions there can be no true happiness." So while life in New Guinea, as elsewhere, is surely not bliss, there is a substantial element in the guide of not being able to allow that the people may really be rewarded by the values that they have developed, since these are not ours. The guide thus encourages the introduction of materialism, consumerism, and the Protestant work ethic. "He has plenty of land that produces well the year around . . . he is a landed gentleman with few economic worries. Work is considered a necessary evil which must be indulged in to a degree if one wishes to live. But there is no point in going farther. In order to help change his thinking the white man has brought in very attractive trade goods, clothing, tools, etc., which he hopes will cause the Papuan to overcome his apathy and get out and work in order to acquire some possessions."

From an earlier guide for Baptist missionaries in India during the period of colonial expansion (*The Christian Conquest of India,* Thoburn 1906), spiritual salvation takes on an even more nakedly ethnocentric meaning. "The gold in California, Alaska, and in the Klondyke, in Australia and South Africa, was kept from the eyes of aboriginal races and of Spaniard and Russian till these regions could come under the control of this one great Protestant race." Now, "the English-speaking race" is "moving forward, through Christian and missionary agencies, to bring the millions of India to share in the same liberty, enlightenment, and civilization to which the religion of Christ has led the Aryans of the West." Of course, the net result of such attitudes was too often in fact complete disaster for the "aboriginal races."

To take a related issue, when Lowe (1982) states the following, he cannot possibly mean by alienation what most of us would mean, a conscious or subconscious discomfort in not belonging.

> Marxism considers . . . consciousness to be superstructural, i.e., conditioned by the economic substructure. The underside of consciousness, Marxism argues, is alienation, the key to being-in-the-world. Consciousness, according to Marxism, is eccentric in refracting the multilevel structure from the top down; superstructural consciousness is always at variance with economic substructure. Alienation is the fate of the human being stretched across the multilevel structure of totality. Only the

unity of Marxist theoretical consciousness and revolutionary *praxis* can overcome alienation.

Surely there are many peoples who *could* better themselves by becoming aware of some economic trap that their culture has fallen into. But does unhappiness and emotional alienation *necessarily* follow from higher order socioeconomic ignorance? Certainly, the usually happy tribal peoples of the Pacific and elsewhere that I have met, innocent of Marxist theoretical consciousness, did not feel alienated from family, friends, or nature. I am not an advocate of the idea of a noble savage (attributed to Rousseau) or any other such simple stereotype. We must keep in mind that there has been fear, suffering, war, rape, and so on among tribal peoples in the Pacific.

But I have seen firsthand that stories of tropical paradise are not based on nothing. It is not simply that lonely explorers and wishful-thinking romantics choose to see these people as blissful in their innocence. Tribal peoples have developed often quite fair political systems and found ways to keep their populations in check and appropriate to their physical and emotional resources.

Admittedly any group may fight with neighbors and spirits now and then when a pig is missing, and in some areas there is nearly constant bravado with hand weapons, fear of neighbors and spirits, and occasional killing. Indeed, it was when I saw firsthand beyond the stereotypes, the needless fears of spirits and neighbors, and the hostilities that this could cause among basically warm, bright, and well-meaning people, that a major organizing theme for this book was suggested to me. But on the whole and when left to themselves, they have had a remarkably firm sense of peace and belonging. If they have been living in a dream world of false consciousness then at least it has included good health, happiness, love, *belonging*, time to relax, diversions, and personal challenges, and they are often remarkably astute and critical.

One meets those various religious and secular missionaries who argue that these people have only imagined that they are happy, and that they will be *truly* happy only when they have television or ideological indoctrination. It is hard to see that they can only be more happy and "connected" by learning that they are alienated and eccentric in refracting the multilevel structure of totality from the top down, but can overcome it through revolutionary *praxis*. Or, again, that they can only be more happy and "connected" when they can be taught shame and convinced that they are sinners but can be saved, or that they are disgustingly backwards but can get loans to catch up.

CHAPTER 10

Intuition in Science and Eastern Disciplines

Intuition attracts those who wish to be spiritual without any bother, because it promises a heaven where the intuitions of others can be ignored.

E. M. Forster

Disciplined Intuition Is Vital in Science

Many aspects of science have been discussed in previous chapters. Here I want to explore the point that traditionally creative science—indeed rationality—has not been simply formal and impersonal method. It has involved the whole individual and a disciplining of the intuitive abilities. The interplay between intuition, disciplined thought, and formal proof is always complex, and it is necessary for one to think about this with regard to the cultivation of our own abilities.

A *creative* modern scientist may depend almost completely on a highly prepared intuition for the genesis of his or her contributions. (Much routine or *normal* science is not necessarily creative and does not involve the sort of discovery that generates broad new perspectives.) By intuition I refer not to metaphysical revelation, nor to philosophical "intuitionalism" (absurd—Bunge 1962). I have in mind only subconscious and/or subverbal processes of the brain that follow covert logical programs as the brain tissues process information. Our brain takes a two-dimensional image from each eye, for example, and combines them into a three-dimensional experience quite beautifully without the help of conscious logic, but the brain cells conduct some covert logical organization of the information. There is no reason to imagine anything occult about this. We might not be able to write out, according to a logic, how we depend on intuition to move each muscle to keep our balance on a bicycle while we concentrate on traffic, to decide in a second that something deserves a laugh and, in musical improvisation, to decide what our fellow player is apt to do with a theme or rhythm in the next few seconds. There is no reason to suspect that any of this is a metaphysical phenomenon.

Intuition often exists in the service of superstition, craziness, and stubborn common sense. The history of science and the West has seen splendid reason and evidence fail to convince when commonsense intuition elevates itself above them. For many thoughtful people intuition has thus come, and for good reasons, to mean the opposite of rationality. It is the shadowy Neanderthal in us. The Hyde in Dr. Jekyll. The fool in the person of honor. It can indeed be an awesome obstacle to any hopes for personal or social progress.

We may think of the tragic individuals who believe what they believe so firmly that nothing can stop them—serene and self-confident, town heroes to the kids, perhaps even the local trustees of common sense; but in a crisis the reason that they use, the evidence that they point to are means to the end of defending and promoting what they see with their guts to be true. They have no doubts, see no need to subject their beliefs to any tests that might falsify them. They seem to know just where they are going, they can charge ahead, and nothing will distract them from their agenda. And if they blunder, we may still want to admire them for their decisiveness and character, and we mumble that anyone can make mistakes. A child may see such a person as a fine one to emulate, someone to look up to, a leader even. Even once we mature and learn to value caution and reason, such persons may still demand a place in our hearts and leave us with mixed feelings of respect and at most pity rather than amusement or disgust. Part of every person who strives to be rational is at war with such nostalgia.

A raw intuition may cloud our attempts at objectivity and mock our dreams of utopia. But there is every reason to believe that intuition is not an *essentially* chaotic force, nor one *necessarily* antithetical to rationality. It can be cultivated and used well or badly. It is a complex, vast, and powerful dimension of our minds that is simply more difficult to think of and to understand and use than are conscious verbal and symbolic reasoning (which are also easily misused, to be sure). If we should all have to stop and think consciously about each muscular movement and each decision in life then we would all die in traffic. Intuition has generated great truths as well as silly superstitions and disastrous decisions. The intuition can be disciplined, as we see in such mundane things as driving a car, debating, or training for military combat, where one begins to think and react like a soldier with one's guts, and not by consciously reasoning out each decision. It becomes disciplined to more sublime potentials in the great musician, the Zen master, the creative mathematician, and so on. It is an important part of human being that we should not neglect.

None of us could play good improvised jazz or *ragas* by raw intuition alone. It takes years of listening, familiarity, thought, and practice to get to the point where in split seconds one may be able to anticipate the other musicians' musical ideas and to have inspired responses to what one has just heard. The musical intuition is highly disciplined within the framework of a tradition.

There is nothing mystical (in the common sense of the word) about this. There should be nothing mystical in asserting that intuition may be an element in every sort of creativity, including creative science. But this is not the layperson's *image of science*, and if one mentions intuition in science it may seem as though one is being mystical or evasive or holding something back. To understand why we do not talk much about intuition, one must understand that the public and impersonal facade of science is *critical* to its proper functioning.

For scientific ideas to mean anything that the community can work with, they must be put into public form even if they *come* from the most cultivated private intuition. The scientist must begin to articulate in "the language of proof" (chapter 11). The formulation of the theory, its correction or refinement through the careful use of data and experiment, the selection of experiments to embody it and to underscore its robustness, may likewise represent intuitions about both the phenomenon itself and also about the state of mind of other scientists. Galileo had to devise particular thought experiments and other methods that would be meaningful to other scientists whose minds were filled with Aristotelian thinking. What arrangement of empirical data, facts, experiments, arguments, calculations, will communicate an insight to others who are searching for truths in the given area of interest?

Most scientists try to be convincing. This may look like advocacy. There is a very fine line here, for *responsible* scientists prepare a strong and honest case to make others *understand* the new idea, not to mislead and not to create believers. They word their case carefully to make it clear that the idea is not a fact that they as experts are vouching for, no matter how convincing this case may be, but it is a suggestion or point of view offered for examination by the scientific community. They should cite the important literature prior to them and make controversies clear so that the reader becomes aware of other opinions. Usually they are writing for an informed audience and only need to belabor certain points, which can create some misunderstanding with a novice or layperson. Sometimes a reasonable and apparently well-supported argument seems so convincing that it is taken by others as fact, but these should know better. The point here is that scientists try to use their intuitions to build public conceptual models that will stand or fall in terms of their own architecture and that others can test and build on.

Scientists try hard to harness their valuable intuition and to validate it and enclose it as much as possible in the "language of proof" so that truths *can* unfold in a logical manner, so that they have a reality in formal empirical reason. Indeed they *must* actually do so at this point, but the power of the public presentation will, and perhaps must, cloak the actual creative process. The public sees new truths apparently emerge purely from technique. But much of the time they see only an illusion.

I discovered a refreshing book, *The Art of Scientific Investigation* by Beveridge, a contributing scientist that discusses realistically, concisely, and in some detail the relative roles of intuition, reason, experiment, and chance in scientific discovery. I can refer the reader (certainly the scientist, but the general reader as well) to this for more information and will try to repeat as little as possible of its discussions and literature citations.

There is no way to know for sure how many *major conceptual* breakthroughs have come from technique alone, but I suspect few. Colleagues in mathematics tell me that insights *never* seem to come from technique. They come from a highly cultivated intuition that is checked, expanded, redirected, or discarded along the way by formal methods. The formalisms are necessary to communicate to others and to oneself, and to prove or disprove the point in a public arena. From personal experience and from talking over the years with scientists who have done important creative work, I think that our situation in science has long been roughly comparable to that in mathematics. We have depended very much on the flash of insight, or perhaps the slow flicker of comprehension, that gets one started thinking in a new direction, or thinking systematically about a mystery; then one uses formal methods to refine or reject the thoughts. The skillful *selection* and *use* of formal methods may also depend on intuition until a final structure is built that will meet "intellectual safety standards."

So, we may depend very much on intuition for ideas about the best calculations or logical arguments to use to follow a hunch or insight. We modify our path as we rush or plod ahead, and we depend on intuitions as well as on formal tools of method and technique, statistical methods, for example, to tell us when to shift direction.

We must decide how to design good experiments, and intuition comes in again. We must decide whether or not to accept the results of an experiment or whether to suspect a contamination or error in the design, and intuition comes in to play again.

Few leading scientists, and probably no creative scientists, could function with technique alone and without keen wits. This is why creative science cannot be reduced to a method or *algorithm*. Even if one advocates falsification, for example, as *the* method or essential feature of science, as Popperians have tried to, then someone must create valid and ingenious tests that can be used to falsify and reject or not a substantial hypothesis. People must be convinced that the test is valid, and so on. How do we decide if a hypothesis has indeed been falsified? Such decisions involve extremely well-informed and careful value judgments — but they are by no means arbitrary or impulsive as the term *value judgment* often implies. And one cannot turn these tasks completely over to a computer using any of the the sorts of formal logic that have been invented so far. What is important is the *attitude* and determination to evaluate as fairly and as carefully as is humanly possible and to use all means at one's disposal. If one tries to reduce that attitude to a presumed "essence" or form, the ritual of "falsification" or whatever, and forgets the spirit in which scientists have always tried to do this, then one risks throwing out more than one can ever gain.

But little if any of that is seen in public. The artistry may not even enter the consciousness of some scientists. If they are not introspective people, it may take a few beers and a deep discussion to make them ponder how they came to do good work. In one poll, 18 percent of the scientists interviewed claimed never to work from intuition. Along with Beveridge (1950) I suspect that many of these, even if they were doing "routine" or "normal" science, may simply not have thought carefully about how their minds do work. Like driving a car, the research enterprise may become automatic and not something one analyzes constantly.

I suppose that many artists forget that they have mastered technique and would imagine on first thought that they just let the ideas flow out and take form intuitively — since that is what many say they do. Some creative scientists similarly may at first think that it is their method that generates the creation — rather than a wedding of method to obscure factors like comprehension, imagination, intuition — since that is what many say they do.

This pretense of a sort is not all bad. It appears to be in fact desirable, esssential, and unavoidable. It keeps standards for data and critical evidence high and is a constant important reminder to scientists of the public and historical character of science. High standards for data have become extremely important in other areas of life, and we should keep our courts in mind. How terrible it would be not to have good rules of evidence.

A mathematician may invent a proof and of course it is *he or she* who has proven the theorem. But one must be able to present one's equations and say "They prove it" — and the equations *should*. This largely is what is meant by *objective*. "The published proof is for everybody," a mathematician colleague confided. "How you *really* did it you save for your students and the initiated who can appreciate it." It might seem illogical to some lawyers or philosphers that the subjective could generate the objective, but in large part this is what happens, though it is not a simple transformation.

This distinction between the domains of the formal and the informal is critical and should not be hard for the mature researcher to live with. The problems only come if one starts to believe the "pretense" and think that formal proof is how fine mathematics or science actually is done. Scholars seeking to emulate fine science may give up on interesting problems if they do not see a clear way to use "the method," experiments, statistics or other specific, supposedly objective techniques. But they may well keep in mind that creative scientists modify and invent methods as necessary and that a great deal can be done with open-mindedness, imagination, careful organization, preparation, a passion to weed out myth and find the truth, a careful weighing of alternatives, the use of all available evidence from different categories ("lateral thinking"), logic, a checking of assumptions, findings, and predictions against reality, and so on. This is empiricism.

Even when I publish a "theoretical" paper with no data or experiments reported in it, laying out the logic of a hypothesis, the hypothesis typically has been backed up by an extensive amount of empiricism. I have checked the original intuitions against critical observations from the real world and modified them if necessary, made predictions, and then gone back to the material world for more observations, checking of predictions, and so on, over and over. Sometimes this takes me years, and thousands of miles of travel and many hours of work, reading, and conversation. There are many reasons why one might not put any of this background material into a theoretical paper, but the point is that valuable theory may be generated by some workers in a process that draws vigorously from empiricism even if the connections are not obvious in print.

Facades of Wisdom East and West Compared

In both the East and the West there are quite different systems of discipline in which human intuition is brought into the service of the individual, producing supposed wisdom. (I am not talking now about metaphysical revelation.) Interestingly, in both East and West the full situation tends to be hidden behind one's ideal image of the final product. In Eastern "no-mind," understanding is said to come because one stops trying (chapters 7, 8). This is true in a sense, but on the other hand it may take much difficult study and preparation and *then* one may give up trying and no-mind may come. A master may claim to be striving to become an idiot, but this is also a sort of misleading pretense. A natural idiot may have no way of dealing with all sorts of passions, fears, and confusions that stand in the way of peace and enlightenment. A natural idiot would still be vulnerable to many of the propaganda images of expansionary ideologies and religions, popular enthusiasms, seductions of mass advertising, manipulations of selfish egos, and self-deception. A natural idiot will not have the highly sophisticated martial arts abilities that advanced monks may cultivate. Freedom from thought is a very sophisticated state, quite different from mere ignorance, indifference, or escapism.

Creative Western scientists also use intuition in a very disciplined manner, but may explain this in an altogether misleading way, perhaps even to themselves. The scientists who are the most creative are not so simply because they have mastered "the method" any more than karate experts can break boards with a blow because they are mindless idiots.

Because of such facades the public, and even aspiring young scientists or Zen monks, may never appreciate that the roots of creativity and mastery of craft are very much the same in art and science or any other demanding human endeavor. In creative science there is an essential and dynamic, and to me quite wonderful, interaction between the public language of proof and the intuition. By being fed with the better examples of thinking and information, and with a deep knowledge of a subject, the intuition becomes *prepared* to generate its flashes of insight and sober, sound judgments (along with some crazy notions, which can come from the same energetic imagination). Formalisms then can serve as a check on the tendency for one to delude oneself; they help to weed out the crazy notions and to spot the unpolished ones. If one is willing to learn from this and does not merely become discouraged, the intuition becomes disciplined and enlarges its scope and abilities.

Zen in the Art of Archery by Herrigel (1971) has become a classic. The bow used is long, elastic, and clumsy. It will not work to apply techniques and expect to be accurate with it—to draw and aim it just so and imagine a geometry of which arrow and target are a part. One does not attempt to master such a bow. One plays with it and lets the arrows fly where they will. One relaxes one's breathing and muscles and becomes comfortable with one's body and with the ways of the bow. Eventually one can begin to hit the target, but it is through an automatic, intuitive feeling for the bow, whereas formal technique failed.

It would be a mistake to assume that one can simply play with a clumsy bamboo bow and become accurate. The play and relaxation depend on a great deal of instruction from a master. After all, one must correct a lifetime of habits and attitudes. There is no pure, unspoiled intuition here that can simply be released.

I often see in good, mature scientists and scholars a virtual reflex alertness for possibly useful information, an automatic scanning of possible criticisms and supporting arguments for ideas, a sorting out of facts into firm and soft, the building of a diffuse but relevant intellectual system or foundation for the comprehension of a subject. Then the insights come. Some are tested in the mind and put on a memory shelf somewhere if they have too many rough corners or are really bad. Others survive and become the subjects of sustained enthusiasm and study. Then, with intuition and logic, the scientist starts trying to figure out priorities—what the critical data would be and how much of it can be gotten, what the most elegant experimental tests would be. These in turn must be gradually detailed technically and conceptually, and accepted or discarded when their form is clear. (Is the study really going to work and tell me anything definitive?) And on and on.

This also takes place over the academic generations in fact, for science is a community effort even though one can sometimes see one individual who is good at every phase of the process. More often, one worker has a nose for theory or thought experiments, another for empirical experimental design, another for gathering comprehensive and clean data, another for calculation, another for comparative studies, another for organizing timely symposia, another for synthesis of several lines of study.

As a practicing scientist I deal daily with real investigators and students having different levels and sorts of natural intellectual ability, different learned intellectual habits, different sorts of courage and drive, different levels of open-mindedness, different sorts of both blind spots and insights, all of which result in different patterns of promise and accomplishment. For example, Galileo's methods give only limited insight into him as a scientist. He had, in addition to brightness, dedication, courage, organization, open-mindedness, an intense interest in seeking truth, and imagination (if he was not pure in each respect, at least he was superior). Such individuals can be misfits in large or mediocre organizations, but nevertheless, if Galileo were a graduate student I would sooner predict from his human qualities that he would make important contributions than from the methods he had mastered in his earliest years. A given genre of methods does not invariably generate a given set of insights in science, as philosophers now comprehend.

Barbara McClintock discovered the important fact that genes will jump about between sites on chromosomes. But her work was so far ahead of its time that it was not adequately grasped by molecular geneticists for about a quarter of a century. How does one develop critical insights so far ahead of great numbers of bright, competitive colleagues? I was fascinated to hear McClintock lecture on how she made her historical, Nobel prize-winning discoveries. With her intimate knowledge of plant biology she had developed a keen eye to spot a mystery. She would intuitively grab knowledge from here and there (a *lateral thinker* before the term had been invented), follow hunches, devise critical experiments, reach back into her treasure chest of basic biology, ponder odd facts of life, emerge again to track down the critical detail that would tell her if she was on the right trail in finding the solution to the mystery. (Keller's biography of her is tellingly titled, *A Feeling for the Organism*.)

After the talk I hung around to see what the audience thought, chatting with graduate students and professors. I was saddened to hear many negative judgments that her work "didn't seem very well organized. She wasn't always clear about her hypotheses. Her approach seemed diffuse."

We had been privileged for her to allow us to glimpse the workings of her creative genius. It was blindingly obvious, and yet many in the audience had been so completely indocrinated in the ritualism of technique that they could not see what she was trying hard to say about how *great* science is *really* done.

Much of a generation had overlooked the significance of her work. Now she had become a living legend. She would try once again to enlighten humankind, this time to tell it by example something about the passion to know, armed with the knowledge of those who came before, fired by the courage to invent, to imagine, to harness the intuition, to defy orthodoxy, to demand sacrifice of herself. "She didn't say much about her statistics," was all one puzzled graduate student seemed to get out of the magic hour.

Western scientific enthusiasts have long tried to replace the intuitive element in science with impersonal technique. And it gets worse as our universities are run more and more like mediocre corporations, and as the academic management system looks for uniform "objective" standards by which to evaluate and predict productivity. How much data do the scientists produce? How many papers do they publish? How many of their colleagues like them? How much grant money do they bring in? Granting agencies and peers weigh the methods in research proposals heavily. In the end there are conspicuous rewards for mastering popular, easily understood techniques and generating lots of data that people will find respectable in terms of what is conventional. While this is not the only path to respectability and security, it may be the most obvious for many workers. It is easy to see how one could be led to neglect the development of large areas of one's potential, in order to meet short-term institutional demands.

James Watson, a discoverer of the structure of DNA, Nobel laureate, and director of the Cold Spring Harbor Laboratories, outraged many of his colleagues with his book *The Double Helix*. For example, "In contrast to the popular conception supported by newspapers and mothers of scientists, a goodly number of scientists are not only narrow-minded and dull, but also just stupid." I think the issue is more complex. I do not think the faculty and graduate students who could not grasp McClintock's message were inherently dull or stupid. But to the degree to which one sees the problem as one of dullness and stupidity, the system of values, perspectives, and institutional demands actually does an enormous amount to encourage it.

Fine scientific thinking demands an interplay of the formal and the intuitive. If one rejects the formal aspects, then one sacrifices the opportunity to augment and train one's thinking with a powerful rational ally. If one rejects the intuitive aspects or takes them for granted, then one sacrifices the opportunity to augment one's thinking with the most powerful, and presently quite irreplaceable, aspect of one's biological endowment.

Sex may be critical in holding the family together. Nevertheless, in our culture we keep our privates covered in public. Some children even assume that their parents are not sexual. Intuition may be critical in science, but we try to keep it covered in public. There are historical reasons for both customs, but in any event the stork does not bring babies.

Humanists may see too little of the healthy side of the scientific mentality, and their intuition and experience tell them that formal logic or the reductionist study of nature alone cannot possibly formulate or answer great questions or lead society in a wholesome direction. Some may have little faith in "scientific" thinking for this reason. They would be *correct*, I would completely agree, *if* the premise were true. But the formalisms are only a portion of fine scientific thinking. They are only the part that scientists must put out front in public. They are the part that is easiest to discuss philosophically or in classrooms where their formal structure can be outlined and analyzed and memorized and quizzed on. A discussion of intuition can get excessively gossipy or may seem superficially mystical—so there is a tendency to avoid it.

Nevertheless, scientific thinking as it has *actually* been done by creative scientists can in principle be valuable to humanists.

The system of higher education and books in the West is developing as one largely to teach and supply *things*: facts, concepts, and marketable skills. It sells these things as goods to consumer students and readers. This is consistent with the prevailing commercialist outlook and value system, and so the situation is unlikely in the main to change. The challenge would be for individuals and small groups interested in humanistic growth to find personal ways to develop latent intuitive abilities, including critical thought habits, lateral thinking, and a greater reflective sensitivity to one's personal energies and creative possibilities. These ways might include not only thoughtful reading, discussion, and writing, but *experiential* explorations of the opportunities to deal constructively and creatively with life, and continuing reflection and re-reflexion on the subtle relationships between self-honesty and the possibility of personal growth.

CHAPTER 11
The Language of Proof

> *If our discovery has a commercial future that is an accident*
> *from which we must not profit. And if radium is to be used in*
> *the treatment of disease, it seems to be impossible for us to take*
> *advantage of that.*
>
> Marie Curie *(in* Marie Curie *by Eve Curie, 1938)*

> *To understand where we now stand, we must realize that we are*
> *living in the midst of a virtual biological gold rush ... where the*
> *rules for decent behavior in more civilized territories are not*
> *easily transferable to the new frontiers.*
>
> James D. Watson, *from "For a few dollars more."*
> Biotechnology Education *1989*

Critical thought is not simply to tear things apart or to be judgmental. At its best it is constructive evaluation that can lead to the development of improved and more useful theories. "Critical thinking is not a characteristic of Western thinking just because it is Western; nor is folk thought uniquely characteristic of non-Western thinking" (Harding 1986), but the most powerful critical thought about "material" systems (with utility even for psychological, social, and ethical systems) did develop *in* the West, out of its philosophical and scientific traditions. At times it has had a tenacious following in the West, as among some advocates of the liberal arts ideal. One might argue that it is vastly better for each of us to go through life with our own tentative, evolving set of beliefs based on critical analysis than to go on defending confining belief systems merely because they have come down to us.

Critical thinking involves a sense of how best to use both evidence and logic in interactive ways, of how to construct robust conceptual models and test assumptions, of how to make careful observations, of what is physically reasonable, of the utility of relative certainty and of tentative hypotheses, of the nuances of probabilistic thinking, of psychological dynamics and pitfalls, of both the values of accurate historical information and its limitations. It is open and *self*-critical. It is constructive and strives for improved theory.

Now, following the discussions in chapters 9 and 10 of the possibility to develop individual potentials, and of misunderstandings concerning the roles of intuition and method in scientific and Eastern disciplines, let us return to some issues raised in chapter 8. What is good "theory" and how did its traditions develop in the West? What is "the language of proof" that began to develop among theorizers in the sixth century B.C.?

Reality in Western Philosophy and Science Is Abstract

I will elaborate first on what it means to say that important concepts are abstractions, because if one is going to understand the history of Western philosophy and science one should reflect on this (keeping in mind how other traditions broadly have dealt with abstraction, with "reality" beyond the senses. chapter 8).

Science Is Essentially Theoretical. Not all scientists today deal directly with the formulation of theory, but Greek and modern science have historically been preoccupied with refining views of the abstract nature of material things, the ideas about things, the reasoned properties of things, diverse abstract "models" of material reality (bear in mind that there are several senses of models—heuristic, predictive, representational, experimental, etc.). Science has long tried to probe (or construct) the abstract nature of the material world systematically, using reason and evidence—it is *both* rational and empirical (though as used in *philosophy* these are opposite terms). Most of the terms in science are not just fancy words to describe the commonplace or even "hiddden truths," they represent abstract constructions—*theories* in the proper sense of "a mental viewing," "a looking at," "a contemplation," which is somewhat different from theory as *speculation* in popular usage. (Similary, a *theorem* is "a proposition that is not self-evident but that can be proved from accepted premises and so is established as a law or principle.")

One should keep in mind that *anything* described by language is theoretical, since the categories and hierarchies of language are ways of, constructions for, looking at things. For ease of discussion, though, I mean here that science deals with formal concepts that are abstract relative to the things that civic discourse treats as directly observable.

Heat is *not* at all the same as temperature, mass is not the same as weight. Temperature and weight are formal quantitative estimates of commonplace sensations. But heat and mass are conspicuously theoretical formal constructions, or ways of viewing matter mentally, and they greatly increase our understanding of thermodynamics and mechanics.

Some scientists may sputter at this point, "But active transport across cell membranes is *real*. I can measure it. I can slow it down, step it up. It is *not* a mental construction, it is right here in front of us." They have a very good concept and don't want it confused with speculation. It is as real as things get in a world where we don't really know the minds of the people who lead us or who decide what is shown on television or what we read in the newspapers, what fashions and life-styles should be in or out. In much the same way, Darwinians get irritated and see word games being played when Creationists say that evolution is "only" a theory.

The point is that heat, mass, active transport, and evolution are meticulous, robust, formal constructions that cannot be seen with the unaided senses or even with any powerful but simple extension of the senses. The little jumps on the dials must be interpreted by a mind that understands that under the tightly controlled conditions of observation, the concept of active transport is indicated. The patterns indicated in the comparisons of the bones and soft tissues of species and in their biochemistry are best explained by the mind as the workings of evolution, but the unaided senses do not see evolution any more than they can see that the sun is much larger than the moon. That meticulous construction, that fact, too is only known from the data, through reason.

At some point it may have become clear that a scientific concept is so robust that the term *theory* may be dropped by convention, but this is not dictated by rules. We still refer to the heliocentric "theory," but that does not mean that it is mere speculation that the earth revolves around the sun.

Nature Becomes Seen as Orderly and Logical. The earliest abstract worlds that people constructed may have been characterized by narratives about, and personality profiles of, gods and spirits. We are most interested here though in the construction of abstract worlds where abstract powers seemed to manifest themselves according to logical necessities that seemed even to determine the relationships between numbers and geometrical forms.

If nature was originally seen to be moved by gods with personalities and passions then, for philosophers, by Aristotle's time it was seen not to be composed and moved by divine whim and intrigue, but moved (or inspired to move) by divine *reason*. There was not a simple progression in the history of such thought, but one can say that as thinkers were becoming more logical in their own thinking, there developed a new anthropomorphism, and schools of philosophers expected that nature, as the most certain man-ifestation of the divine, must be rational, logical, and deliberate in its "es-sential" nature. Even if it was a soul that accounted for a certain phenome-non, then it must be a soul with a simple and rational nature, not the impulsive scheming, affection, or rage of an Olympian.

In *Metaphysics* Aristotle allowed, "It belongs to the student of nature to study even soul in a certain sense, i.e., so much of it as is not independent of matter." Scientists would not try to study soul today because it was found to be an unnecessary and unproductive concept in understanding nature. So Aristotle is out of step with modern science when he assumes that meta-physical causes are active in nature, but the *motivation* to discuss system-atically and critically and to explain things in the most simple and yet accu-rate generalities possible is there in Aristotle nevertheless. He had the important attitude that nature had an orderly and basically simple, imper-sonal (even if divine) character. This could be determined by "the language of proof." He had the attitude that through systematic study and analysis, be it largely empirical or largely deductive, rather than through divination, the nature of things could be understood.[1]

Extending the Idea of Proof Beyond Geometry and Mathematics. The Greeks did not invent mathematics, or even geometry, or even mathematical *proof.* There were mathematical problem sets and there is direct and indi-rect evidence of formal proofs long before Thales or Plato (from India, Babylon, and Egypt), and no scholar any longer advocates the "Greek mir-acle." It now seems that the Babylonians knew the theorem of Pythogoras at least a millenium before Pythogoras lived in the mid-sixth century B.C. What is most interesting is how Greeks transferred the idea of mathematical/ geometrical proof to the building of abstract models of material reality.[2]

The Greeks excelled at a type of reasoning known as abstract geometry, until recently taught as grand training for the mind, that tradition says was invented primarily by Thales (in the early sixth century B.C.).[3] It is particu-larly good for the proper exercise of deductive reasoning, which allows new statements to be generated from accepted facts. Thus, there is the possibility of step by step reaching mentally beyond what one can sense and measure directly. With this powerful reasoning tool, and thought habits modeled af-ter it, the Greeks could hope to explore with some precision the supposed Truths beyond the senses that they and others so intensely anticipated.

Even though we live upon it, the earth beyond one's travels is an abstrac-tion, and in the earliest days there was much speculation about its form and the form of the heavens. Babylonians, Egyptians, and Hebrews saw, in vari-ous versions, the heavens as a vaulted roof, or firmament, that supported the stars and gods, and the earth as a flat, bulged, or bowed floor beneath this. To these local beliefs Greek philosophers added such ideas as that the world was a cylinder or even a sphere within a sphere or spheres that supported the heavenly bodies. Could one decide if one belief among all these was true and the others wrong?

Aristotle observed, for example, as part of a series of integrated observations, that the shadow of the earth on the moon during an eclipse is round. The earth *must* be round. Other empiricists used geometry. They extended the certain logic of geometry to the heavens. They began actually *to measure* nature to find out what the Truth is. Many readers will know some geometrical Boy Scout tricks for measuring the height of a tree or the width of a river using only a small stick, knowing the length of one's pace, and such. Eratosthenes actually might have made a fairly accurate *measurement* of the earth's circumference by measuring shadows and angles at different points in Egypt. His methods were good, but we aren't certain about the distance units that Greeks used. Aristarchus attempted pretty reasonable measurements of the diameter of the moon, and perhaps as important, he proved that the sun must be many times farther from the earth than the moon is. He also *suggested* that the earth rotates about the sun, though he did not present his evidence for this. Hipparchus's attempted measurement of the distance from the earth to the moon was rather reasonable. Astonishingly, by comparing older data and being able to evaluate their reliability, he also discovered the precession of the equinoxes (such that Polaris cannot always be the North Star) and he and his colleagues correctly estimated the cycle to take about 26,000 years.

These sorts of studies differentiate Greek astronomy from other intensive studies of the heavens. Others apparently did not work so hard and ably to figure out the actual shape of the earth and heavens. They were content to keep careful records, make calculations, and think about correlations between events, or beyond these they were content to speculate or to cling to traditional beliefs, or to seek inspiration in dreams or from prophets.

Aristotle dismissed the mythologists such as Hesiod, but paid careful attention even to those philosophers with whom he disagreed, " . . . into the subtleties of the mythologists it is not worth our while to inquire seriously; those who use the language of proof we must cross-examine. . . . " (*Metaphysics*). Geometry carries the powerful idea of the possibility of the public *proof* of a proposition that attentive individuals must agree on and that then may be said to be true for all persons everywhere. Mathematical insights may, to be sure, *originate* through the cultivated intuition, and there is much room for art in this respect, but they must be *demonstrable* through formal logic that others can follow and try to poke holes in. Thinking based on a model of critical logic and proof offers the hope to build a solid bridge across a chasm of ignorance, confusion, and illusion that separates one from truths or Truth that one suspects must exist on the other side. Indeed, one keeps finding truths and this sustains faith in the approach. (But there are no guarantees of unconditional Truth. Recall after centuries the discovery of a non-Euclidean geometry that revealed that Euclidean Truths are conditional.)

Disputing Truths. The claim of absolute Truth based on proof has obvious power implications and makes an attractive prize or a compelling target. It seems that claims of "one Truth" can provoke strong philosophical challenges. Faith in the idea that some form of absolute Truth exists can be manipulated by bureaucratic or authoritarian power structures to gain control of the population through the deception, and perhaps self-deception, of rationally sanctified absolute moral authority.

Tribal democracies must negotiate between differences in perspective and this can get clumsy as groups grow larger. A language of proof could aid the negotiation process, help to weed out flimflam and spur people to sharpen their conjectures. But then it could also become misused and co-opted by power-seeking forces. It may also seem to some to disqualify the passions. Demystification can be a double-edged sword. I cannot help but think of the struggle in *The Bacchae* between Dionysus, with the unwise energy of emotion on his side, and young Pentheus, trying to be rational and to impose his imperfect rationality, in fact merely his will, on others.

Of possible interest here, Plato and Socrates were suspected of being antidemocratic and even suffered for this in time. Surely their logic has since been used repeatedly to make monopolistic claims to truth and/or spirituality. (See Popper 1950, Stone 1988.) Aristotle was accused of sacrilege and forced to flee to Euboea following the death of Alexander.

Aristotle, in staking his own claims to Truth, seemed in no small part to be reacting against the increasingly popular idea, promoted by Pythagoras and Plato, that numbers and ideas are fundamental Reality. He got at once to this issue in *Metaphysics* and closed this book with the conclusion that numbers and ideas cannot be physical causes or substance.

Somewhat similarly, the often materialistic, logical, and methodical Mohists and logicians in fifth-and fourth-century B.C. China developed interesting small movements in reaction apparently as scholarly Confucianism began to transform into a state religion. Taoism had its origins in shamanism, and much like Pythagoreanism it attempted to develop a systematic "scientific" aspect. It was looking for "ethical" guidance in nature, somewhat as the Aristotelians and natural theologians would in the West. It reacted to Confucianism, which was growing religious and becoming official doctrine. Taoism looked for meaning beyond rituals and the social order, in nature and in individual experience. Relative to sociopolitical Confucianism this face of Taoism was "materialistic."

In India an early materialistic enterprise is known (it was criticized in the Upanishads). This seems to have developed into the Cārvāka school around 600 B.C. This was a major politically ineffective reaction against the superstitions and fears reportedly promoted by the brahmins to consolidate their power. Late Cārvāka texts argue, for example, "Hence it is only as a means of livelihood that *brahmins* have established here all these ceremonies for the dead—there is no other fruit anywhere" (*Sarvadarsanasamgraha*).

Even Buddha had been reacting against the grip on people's minds and lives of institutionalized ritual and supernatural claims; but of course his psychological teachings became co-opted by several religious systems. In some, he was even deified, and his psychological teachings were woven in with supernatural cosmologies and devotions.

Scientific Shamans. Authors who give the impression that the philosophers of Greece made a complete break in their thinking between the natural and the supernatural are misleading. Indeed, Dodds and others (Luck 1985) have emphasized that Orpheus, Pythagoras, and Empedocles in their times were seen essentially as *shamans*, with magical powers and the power of prophecy, and many scholars accept that they are indeed best thought of as true shamans. Pythagoras then, would have been a shaman who studied numerology and "divine geometry." Today we classify him as a philosopher because of his influence on Plato and because he was one of the first Greeks known to experiment with the language of proof.

Similarly, the line between alchemy, philosophy, and science was long blurred (e.g., Jacob 1988, Ronan 1982).

Metaphysical notions such as occult numerology were important in early systematic thinking. The Pythagoreans were extremely important to the history of science in their promotion of an interest in numerology and proof in the abstract world. By reducing reality to numbers and forms they helped set the stage for the later emphasis on mathematics, quantification, and abstract models so important to the advance of modern science. These essentials of modern science—an intense interest in abstract models, the notion of the possibility of the *proof* of ideas in the abstract world, thereby the possibility of Truth, a sense of the importance of geometrical and analytical thinking in probing the world of abstractions—were already becoming established, then, in antiquity.

Truth from the Study of Matter. In *Metaphysics*, Aristotle says Thales was the first (Greek) to explain the world in terms of "natural" causes rather than by myth and the supernatural. According to this school one should begin the search for wisdom by starting with what we know best, a study of the permanent substances (physical basis) of things, though, "on the number and nature of such principles they do not all agree" (*Metaphysics*). This makes sense for a mind that also invented Greek geometry, where one begins an analysis with simple facts that one believes one can be sure of.

Most modern researchers typically pick as simple a system as possible and then begin to study its properties in detail in the tradition of Galileo and Newton. This is a valuable part of *methodological reductionism* (though it is not the only valid way to do research). It is not hard to imagine Thales's thinking as being along similar lines. "Let's start with an analysis of something very simple and see what else we can explain once we understand it."

Once Greeks decided to turn away from the passions and exploits of the gods and to begin thinking about what-we-know-best, Thales, Anaximander, and Aristotle might have gone in other directions with the language of proof rather than to concern themselves with matter and formal logic. They might have bypassed matter in favor of numbers and forms, as Pythagoras and Plato did. They might have decided to examine systematically the realities of the strivings and frustrations of the individual human psyche, as Buddha had done earlier. They might have decided to probe the reality of the social order and of the individual's relationship to it, as Confucius had so brilliantly done. The main point is that the epochal systematic program to discover the abstract and general properties of air, fire, water, and earth and such as a way to wisdom was somewhat roundabout to the central humanistic concerns, but it was very important for the eventual maturation of modern natural science.

Greek science continued but declined in creative vigor under the rule of pragmatic Romans. Under the Christian factions who won the early internal theopolitical battles and eventually gained control of the official thinking of Europe, conceptual science was essentially suppressed among the highly educated. This was in significant part because of its various associations with the pagan culture that those Christians were determined to eliminate once they gained political power. In particular, the empirical approach took too much interest in the nature of the world of here-and-now for their tastes and beliefs, rather than the world of the soul and the afterlife. Though much of Greek philosophy was concerned with abstractions, to the Christian mind empiricism was *materialistic*, since it was concerned with physical matter and things that they associated with pagan Dionysian earth spirits, the gaining of knowledge through the senses and such. So out went science, along with things bad, good, and harmless, with the carnage of the Colosseum, orgies, the library of Alexandria, decent sanitation systems, all homosexual and most heterosexual acts, instrumental music, the Olympic games in honor of Zeus, and nude statues.

A somewhat parallel negatively toward "materialist" science by the ascetic religious establishment, along with conditions arising from the caste system, produced an inhibition of, even a forgetting of the history of, Indian science after an impressive early flowering (Chattopadhyaya 1986).

The General Substrates for Critical Thought

It is traditional to look for the origins of science in the knowledge of mathematics, astronomy, and medicine of the most ancient civilizations. But the substrates for science must in fact go back before agriculture and monumental architecture, to the dawn of the species.

Classification into meaningful categories is fundamental to science, and even other species make pragmatic classifications of nature, to identify food and nonfood species, good and bad habitats and neighbors, and so on. Many so-called preliterate or primitive, peoples today have an admirable knowledge of edible and medicinal plants. Indeed, my own work with the Seri Indians began as part of an effort to verify whether their vocabulary for sea turtle biology was based on scientifically valid ecology and behavior. In the end the Seri even led us to new scientific insights, such as turtle populations that hibernate on the floor of the sea. Such knowledge implies traditions of careful observation, debate, classification, and forms of experimentation. Imagine a tribe entering a new area and making careful observations on the natural history of plants and animals, eventually sorting many similar-appearing creatures into distinct "species" with distinct properties.

The discovery of new useful plant species and their classification into reliable categories might have originated by random experimental nibbling. But thoughtful and extensive observations of *animal* feeding behavior can provide a lot of clues about the food and medicinal values of plants. Beyond that, people still must experiment to find how humans react to one substance or another, to learn to remove poisons with heat or leaching. They must negotiate over where, when, and how to go hunting and gathering as resources fluctuate from year to year. All of this implies episodes of critical discussion, however practical the issues.

The concept of measurement is very old, with distances represented in moons, suns, sleeps, and the like. Trade items were long counted and measured.

Among "primitive peoples" there have always been relatively strong critical thinkers interested in general philosophical issues, as well as (in our terms) in practical ones, as for example Paul Radin (1927) long argued in *Primitive Man as Philosopher*.

Alert observation, measurement, and critical thought aims at taking the practically infinite variety of information in nature and arranging it into a more or less predictable framework. In a general sense some humans have always been philosophers, scientists, and technologists, so one must probe further to understand the very special nature of *modern science*, its very restricted origin, and its impact in strengthening critical thought.

Monism and Calendars

It is intriguing that philosophers and scientists for so long held out the fervent belief that reality should be underlain by a single universal principle, system, or force of necessity, though the evidence has consistently been against it. Recall, for example, how the explanatory reductionists in physics for so long were wrongly convinced that *all* of physics could be reduced to the system of Newtonian mechanics (Einstein and Infeld 1966). Recall idealist assumptions of *the* scientific method. Pursuit of various monistic mirages has been a powerful and valuable motivating force in science that has helped us to stumble upon and see common elements among material systems.

I would agree with those who argue that monistic thinking was used to validate authority as centralized states developed—universal explanation from a single point of view. The influence of *monotheism* (established by the time of the Upanishads among Indo-European cultures), a separate and complex if related issue, must also be considered. But I am not convinced that such lines of thought will completely explain the origin of monistic convictions.

People have long worked to place celestial events within a single time system, for political reasons, even though the cycles are in fact independent of each other. Recall the construction of a zodiac and calculation of the moon and sun's movements through it. Could the relative success in reducing heavenly movements to a single system of reference have reinforced faith in an underlying single universal force of necessity, a single universal system or power of causation?

Heavenly bodies have always had spiritual meaning for people everywhere. Presumed gods and spirits were always hard to predict and understand and negotiate with—one day being a friend and the next day making one suffer. Coyote was god, creator, and *trickster* to many Amerindians (Radin 1972) and even the Bible tells one not to presume to know the Jewish God—sometimes horribly punishing and sometimes completely nonpolicing. So people speculate and make offerings and pray and try to shift the odds in their own favor, and it is reasonable to imagine early humans in intellectually active cultures trying to gain precise knowledge of the gods of nature, including those in the sky.

Anyone alert to the sky would notice that natural daily and seasonal rhythms of plants, animals, and weather, were tied to the largely simple movement patterns of the "sky creatures": the sun, stars, moon (possibly, people may long have wondered, the planets?).

Sky watching long had prominent divinational functions, and paleoastronomical knowledge sometimes went far beyond what would have been required for agricultural purposes. Eclipses, long-term cycles of the moon, and conjunctions of planets may be of profound interest to people who see these as powerful gods, but they have no important ecological implications. It seems probable that some ancients were largely motivated to make systematic observations by a desire to better understand the behavior of the presumed powerful forces in their universe. We see a "practical" leaning when the Babylonian astronomers/astrologers recommended, for example, "When Jupiter goes with Venus, the prayer of the land will reach the heart of the people. Merodach and Sarpanitum will hear the prayer of thy people and will have mercy on thy people" (Taylor 1949). Which is perhaps to say, "Last time we had a good response from our offerings to Merodach and Sarpanitum was when Jupiter was with Venus." Even within recent centuries in Islam and Europe, respected astronomers were supported financially because the rich and powerful hoped for astrological insights.

Ancient peoples surely had curiosity (curiosity is a typical mammalian trait) and a conceptual framework for observation and interpretation and even debated their techniques. This is not to say that they were necessarily searching systematically for Truth. Their astronomy may have been more like a descriptive natural history study of the sky, and *quantitative measurements* eventually became one of the most appropriate ways to record the data. At best such is an attempt to develop religious technique, or to organize information. It is not science in the modern, noncolloquial sense. (People will say, for example, that such and such is "down to a science." By this they mean only that the techniques are quite effective. This colloquial usage is worse than useless for historical analysis.) Exactness and quantification did not of themselves give birth to modern conceptual science, and when one ponders the matter carefully it is hard to see how they could have. Similarly, written language was important in the development of science—perhaps necessary, but not sufficient. But such activities were certainly significant parts of the origins of modern science.

Calendar systems may not be so important merely for knowing that it is time to do certain ecological things, such as hunting, planting, rotating crops, etc., as they are for synchronizing the schedules of activity of communities, as anthropologists such as Hogbin, Leach, Turton, and Urton have argued (Urton 1981, Krupp 1983). "We must stop arguing and schedule 15 days of preparations to do such and such on the day when a certain star will

return in the spring"—that sort of thing. Then tradition becomes established. The more complex the community, the more difficult the logistical
and political challenges of coordinating community functions to seasonal
and longer cycles of labor. Some argue that calendars denote civilization.

> The Chumash were settled but hardly agricultural, and their shaman-priests
> kept track of celestial cycles, counted out the days in a lunar calendar, and
> established the time for important public events like the harvest festival
> and the winter solstice ceremony. Many people gathered for these holidays.
> Their impact upon the social life and the economy of the community was
> great. They provided structure, stability, and equilibrium.

> This then is the real purpose of the astronomical calendar. . . . Small farms
> and backyard gardens . . . can succeed without such institutionalized
> timekeeping. Folk traditions, nature lore, and informal skywatching provide
> ample cues to the appropriate times for planting and harvest. It is when
> farming—or any other activity—grows more complex that the calendar
> assumes greater importance. This is a reflection of society's increased
> complexity—not a legacy of discovering how to get crops to grow. (Krupp
> 1983)

Politically attractive fixed calendars based simply on *counting ahead*
days or moons slip out of phase with the seasons. From an ecological perspective, an inaccurate calendar could sometimes be worse than no fixed
calendar at all, as Turton and Ruggles (1978) have argued.

Much of the world today synchronizes its social intercourse by agreement
on an artificially single but useful and workably accurate calendar system
and has resolved most of the political problems that had occured without it.
This *Gregorian* calendar is much better for our needs than the systems civilizations were getting by with thousands of years ago. (Unless one worships
the moon. And keep in mind that before printing it was very practical to
mark time by the phases of the moon.) So it takes considerable imagination
to picture what it must have been like intellectually and politically without
the construction of one or another standardized, predictive, system for time.
The problem is that the observed cycles of the moon and the sun and the
stars slip slightly but progressively out of phase with each other, each following their own time courses. The lunar phases even vary from month to
month (29.26 to 29.80 days), since the moon's orbit is not a perfect circle.
So predictive solar or lunar calendars involve invented abstract systems,
which for political reasons must be granted more reality than they actually
have in astronomical terms. It has been common to use the lunar cycles, the
daily cycle, and the annual cycle separately, since they are in fact not single
natural parts of one time system; this was obvious to ancient people.

A lunation—a month—is a convenient measure of time, visually marked by the phases of the moon, and the oldest, most widely used calendars were and are based on this. Orthodox Jews and Muslims may still use lunar calendars. But there are only about 354 days in twelve lunar cycles, so lunar years slip out of phase with the seasons (based on the *revolution* of the earth) even more quickly than years based on counting ahead the daily *rotations* of the earth. Once one has determined some underlying mathematics, one can experiment with systems to intercalate (as we do with leap years to correct our solar calendar) to somewhat correct the lunar calendar and keep it from slipping hopelessly out of phase with the seasons. Previously, going back at least to the Babylonians, the priests would simply get together and add a month every few years to start the twelve-month cycle over in synchrony with the seasons. Then they began to add a month in versions of a nineteen-year (Metonic) cycle on the third, sixth, eighth, eleventh, fourteenth, seventeenth, and nineteenth years once they were convinced this was a significant advance over ad hoc decisions and competing *systems* of intercalation to keep the moon and sun within a single framework.

Before such algorithms, in all cultures *someone* had to make a decision. Sometimes priests were bribed by people who stood to gain if festivals or other events were or were not put off a month for a given year. Imagine how badly some merchants depending on immediate profits would take it today if someone suddenly decided to insert a month between October and November, thereby delaying the Christmas buying rush. A heavy debtor not involved in sales might, though, cheer the action. Other times towns or sects would fight over who got to decide which days would be holy and which would not. It should be no surprise that the *Encyclopaedia of Religion and Ethics* (Hastings 1961) devotes 80 pages to *calendar* (63 to *God*). Calendars have been heated religio-political issues.

Europeans may like to pride themselves on having "discovered" a single "stable" system for scheduling and recording events, but the system is quite artificial (and not completely accurate). For example, like the Egyptian civil calendar, our months no longer bear *any* relationship to the lunar cycles; and Easter wanders senselessly. The Jewish (or Quartodeciman) Christians originally celebrated Easter on Passover, the night of the calculated first full moon after the vernal equinox (14 Nisan). That made sense, since Jesus was arrested on Passover. In a way he was identified with the first-born sacrificial Paschal lamb. There was confusion though, since the later Romanized, gentile Christians gave most religious significance to the days of the week and wanted to celebrate the Friday death and Sunday resurrection.

Then again there were disputes among both factions because the Jewish system for calculating the vernal equinox was inaccurate. Some followed

Jewish religio-political decisions on the dates and others followed newer systems of calculation (also inaccurate). When the gentile Christian tradition came to outnumber the Jewish Christian tradition, the Quartodecimans were even excommunicated by Pope Victor; but political power had not become consolidated and the Quartodeciman tradition lingered on. The Council of Nicea tried to end such conflicts in 325 and it set the date for Easter on the first Sunday following the first full moon after the vernal equinox (Passover). Since the equinox could not be accurately predicted, tradition fixed it at March 21.

So Easter slipped out of phase with both Passover and the full moon. Fixing its date solved deep political problems, but it led to the detachment of Easter from its historical relationship to Passover and the heavens. Today the date shifts confusingly and for no valid traditional or astronomical reason.

The Eastern Orthodox church uses a pre-Gregorian system to calculate the date for Easter; and so even today not all Christians celebrate Easter on the same day in a given year.

To return to the major point, the coordination of diverse work tasks, as well as religious functions, are political issues that long created pressures in complex societies to develop accurate astronomical knowledge and stable calendar systems. There were calendars that attempted to fit the moon and sun to a single system going back at least to the third and fourth millennia B.C. among the Mesopotamians and Egyptians. These involved crude experiments with systems for intercalation. At least by late Vedic times it is clear that Indians and Chinese were also experimenting with single systems for heavenly events. The base for *accurate* systems can be dated to the reign of the Babylonian Nabonassar (747 B.C.) when precise celestial records began to be kept, implying also the development of a single conceptual system for doing so. Then over decades and centuries one could begin to calculate more accurate patterns from the data as they grew, whether lunar, solar, stellar, or planetary.

One significance of advances in astronomy, I suspect, was to contribute to an atmosphere in which critical discussion of a systematic (and almost mathematical) character was forced (even politically) upon metaphysicians, philosophers, theologians, and shamans interested intensely in the nature of reality and ethics. I suspect it would have upped the standards for precise debate in some circles, and intensified interest in abstract physical and rational truths. As one could represent the heavenly bodies as moving with regard to a reference system, as across a zodiac, and could better predict movements in terms of this reference system, they may have been forced to wonder, does this fact imply some underlying necessity that has not animal properties, but geometrical or mathematical abstract regularity? Does it imply a universal "logical" or numerological rulership of the sky creatures and earthly cycles? Even slow progress in constructing a framework for study and prediction of the heavens could have helped divert interest from the dream world and from speculation about the whims of local gods, to include critical thinking about possible abstract rational universals.

Another important implication of civic agreement here may have been to encourage monistic thinking. If one begins to think of the seasons, etc., in terms of the useful illusion that the computed cycle of the moon (or sun) is a True source of absolute reference, a universal validating standard, then a monistic habit of mind has been acquired or reinforced. An *artificial* system for conceptual order may create the illusion that there is somewhere a single *natural* system of order.

The Modernization of Science, Critical Thought, and Western Religion

From Theism to Deism. From the times of Galileo, Newton, Kant, and Darwin it was shown that simple physical mechanisms can easily explain the obvious features of the natural world. Spirits and final causes were not ruled out by proof, but were left unneeded and unemployed. They were indeed distracting and misleading ideas, so there was no reason for scientifically informed people to probe them. It was not until this relatively late time period that Descartes and others began to formalize a reasonably tidy division between the natural and the supernatural. Today we would define as supernatural supposed phenomena that clearly could not be explained by known simple physical, mechanical, forces.

Often we do not know what these natural forces are, but only that they can be described by simple mathematical formulas—there is no trace of caprice or creature passion. In a sense they are the old unseen metaphysical concepts of the Greeks reconceptualized into unseen "clockwork." This mechanization of nature was discovered largely by devout Christians who felt that their careful observations were revealing God's ways and plan.

The clockwork world that emerged from scientific study has led mainstream European religionists of the last few hundred years to develop *deistic* views, in which God is the clockmaker. There were no patterns of daily divine help or intervention. Jehovah became an otiose deity. Many religionists see this as a very positive step, resulting in a more benign diety that is not busy personally destroying things and causing pain and misery with plagues, earthquakes, floods, lightning, and other "acts of God." How much more beautiful, believers may think, that a church is blown away by lightning because of its height, rather than because God, for obscure reasons, is making good people suffer while bad people are spared. In this *reconceptualization* of God by theologians, the methods of science are probably not suitable to detect or refute the existence of such a deity, since they could not differentiate deistically generated mechanical clockwork from a-theistically generated clockwork. But this is not a historical shortcoming of science. Rather, Western religionists, needing to deal with scientific advances, have reconceptualized God so that *by their thoughtful definition* this issue has been placed outside the scope of science. Thus, it creates a serious misunderstanding to claim merely that science has been myopic here.

A Mechanical View of Nature Emerged from Theological Concerns. The claim by some extreme religious critics of science that atheists simply invented a mechanical view of the world in order to hide God from our eyes completely ignores history. Science may have made atheism an easier position to hold, as apparent miracles of daily life, like wind and rain, became understandable as mere physical happenings, but by no means did atheists create science. Organized religion and common sense were often opposed to scientific explanations, but it does not follow that atheists were out to eliminate God from nature. Historically, a mechanical view of nature emerged not by intention, but as a surprising conclusion to the painstaking studies and debates *of an entire civilization*, including pagans, Moslems, Jews, and Christians (and recently some atheists), over many generations and political regimes.

Such claims create confusion about the nature of science, as though science as a community effort is only capable of seeing what it already understands. Plenty of examples support this, and plenty of individuals are myopic and narrow. But the range of personalities, motives, and courage in science has fortunately been quite diverse. Some scientists may see only mechanical phenomena, but others will take an interest in things that appear to be real, even if they do not have a mechanical framework ready to explain them.

For example, scientists have been intensely interested in mysterious diseases that appeared to be real, empirical phenomena, even when their mechanisms have not been clear. Patterns in the anatomy, embryology, biochemistry, and biogeography of organisms overwhelmingly point to the evolution of species from common ancestors. This is of obvious scientific interest even if we cannot reproduce the past history of life in the laboratory; and such patterns were of interest long before Darwin proposed natural selection to explain them. (Pre-Darwinian Christian biologists tried to explain the patterns in theological terms, but their models lacked predictive power.) Astronomers and physicists were interested in evidence of pulsars before that evidence could be explained.

On the other hand, ghosts are of little scientific interest anymore because it is not clear that they even exist outside of the imagination. The evidence itself proved to be not good. There *were* scientific investigations of ghosts over the years, but the only conclusion from such research was that people are vulnerable on such issues. They imagine things, play jokes on each other, and many frauds have used all their ingenuity and guile to perpetuate elaborate deceptions.

The unexpected and difficult to explain in nature will always be of interest to some scientists who have high levels of curiosity, because the unexpected may be the breakthrough that leads to new ways of thinking. If there were any indication that gravity involved the changing disposition of a deity then this would be properly measured and reported. One cannot keep such secrets, and religions with strong theological schools have come to appreciate this fact. Whatever happened at Fatima, Portugal, happened in the minds of some individuals; if the sun had actually, objectively, fallen, there is no reason why scientific instruments all over the world should not have detected it. Roman Catholic theologians well appreciate this. If any deity is hidden, it hides itself; it is not the methods of science that hide it. Those methods are merely the diverse tools of human beings who have on average proved to be alert, curious, open-minded, energetic, and resourceful.

Attitude in Science. Karl Popper (the reluctant patron saint of today's methodologists) stresses, though this seems to be little appreciated even among many scientists, that not just formal methods but also attitudes and values are critical for good science and cannot be taken for granted. For example, he mentions that respect for predecessors, for those who went before is "incredibly important" in science, and that the Greeks had this. He also points to the fact that the Greeks criticized each other's ideas and, essentially, tried to falsify ideas. Moreover, "the ethics of science has to be founded in a tradition, and I think that the loss of this tradition would mean the real end of our civilization. Among the various elements constituting the ethics of science are the respect for truth and the respect for the impersonal character of critical discussion" (Krebs and Shelley 1975).

As a practicing scientist I also see critical factors that express a kinship between Greek efforts and ours. This spirit of inquiry, optimism about the possibility of finding Truth in the world of abstractions, appreciation for the power of systematic thought and debate in the language of proof, the idea of historical struggle in thinking (so that one remembers and with humility respects others for their contributions, even if they were incorrect), and the belief that one's own work will be bettered at some time in the future, are fundamental debts that modern science and scholarship owe to the Greeks.

Science and Ethical Aspirations. The outstanding pioneers of a new critical thought were Leonardo da Vinci (1452–1519), a keen student of the senses and the material world, and Galileo Galilei (1564–1642), who had a facility for abstract concepts and for the use of experiments in refining them. Economic and social changes in Italy had created elements supportive of their efforts among craftspeople and merchants (in contrast to the traditional forces of throne and church).

Thinkers of their day, certain pioneers at least, were extremely careful—this is very important. They were meticulous in reporting their observations and in their reasoning. They had to be to interact as a network of seekers of truth in times of uncertain support. Much of Europe had discovered that mere intuition and belief, supported by logic, resulted mostly in political conflict, or war—"Might makes right" in effect. Too many religious factions, wars, and inquisitions in the name of the Prince of Peace were bound to be discouraging (much as today we are proud of the up side of technological development and disturbed by its down side—we long for better). There was a widespread perception of corruption in the religious bureaucracy and many were skeptical of offical explanations. So the pioneers of science explored dynamic interactions of logic and observation—the best of both—to create in effect an intellectual system of checks and balances. Using those two tools in careful and mutually critical concert, they were to build powerful new systems of thought.

The pioneers had to be very careful, in order to gain an audience and win acceptance, since they lived in a culture with factions still very skeptical of the senses, for good empirical as well as religious reasons.

The development of methods was extremely important. But one should not concentrate excessively on their methods, as is fashionable in our technological age. Character and attitude were also important. There was a bold open-mindedness. No one can be 100 percent bold, 100 percent open-minded, or 100 percent objective. But they were not afraid to travel cautiously into the intellectual unknown, beyond conventional wisdom, beyond their peers, beyond common sense, beyond the intellectual authority of giants such as Plato, Aristotle, Augustine, and Aquinas, to look at the world with fresh eyes. Thinkers like Galileo and Descartes would write in the vernacular, reaching out to all; they were not simply interested in advancing within their professional circles.

We would all like at least some degree of certainty in the important things in our lives. But the domination of throne and church brought neither certainty nor anything even close. Each prayer, each candle lit, each donation, the following of each instruction was a gamble. Who could guarantee that it would bring any results? Effective paths through the complex world to salvation were always open to argument. A multitude of philosophical sects proliferated within the Church (and this is why it has been called Catholic) prior to, as well as during and following the Inquisition. The Bible made different sorts of sense to different sorts of people. The philosophical options, and hence the general level of philosophical uncertainty, increased even further with the proliferation of Protestant groups. Soldiers went into the frequent wars believing that God was on *their* side, but knowing that the enemy believed that *they* were blessed too; yet someone always lost. Large clergies worked hard to explain over and over to each generation why prayers were not answered, why good people had misfortunes and died young, why bad people rose to power, why life must be such a hopeless struggle for so many.

If the common people wanted to eliminate technical science today they could not do it, because its economic constituency is too strong. The classes with power realize clearly that technical science is critical to the development of new technologies and to today's economic competitiveness, capitalist or socialist. The intellectual and moral benefits of conceptual science remain real, but we have come to take them for granted and fail to think of them as such. Since they are hard to see and appreciate they now have only a tiny constituency.

In its early days, modern science had a powerful constituency, since it was tied to the development of new technologies for war and commerce, and this is the ground where the economic determinists have placed their claims and done their productive digging (e.g. Hunt 1986). Scientific thought also helped to validate the more empirical outlooks of craftspeople and merchants, relative to the old power structure. There was also the search for natural laws that might give moral guidance as European life grew too complex to be guided entirely by the Bible. The new thought also had a large and diverse constituency that saw in it a better way *to think*, about nature and about being human. Many Europeans were not altogether satisfied with the strife and quality of human life that the thinking and institutions of the past had produced. The promotion of critical thought was in everyone's *self-interest*. This is one important reason why science came to have such a large impact on the overall evolution of Western thought. *It did not remain an esoteric matter for professors, investors, and merchants.* It profoundly influenced the legal system, approaches to solving public problems, ideas of fairness, the arts, religion, the development of public institutions from public libraries to schools, habits of speaking one's mind publicly, and politics.

Nontheoreticians in Science. There were diverse dimensions to the reinvigorated empiricism. Leonardo was not quite a mainstream scientist—he was not so concerned with theory and abstractions as were Galileo and the others. But it is clear from his notebooks that he was an organized, keen observer and critical, fearless, open-minded, and systematic thinker. His keen eye and mind provided a solid knowledge and understanding of material things, from geology, color, and water to the way clothes drape, which in large part enabled him to make such wonderful realistic drawings and paintings.

Like Leonardo, some modern scientists are not much concerned with basic theory, but, with an eye for the organization of nature, they work to find out how phenomena interconnect. They might figure out that fish spawn when certain types of food are most abundant, or that a rising barometer in June brings certain kinds of weather, to take simple examples. Such persons are vitally important in today's scientific community, and even though Leonardo did not cause an identifiable major shift in our abstract concepts, we can see him as an important symbol or example of Renaissance mind.

Leonardo was a Christian but he was a new sort of Christian. The questions in his notebooks were largely not about Christian issues, but about the material world of the senses and a person's daily experiences. He was often able to discuss the world in terms that are still meaningful, objective, and analytical. He wrote notes and thoughts on shadows and colors, on water, on the moon, and on the types of cracks that form in buildings. Even when touching on morals, he leaned on his own critical wits.

He puzzled that fossil shells in the *mountains* were of *marine* species. How did they get there? Why are they all in layers? Shells sink in water, so how could the Flood (which he seems to have believed in) have carried them upwards? Rain washes things downhill, so rain could not have carried them. Perhaps the shells crawled up. But these species cannot crawl and. . . . And so on. For many of us it is second nature, common sense even, to think this way, but this was not always so. Scientific thought has greatly raised the standards of the most cautious practitioners of common sense in our society. How many others had looked at those layers of fossils and been satisfied with simple noncritical answers as to how they had gotten there? Outstanding persons like Leonardo had to demonstrate the power of such critical ways of wondering, observing, thinking, and asking questions.

It is the theoretical tradition that most counts when discussing the history of scientific concepts. But before getting back to that, I want to make the point that "nontheoreticians" in the Renaissance were rediscovering how to trust the senses and were dealing actively with the objective world without waiting for theory and philosophy to catch up.

An Interconnected System of Abstractions. Returning to theory: What did Galileo do when he invented modern physics? The faith, true or not, that explanations must be in harmony with one another can motivate a very careful and critical look at basic assumptions. If the earth is spinning, why do things not fly off? Why do projectiles not curve to the right or left?

If you throw a baseball, why does it continue to move once it leaves the hand? We could say that it is because that is what things do, or that God makes it continue. Neither response leads us very close to understanding mechanisms. These are nonanswers from a Western scientific point of view and will not satisfy someone who is really curious to know.

If you watch a ball sitting on the ground for a year or more, it will not fly, but if you pick it up and drop it, it moves to the earth. Aristotle weighed such facts and concluded that *the nature* of objects such as balls is to remain motionless or simply to rush straight to the middle of the earth, but that it is not in a ball's nature, anymore than in a dog's nature, to fly. So their natures do not explain why balls continue to move once they leave a throwing hand. Rather, Aristotle proposed, the power of motion of the swinging arm is transferred to the air, which then pushes the ball along. It is common sense that it takes force to move something.

A few thinkers questioned this theory of motion. Why then does a top continue to spin? Surely the air is not pushing the top around and around. (For more see Shapere 1974.) One might speculate that perhaps one transfers some power from the hand directly to the ball or top. The Byzantine Philoponus, in the sixth, and the Parisian Buridan, in the fourteenth centuries, made such suggestions. Indeed, Galileo himself used such thinking in his early writings while at the University of Pisa. These are versions of the Aristotelian theory in that an "impetus," an "incorporeal motive force," a "power of movement" is transferred from the arm and that "it" moves the ball. (In modern physics the flying ball is not "being moved" by a force or anything else.) What is this power? Does it push or pull? How does it know how to make a ball fly outwards and a top instead spin in a circle? Is it on the surface of the ball or inside? A power or force was easy to propose, but one still did not know much about it.

Such issues did not exist in isolation. They were critical parts of tightly interconnected world views, and tampering with one element meant having to explain how others were also affected. Galileo began with the conviction that the earth *is spinning* and this got him into a whole set of questions about why objects do not shake and slosh about on earth and fly off into space. He had the scientific faith that things must "add up," that physics would have to make sense for a spinning as for a motionless earth. The comprehensive astronomer would also have to become a physicist to figure out if the double motion of the earth theory could be true and not just a mathematical construction. That took him right to why balls fall when one drops them and fly when one tosses them.

Galileo's actual thinking, conscious or subconscious, remains obscure. But judging from what we do know and from how creative scientists think today, it is reasonable to speculate that since he was deeply attracted to the idea of a spinning earth on empirical/logical grounds, that led him to look very critically at Aristotelian physics. Perhaps *it*, rather than Copernicus, was crazy. Particular problems with Aristotle had already been discovered, but perhaps there were even more fundamental problems. In any event there was a shift in Galileo's thinking, and "the power of movement" became a notion more mathematical and descriptive and more free of metaphysical connotations. This was a major step in the evolution of modern physics. Galileo begain to work and write about physics virtually as though he had discovered the concept of *inertia* in the post-Newtonian sense. Some scholars believe that he did and concealed it for various reasons having to do with the politics of scientific presentation or religion.

Today we appreciate that once in motion a baseball keeps moving not because it is pushed or pulled along, but because of its mass or inertia, and it is slowed and stopped by gravity and the friction of the air or earth. In contrast to the Aristotelian view based on common sense, where the natural state is rest and so some sort of power must exist to move the ball, inertia is no "force" at all; it is a mere abstract feature of mass. Indeed, mass and inertia are the same to the physicist—the resistance to change in velocity. This can be a difficult idea to grasp at first, since in the Aristotelian common sense that still lingers, *something* must be *moving* the ball.

It seems likely that Galileo intuited that there could be a simple coherent system for thinking of motion that would not conflict with the idea of a spinning earth. To avoid unnecessary complications he simplifed the Aristotelian conceptions of the nature of things to almost pure mathematics, as Newton and his generation would later do so effectively. He might have said to himself, much as a modern scientist would, "I can explain this problem simply by the way these components actually behave in measurable and repeatable ways. Maybe there is a soul in them, who the devil knows? If I just ignore that sort of question then I can go on, but if I get into speculation on the substance of things then I have to explain why Parixo must have been wrong about the dislike of air for saltwater after a good meal of heat. I guess I will worry about the nature of substances and spirits as little as possible, and leave it to someone else. I'm going to assume that any spirits only have natures that I can see and measure. If one belches, then I will write it up and get another publication, ha ha. But by just thinking of the natures of these things as mathematical equations then, wow, I can next explain why it is that . . ."

This verbal cartoon illustrates again an important feature of modern scientific thinking. One tries to keep assumptions to an absolute minimum. If something can possibly be explained in very simple terms, then check that idea out very carefully before introducing any unknowns. If you do an analysis of blood and can only explain 90 percent of the components, check for a leak in the equipment before jumping to the conclusion that there is a mystery substance that makes up 10 percent of the blood. Scientists like Leonardo, Galileo, and Gilbert may have believed that there were psychic forces in nature (Westfall 1971), but they did not let this sidetrack them much from dealing with what they could see and measure and safely infer on the basis of evidence.

Even though people such as Galileo were religious themselves, their insights were the beginings of a comprehensive nonmetaphysical theoretical science.

The Cartesian Dualism Makes a Truce Between Science and Religion. No one individual could have been responsible for a great intellectual and social movement taking place over a continent. Philosophers of science usually see the philosopher Descartes (1596–1650) as the main founder of the "materialistic" philosophy. He reasoned that reality is divided into mind and matter, and this (to us simple) move took all the souls and psychic forces out of nature and put them in a spiritual world of the mind and God. So he was not materialistic in the common sense of the word, since he was by no means religiously atheistic (and his values were not necessarily money, power, and so on). His philosophy of materialism pictured a purely mechanical world where things move about by physical necessity. The famous Cartesian dualism was formulated in terms that much of Europe was ready for (Jacob 1988). *Discourse on Method* (1637) was published only four years after the trial of Galileo before the Inquisition. Descartes was a person of and for his unique times. He articulated a sentiment that many scientist-philosophers had for some time been been expressing in their research attitudes—the daily workings of nature may well take place without psychic forces, and any world of the human soul and God are a separate issue.

Many details of his materialism and the way he came to them seem bizarre today and one may find it difficult to think of him as a scientist at all, but it seems he did much for the outlook of natural philosophy. The working attitude (as distinct from the personal beliefs) of men such as da Vinci, Copernicus, Galileo, Gilbert, and Kepler in a changing yet *very much* Christian society produced new findings and ideas that virtually called for a philosophy that would keep God and yet make room for a new simple way of looking at nature.

The modern atheist may reject God for *philosophical* reasons (e.g., Herrick 1985, Johnson 1981, Thrower 1971) or from *disillusionment* with miracles or prayer, but Descartes and others showed that an a-theistic *view of nature* does not necessarily lead to religious atheism, and most religionists today have become quite comfortable with one version or another of the Cartesian dualism.

Important ideas are always controversial; I do not mean to imply that Descartes's ideas became popular with educated people overnight. Indeed, they remain controversial today. But attitudes in the West in time came largely to reflect the Cartesian dualism. The practical effect of deism was enormous, since it allowed Europeans to continue the research and discussion necessary to the progress of their thinking and awareness, while keeping relative peace between materialistic science and religion. On the other hand, it has helped science deny its debts to and roots in ethics and morality.

Some modern scientists have a variety of personal supernatural beliefs, hopes, or suspicions, but if they are competent they know how to keep them out of their critical methods. Pasteur may have been motivated from his religious beliefs to disagree with the advocates of spontaneous generation. The Bible says that creatures were created by God and once; it makes no mention of spontaneous generation. But this belief led to a series of good experiments that were purely scientific and sound in method and thinking. His research was quite respectable and had no similarity to today's "Bible science," despite the similarity of his religious beliefs to Creationism. Pasteur worked in "the language of proof." He was systematic and used critical evidence and experiment in a proven manner. It was his solid approach to research that made him a contributing scientist, rather than his personal motivations or beliefs, or the origins of his insights.

A Lexicon in Disarray?

What is the status of the "language of proof" *for a citizen*, for the liberal arts? What is the status of its trustees and teachers, the scientists and scholars in the rising system of intellectual advocacy and specialization, the multiversity, and science organized on the corporate model?

Massive industrial societies cannot afford to wait for the odd creative individual to come along and take an interest in this issue or that. So career structures were invented to attract and keep busy massive numbers of people in science and scholarship, and to *certify* expertise as well. There have always been specialists, but specialization has virtually become a moral imperative, and our universities gardens of career ladders.

Skill and expertise are wonderful things. They can be a source of *personal* pride and self-esteem. They can sustain one through periods of misfortune or self-doubt. They reflect organization and self-discipline that can be useful in many aspects of one's life. Acquiring them can be an enriching experience. Moreover, society as a whole does benefit from having skilled and expert people available. It is certainly not my intention to criticize specialization per se.

I enjoy, admire, and respect people who pour their lives into studying and talking about bees or painting or ancient pottery or the history of Hawaii. If one expects to make much real progress when working on difficult projects then one must devote oneself eagerly and even passionately and learn to follow through. Directing obsession to gentle idealistic or creative activities is one of the finer human accomplishments, just as employing it for harmful or fanatical activities is one of the most regretful errors of character and judgment.

On the other hand, even in personal "practical" terms there are dangers in specialization, especially if it is imposed by the career structure rather than sincere passion. The world is changing rapidly and if one pours one's whole being into a mold of marketable skill, then one risks becoming obsolete when the market changes. We are a competitive society and one is not likely to remain the expert in one area for more than a decade or two.

Moreover, while society needs experts, these have proved to be *not enough*. We have tracked many of our best minds into specialization and now find a crying need for persons who can develop broad overviews and judgments from more than an ideological perspective. We need broad perspectives and wisdom not only at the top, but at every level, certainly at the level of the middle management and professional classes.

Some would argue that we are evolving into classes of functionaries, that there are too many "dial-a-scientists" and other academics who focus their creative energies and potential individuality onto nearly any narrow problem that some party will pay to have solved or that some group will give honors for. It is true that nearly anyone can earn a Ph.D. who is persistent and willing to master the specific ideas, facts, and techniques that have been worked out for a given field of study. Historical awareness, vision, or independent commitment is no longer expected. This is the Age of the Expert. One must avoid even the appearance of being a jack-of-all-trades, master of none. Character in our society has become a fuzzy issue. Often, what is promoted as character is the sort of *grit* people believe high school football coaches teach. It may have little to do with truth or individuality.

Creative talent is still highly desired (though largely when it can be focused on the program goals desired by the management structure), but we will always settle for technical competence in training graduate students. Albert Szent-Gyorgyi put it, "Research is to see what everybody else has seen, and to think what nobody else has thought." We would like that in students and colleagues, but we will settle for competence. So the system too often teaches technique, since it is not established how to *teach* inspiration and creativity at a late point in life (though see Beveridge, *The Art of Scientific Investigation*).

Franz Boas, a father of modern anthropology, was originally a physicist. Pasteur had been a chemist. Freeman Dyson pointed out to me that two of the recent important astronomers began as a lens grinder and a mule driver. Brains counted more than credentials.

Since the post–World War II surge of industrialization and the hyperorganization of society, the opportunities to break in or out of disciplinary boundaries have been vanishing especially fast for reasons relating to management philosophy, economics, public perceptions, and increasingly parochial outlooks within disciplines. It was once common advice to change research areas every five or ten years in order to remain intellectually fresh and creative. Now many social and economic pressures weigh against such advice; I have not heard this discussed in years.

Much of this has coincided with the reshaping of the university into what former University of California President Clark Kerr called the "multiversity," and the entrenchment of what R. M. Hutchins called "departmentalism". Others refer to the "balkanization of knowledge." A campus may be a collection of competing departments sharing some common facilities such as a library and computer center, but the idea of a community of scholars is merely words at most campuses. Former University of Minnesota President Kenneth Keller said that most interactions that administrators see these days have become disputes over turf rather than meaningful discussions of institutional goals and standards, educational coherence, or the general state of our civilization. A dean claims he works hard to keep departments "thin-walled," but he is only partly successful. He claims that in academics, and in our society in general, the art of long serious discussion commonly has been lost in favor of emotional heat or emotionless banter.[4] One hears similar stories around the country. One certainly also sees for oneself.

Knowledge is power is money. After World War I and especially World War II, society began to mine and manage its knowledge resources as never before, as Clark Kerr detailed in *The Uses of The University*. The ivory tower remains only as a myth. Institutions of higher learning reflect the pedestrian routines and organizational outlook of society at large and have done relatively little as "communities of scholars" to try to offer leadership to the times. Any coherent sense of mission has been reduced to a dysfunctional, vestigial rhetoric, preserved largely as ceremonial adornment.

In reflecting on the increasing specialization, on this proliferation of narrow "experts" who must stay sharply tuned like the keys on a piano, waiting to be used, fearing their potential obsolescence and disposability, I sometimes recall the Chaplin film *Modern Times*. The little hero struggles at his task to continue moving fast enough to keep his job. But the assembly line speeds up and the situation becomes comically and crushingly hopeless. And one's life may become such an obscure fragment of the whole that one may wonder, what *does* matter? And how long can the whole thing keep going without facing itself squarely?

I am not certain that science as a whole has been done in so badly yet. I know many fine and concerned scientists, and society may yet face itself and pursue corrections. But as an informed and concerned insider, it would be irresponsible not to ponder this serious question and explore such comparisons.

Are we modular men and women? Have we made specialization our creed and the assembly line our altar? Is this the "Year of our Ford 19 ... ," as Aldous Huxley quipped in *Brave New World*? We seem to have moved in this direction.

Management philosophy certainly tries to steer us in these directions. But there is no science of management, only beliefs, and the great experiment in these beliefs has been looking sour even to business itself. Walter Kiechel wrote in *Fortune* magazine ("The Real World: Corporate Strategists Under Fire," December 27, 1982):

> And if the apostles [of corporate strategy] have have fallen on slightly hard times, the object of their devotion—the concept of strategy itself—is positively bedraggled. What went wrong? Where did the dream fail?

Many argue that management has not given enough attention to "messy" issues like talent, values, communication, the corporate culture, and has given too much emphasis to apparently "clean" issues like structure and strategy. Kiechel concludes, "In other words, it isn't going to be nearly as simple as the dreamers of the dream once thought—hoped—it would be."

As I write, American businesses are going bankrupt at record rates, while European and especially Japanese businesses are doing better, especially considering their much more limited resource bases. Management philosophy has been looking bad not only in business but in the military and academics as well. As I write, there is deep concern because our system has managed to equip our troops with many billions of dollars worth of badly designed and badly built weapons. But no politically comprehensible alternative has emerged and our culture slides along nervously, trying to be optimistic.

Obviously, I have deep concerns about *all* this, but in this discussion I wonder especially what we have been doing to this powerful thing called scientific thinking that our ancestors worked so hard to carve out. It was a bit strong for me when the much honored biochemist Erwin Chargaff wrote, in *Heraclitean Fire*:

> What have the universities done to science? They have bled it for overhead; they have cheapened and vulgarized it to the point of nonrecognition; they have made it into a public-relations "gimmick." If the products of this kind of education often still are so good, it testifies only to the resilience of young minds. But many are damaged irreversibly.

Some of my colleagues branded such views as elitist, seeing a lack of respect for the ordinary man or woman in science who is just trying hard to make a living and do something status-worthy and, they hope, interesting and useful with their lives. But many found that they in any event agreed with his general thesis that our management of economic and political pressures creates researchers who must work too fast to think about it all, who become so self-interested that they cannot afford to think about it all, and who must become so specialized that they *cannot* think about it all; that it is profoundly disturbing to see the world's problems growing ever greater as its potentially best minds grow ever smaller.[5]

It saddens me that William Gilman, *Popular Science* associate editor and chemist (1965), concluded his survey of more than 100 laboratories with a very negative reaction to "big science."

> ... the unhealthy corpulence stumbling under its overload of incompetents, predators, presumptuous politicians, pursuers of the bitch goddess; the recklessness, arrogance, and petulant demands of a self-annointed aristocracy; the social irresponsibility for which it is damned twice over: for the things it does and the things it fails to do. What can be set in the opposing column? Simply refusal to allow the pursuit of knowledge to degenerate into this.

I have seen disturbing changes take place even more recently. And as sad, I see enterprising revisionist historians, philosophers, and sociologists starting to move in even before the body is cold, to sell to an eager market, to argue not that we have been mismanaging a trust but that science is not changing at all and that previous generations were given completely to distorting their motives and circumstances. They go far beyond debunking children's myths that scientists were once superhumanly pure, selfless, idealistic, and brilliant. They go to the other extreme and generate new children's stories. I watch and reflect that if the emerging generation really needs the emerging mythology, that science was built *simply* on narrow self-interest by people *simply* having fun or pushing their personal prejudices onto history, seeking only glory, then perhaps our era of history *is* as grim as Chargaff and others have painted it. We may be entering an era in which we can only have knowledge without wisdom.

In presenting his remarkable studies of anatomy to us, Leonardo da Vinci underscored the traits of character required.

> And if you should have love for such things you might be prevented by loathing, and if that did not prevent you, you might be deterred by the fear of living in the nighthours in the company of those corpses, quartered and flayed and horrible to see. And if this did not prevent you, perhaps you might not be able to draw so well as is necessary for such a demonstration; or, if you had the skill in drawing, it might not be combined with

knowledge of perspective; and if it were so, you might not understand the methods of geometrical demonstration and the method of the calculation of forces and of the strength of the muscles; patience also may be wanting, so that you lack perseverance. As to whether all these things were found in me or not, the . . . books composed by me will give verdict Yes or No. In these I have been hindered neither by avarice nor negligence, but simply by want of time. Farewell. (*Notebooks*, No. 796)

His scientific method was essentially a passion for truth, the skepticism to question, the intelligence and patience to devise organization, the honesty to verify, the determination to make progress, the courage to break new ground and face criticism, dedication to his cause, humility, carefulness.

I see a rapidly growing number of career-oriented researchers who show naked disbelief and even contempt when such matters are discussed. They see this as "mystification"[6] and believe that science surely *must* be impersonal technique that can be memorized and mastered like prayer and brandished like cold steel, without painful self-examination, without exposure of vulnerability.[7]

It does appear that a shift is underway in the balance of personalities of people in the academy. One still sees the "Craftsman," but more frequently the "Jungle Fighter," the "Company Man," and the "Gamesman" described for corporations in Michael Macoby's *The Gamesman*. Of course many of the best scholars have long practiced *forms* of advocacy, and many of the greats have mostly been having fun exploring ideas. But this was often enough balanced within the context of an almost religious personal devotion to ideals of truth and objectivity rather than to offices and techniques— something essentially inconceivable to many careerist academics.

Some see this shift as the result of accelerating competitiveness in society and in the academy (e.g., Eiduson 1962). The generations entering more competitive job markets cannot afford to be otherwise, it is argued. Intellectual styles that signify social respectability must come first. So one must use what are perceived to be respectable techniques and concentrate on a narrow career track, even if the questions addressed are of limited scope in the greater scheme of things.

The most extreme of the competitive academic careerists, perhaps unsure at heart even of the techniques that they have learned, try to justify the contribution of their research styles to human betterment on the basis of a metaphysical progressivism, an intellectual version of social Darwinism. On this, even the most extreme and aggressive economic and biological determinists unite against recorded history. Truth and progress will somehow or other emerge from conflict and naked advocacy. They claim that mere competition for money, status, or sex has always driven all advancements. Galileo, Newton, and the others are tried in absentia and are declared scoundrels by people who are clearly not their peers. If one cannot be great by one's own creations, then perhaps one can attract attention by belittling giants.

Much as Pythagoras believed that everything can be reduced to numbers, they believe that civilization and creative energy can be reduced to competition and conflict. But so too, one must respond, can *anything* be reduced to numbers or to atoms, music reduced to notes. When questioned closely, they offer not much more than a biological version of numerology or atomism. Theirs seems to be the commonsense, unquestioned faith of a highly competitive class. The forces that they see as obvious today must, they believe, be essential, universal, and eternal. They may be sincere enough, and pleasant enough as individuals. We may like and even admire the highly competitive Chimbus of New Guinea too. But we should not accept their metaphysical beliefs simply because they are powerful in their land.

The poet René Daumal in his 1940 "non-poem," "Holy War," wrote of the war of the spirit, within oneself.

> [This will not] be a philosophical discourse. For to be a philosopher, to love the truth more than oneself, one must have died to self-deception, one must have killed the treacherous smugness of dream and cozy fantasy. And that is the aim and the end of the war; and the war has hardly begun, there are still traitors to unmask.
>
> Nor will it be a work of learning. For to be learned, to see and love things as they are, one must be oneself, and love to see oneself as one is. One must have broken the deceiving mirrors, one must have slain with a pitiless look the insinuating phantoms. And that is the aim and the end of the war, and the war has hardly begun; there are still masks to tear off. (Translation by D. M. Dooling, *Parabola* 7(4): 11–13.)

One can, though, still find scientists and scholars who see the broad issue of truth as being in one's self-interest and who are eager to look critically and deep within *themselves* and their societies.

It is interesting that some personality types are attracted to science for its power and the apparent certainty that the hard, physical, sciences may appear at first to offer. Here there is no conflict between science and authoritarianism, since such persons may focus on the apparent authority in well-established areas of science and can get by for quite awhile without having to confront, even in their own work, the self-skepticism and quest for objectivity that made the relative certainty possible in the first place.

Not all of these problems are the result of *inevitable* forces and trends in sociology, despite the claims of many analysts. There are a lot of good, eager young scientists and scholars who are forced to work in production units, managed poorly in terms of interpersonal dynamics (and academic goals), like mediocre corporations. They are in situations that can too often bring out the worst in people and leave little room for the best in them. It is naive or self-serving, somehow grotesque, to lay all the blame on individual scientists, or on inevitable forces, while management philosophy and practices and social goals are so conspicuously misguided and even shabby in our society.

Yet for all its problems, and indeed dangers, we dare not abandon science carelessly. To its most severe social critics I will again emphasize that science is not just a source of growing power that even those within it may not be able to control. Scientific thinking also has been a large part of our modest progress toward being decent and humane. Understood, it can be a contributor to wisdom. Justinian ended open thought, and this contributed to the negative face of the Dark Ages in Europe. We should have learned from this all too well. Oedipus felt betrayed by his eyes and put them out. But it was his inability to pursue clues and to *use his eyes with imagination and open-mindedness* that were at the root of his mistakes. Science has been very helpful. It is still today very much our eyes, and there is a very long way left to go.

It is too soon to say what multiversity science and academics may become in the future or if they will bear much resemblance to the project of the greats. Science grew out of religious and humane concerns and for a long time carried with it, bound tightly to the language of proof, an implicit and explicit devotion to truth and a willingness to try to approach objectivity, in some ways in the tradition of the shaman and the wiseperson. For some time it participated in the liberal arts ideal. Can scholarship afford to abandon this? We cannot foresee the answer clearly, but my observations within the academy make me less than cheerful about the prospects if there is complete abandonment.

Some Contemporary Disappointment with the System That the Language of Proof Was Used to Build

Runaway Technology. Science very much was part of a liberating movement in history. Yet many humanists are opposed to it today (as discussed in other chapters). This is partly because they see runaway technology changing our lives at a rapid pace and of its own momentum, and taking many decisions out of our hands (Ellul 1964, Lapham 1985). I would argue that we should keep distinct the differences between self-serving, ideological, and commercialized science, runaway technology, genuinely beneficial technology, and scientific thought, and keep in mind the indispensable benefits that the last has brought to our aspirations to be treated fairly and to treat others fairly. At least, it puts the goal potentially closer.

Scientism. Critics may nevertheless complain that scientific *thinking* is *scientism*—the lifeless if popular idea that *only* scientific research gives understanding of the world. I have also been critical of scientism (e.g. chapter 4, note 2). This is not critical scientific thought, it is self-serving ideology. Poetry, literature, art, and self-discovery, for example, need not be merely meaningless diversions. They can clearly enrich our understanding and connectedness and help to formulate values. What one can learn about oneself and life from thoughtful art, literature, backpacking, love, or travel cannot be squeezed out of a memorized equation. We must certainly do more with powerful scientific thinking than make it into a tool of the powerful or into sterile ideology; but if we are to use it creatively, we must begin by understanding its potentials and misuses.

Technocracy. There is the concern that scientists and other technical people are the ones who in effect hold power in Western society. Their beliefs and self-interests direct "the System" in the larger, historical sense, since even the rich may believe what the technocrats are telling themselves. (Who gets to define reality?) But some class or other is always mostly running things in a large society. The key issues, it seems, are how to minimize totalitarian tendencies, maximize participation, and insert values of equitable virtue and wisdom into the power structure, whatever class may be in whatever degree of control.

Professor Holton (1986) of Harvard quotes Cardinal Newman, "Not to know the general disposition of things is the state of slaves." The new slaves, Holton muses, *voluntarily* turn away from their freedom and renounce "self-government on the hard issues that determine the fate of a people." In essence, is the public, needing technical science to drive the particular socioeconomic system, selling itself into a new form of servitude?

The Two Cultures. Many argue that the Western humanists have largely cut themselves off from a realistic understanding of science in an age of technology and science. They seem to be quite busy, fighting battles in a multitude of specialties. Understanding little of science, some may become vulnerable to the gratuitous charge that they are being unscientific. They may allow their critics and managers to demand cartoonish modes of science from them. They are told that they must grow Kentucky bluegrass lawns like the rich folks if they want to stay in the neighborhood. Some retreat into the tidy pursuit of esoteric detail and others abandon the humanity of their subject matter to embrace uncritically forms of scientism, reductionism, and quantification that have the surface respectability to win them professional advancement. Others attack—though not only ideological reductionism, technocracy, scientism or runaway technology, but science generally.

Scientists for their part may take the condescending attitude that the humanists "are being paid to teach values." So they feel they can largely avoid deep study or thinking on those issues that should matter to everyone and that so much involve their own enterprise. They can hardly enter the demanding dialogue, since too often they know little of the subject matter of the humanities, let alone of what goes on or does not in the debates within the humanities. They may become isolated within ideological litanies, such as the familiar rhetoric of ideological reductionism and progressivism.

The words of C. P. Snow on the division of the sciences and the humanities fit well here: "The division of our culture is making us more obtuse than we need be," though, "changes in education will not, by themselves, solve our problems: but without those changes we shan't even realize what the problems are." There are natural grounds for much seminal interaction between science and the humanities, but much of this potential goes unseen. I have touched in this book on multiple issues that should be of common concern. Did the "Two Cultures" gap end a dialogue that would have been necessary to develop the agenda that optimistic Enlightenment thinkers began to work out?

The "Postmodernists of Reaction." The spirit of the Enlightenment has not brought utopia within grasp. Admittedly, the problem of institutionalized poverty and disenfranchisement remains today. And, societies have effective if subtle mechanisms for delegitimizing ideas that challenge them. And, we have seen how simply horrible modern technological warfare can be. And, enormous and growing numbers of people balance on a shrinking and uncertain resource base, not knowing how to develop their modes of production, to manage massive economies or their wastes, or to define values accordingly. And the spirit of the Enlightenment was largely invested in public programs, government agencies, and academic departments that turned their vistas inward to their own career structures and funding needs and often forgot their historical missions.[8] And, society forgot why it had set up many of these expensive programs, and even forgot the origins of many of its favorite slogans.

But there are some bright spots, and over much of the world a man today does not have to train in physical combat or wear a sword or make his wife wear a chastity belt, or quarter the king's soldiers, or worry about being tortured or burned at the stake as a witch if he has his own ideas, and so on. And a woman. . . . There have been some advances.

Several appealing romantic versions of "the good old days" before the Enlightenment are being peddled in our troubled times. But the days of old often were peopled (on a per capita basis) with as much or more violence, institutionalized poverty, injustice, war, infidelity, crime, prostitution, immorality, and distress as we can find in our own time. And much of life had this unwholesome texture even with often very brutal and repressive laws, and with the fear of eternal damnation.

If one finds our own era discouraging at times, one should not condemn the entire Enlightenment or all of science or rationality as a failure. This is not a realistic way to respond to today's problems. Let us be selective. At least the Enlightenment could articulate a tenable, humane agenda from which came a great deal of the good that many of us now take for granted and aspire to make more available. If pressed, the critics of modernism and the Enlightenment, especially those Foster (1985) calls the "postmodernists of reaction," whose solution is a return to supposed "tradition," including many neoconservatives, gratuitously reassure us that they are not talking about leading us back to the Inquisition, but only returning to the reputedly wholesome aspects of the pre-Enlightenment, pre-secular-humanist days.[9] But they take it mostly on faith that changing our laws, education, and values to their own tastes and definition of "tradition" (usually one or another myth of harmonious Christian, Jewish, or Islamic hegemony) would give us rainbows without rain. The Age of Faith was an epic experiment that produced some beautiful art and castles and some enduring social institutions but in many ways did not go so well for serfs and others. There was enough rain with every rainbow. Witch hunts are like unintended pregnancies—they may not be on the agenda, but they are risks that with a certain logic may follow certain enthusiasms. I believe that we must open ourselves, judiciously, to ideas from the past as well as from other cultures. But a return to the mythical "good old days" is no practical or humanistic alternative to ills of the world that society has used (or misused) the language of proof to build.

Managers of Science Dictate the Terms of Social Discourse. This is perhaps an extension of the "technocracy" concern from the more obvious political level to deeper influences of the scientific establishment on consciousness. The managers and funders of science (technical people or not) can define to a large extent what is "respectable," "interesting," "good," or "sound" science in terms of subject matter and areas of concern, methods, and social relevance. Since science in our age greatly defines how the effective members of society conduct discourse, science managers tend to shape social discourse. So science has great power to help construct social perceptual realities and agendas for attention as well as to construct material/economic realities (e.g., Aronowitz 1988).

It would seem to be to the advantage of a bureaucratic system (capitalist or socialist) if social problems were seen as the effects of simple causes that lend themselves to solution by technical inventions, for example. But, in fact, many problems may have *complex* causes that would be difficult to deal with politically. Drug and alcohol abuse, for example, have complex causes that include how well families and society effectively nurture children and teach them to deal with their feelings. They involve the pressures of adult life-styles and patterns of personal interactions. This is scientifically completely clear in general terms. Yet it is convenient for bureaucracy and the power structure in science to focus reductionistically on simple models of causation that they can appear to be dealing with. Thus, abuse is caused by chemical action on brain cells. Implication?—Focus major efforts and public attention on funding of brain chemistry research and on the elimination of the drug traffic; stress the simple elements in causation, and hold out hope for simple solutions; relegate analysis of family dynamics and life-styles, and of their relationships to broader socioeconomic patterns, to "less substantial" items for discourse (and action). Yet in fact these last are clearly interesting, perhaps even more interesting than the allegedly "more scientific" studies on brain chemistry. They would deserve priority for detailed analysis and discourse if the pure pursuit of knowledge of alcohol/drug abuse were the only issue. (Anthropologists refer to the "medicalization" of social issues.)

If the scientific establishment were merely directed to the pursuit of the truth, the whole truth, and nothing but the truth, it would give more priority to holistic analysis and synthesis, on a variety of issues. But it is highly selective; and its reductionistic vogues are more apt to yield commercial products, and are in greater conformity with the outlooks that agencies in bureaucracies find help them to stay well funded and competitive with other agencies. It is politically easier to find funds for test tubes and guns than for family education programs.

In or out of science, one needs to be sensitive to the potential for such value-and perception-shaping dynamics.

A Feminist Observation of Science

Radical and bourgeois feminists alike have made *blanket* denunciations of science and scientific thinking because they see it used in attempts to validate patriarchal social norms, or as vital to power structures in which they do not share equity.

Other feminist scholars have examined science in more critical detail. Harding (1986) observes that only relatively late in history did scientists become heavily tied to career ladders within the establishment, and begin to propagate self-serving myths of value-free objectivity. Prior to that, "by the seventeenth century the characteristics of experimental observation were among the central features of a self-conscious political movement." She fo-cuses on the New Science Movement in Puritan England during the 1640s, and 1650s with "radical social goals." She notes too that "science for the people" was Galileo's phrase.[10]

"The six traits and goals of the New Science Movement ... bear an eerie resemblance to those often stated for feminist inquiry. First, feminism's suc-cessor science projects challenge authoritarian attitudes and emphasize personal experience as a source of knowledge; feminism supports the self-confidence of the individual member of subjugated groups heretofore not regarded as social individuals; and political emancipation is central to its purposes of inquiry."[11] A feminist science agenda believes that it is both de-sirable and possible to "redefine political and social progress in ways that reveal the social hierarchies of racism, classicism, sexism and culture-centrism not to be natural. ... Can it be that feminism and similarly es-tranged inquiries are the true heirs of the creation of Copernicus, Galileo, and Newton?"

So, she argues that if one sees past the myth that science is value-free ob-jectivity, it can become for people outside the power structure a personally empowering tool, rather than a disenfranchising force. Her aim is more po-litical and activist than mine, but we have recognized some of the same per-sonally validating and empowering potentials in scientific thinking that has been stripped of its conventional ideologies.

NOTES

1. The early philosophers seemed able to discuss these divine natural forces fearlessly, that is, without the fear of offending gods or of being punished by them. See chapter 1, note 2.

2. One is often not talking about the immediate emergence of a familiar materialism, or the natural versus supernatural in the modern sense, when one discusses the origins of science. There was very little in antiquity of anything like complete a-theism in the modern sense (e.g., Drachmann 1922). Theism was belief in the "official" gods of pagan mythology, such as Zeus and Hera, so a-theism mostly signified without-the-pagan-gods. In Roman times Christians were atheists in this sense. There long were doubters of various sorts, but there was little in Europe that could be called atheism in our sense until the eighteenth century. Since physical, mechan-ical forces in our sense had not yet been conceptualized, a modern atheism was barely possible and hence quite rare, and the subject is complex (as Drachmann demonstrates). The physical world of the mainstream Greek philosophers was filled, if not with the gods of Olympus or the local spirits of the rural and middle-class people, at least with all sorts of things that we would call supernatural today in this context, like reason, purpose, and souls, and even moral qualities, that move nonhuman things about and shape them.

3. Mathematicians argue over whether or not we should have deemphasized Euclidian ge-ometry in education. Some claim that other forms of mathematics teach logical thinking just as well, or that computer use also teaches one how to be rigorous and systematic in one's think-ing. The advocates of Euclidian geometry claim this is not true, that it is not so easy at a critical young age to grasp *the nature of proof* from other forms of mathematics, and that these others are too easily taught as computation and problem-solving devices. Moreover there is a visual

component to Euclidian geometry that can help visually oriented people too into habits of proof and rigorous thought.

Geometry was valuable to me because it helped me toward realizing that even if something seems intuitively obvious it may not be true and that some additional means of proof (or disproof) is desirable and can be relatively simple. As a very simple example, it may seem obvious that the opposite angles formed by two crossed lines are equal. But they could differ ever so slightly. Someone else could claim that they do not look equal to him, and I could only say that he is nuts. Then he could say that I am nuts, and so on. We could measure, but one could claim that slight differences are beyond measuring power, or that we need to measure all possible angles. Once one proves geometrically that they must be equal, it becomes clear that this way of making a judgment is superior to intuition alone, or even measurement. To take a more advanced example, it may not seem intuitively obvious that the interior angles of a triangle of any shape should add up to 180 degrees. But we can prove this more satisfactorily by geometrical thinking than by even measuring all possible combinations of angles and adding them up. Geometrical proof leads to many conclusions that are not necessarily intuitive at first.

Intuitions need testing because sometimes they are correct and sometimes they are wrong. When one learns how to learn from being wrong, what sorts of important things one was tending to overlook, what sorts of careless jumps in drawing comparisons one was prone to make, then the intuitive power can improve greatly. Systematic thought is very powerful and can help to check and improve the intuition, especially if one is reflective. Abstract geometry is at the very least a superb vehicle for such learning. I think we should go back to teaching it more widely in grade school even if it has little "practical" value.

4. Why should the art of serious discussion have become so disfigured? There are many possible reasons. Our society in general has become increasingly legalistic and adversarial. Economic, political, ethnic, professional, religious, and sexual factions have become politically self-conscious and demanding. This has increased the adversarial tone in public life. We have at least temporarily lost much of our sense of purpose as a people in the wake of the Kennedy assasinations, Vietnam, Watergate, the Iranian embassy capture, the Iran-Contra scandal, economic decline. Then, too, if we are becoming the other-directed persons that David Riesman predicted in *The Lonely Crowd*, we will see serious matters as being out of our individual hands. Social acceptability and position will become of foremost importance, and confrontation and banter may indeed be sufficient to ensure these. And possibly of great importance is the amount of time many of those with whom we would otherwise interact and become involved spend their time passively, watching television or at other diverting attractions. These may absorb time, encourage passivity, create an illusion of participation, and condition the attention span to small blocks of time. We have become terribly busy just *reacting* to things.

We do not know the final answer to this great question, though.

5. My language is quite restrained compared to that used by Jaroslav Pelikan, former dean of the graduate school at Yale, in elaborating on his commissioned book (Pelikan 1984), to some 100 university presidents, graduate deans, and foundation officials at a conference sponsored by the Carnegie Foundation for the Advancement of Teaching, the Institute for Advanced Studies, and Princeton (*Chronicle of Higher Education*, December 14, 1983). Apropos of specialization, he notes that, "the difference between good scholarship and great scholarship is, as often as not, the general preparation of the scholar in fields other than the field of specialization. It is general preparation that makes possible that extra leap of imagination and analogy by which scholarship moves ahead." But the response to worsening economic times has been for even greater specialization, a trend that gives him "nightmares."

Pelikan sees other factors also leading to a decline in the quality of young scholars. Despite the fact that he sees some few very bright young people entering scholarship, he sees an overall bleak future if some action is not taken. "I am deeply disturbed by the question of who our scholarly posterity are to be. How can I communicate to Yale juniors and seniors the excitement and fulfillment I have found in a life of scholarship, without leading them down the road to frustration, disappointment, and tragedy? . . . I have come to believe reluctantly but ineluctably, that the very survival of scholarship is at stake today."

As a possible correction to the trend, he suggests that graduate schools begin to admit students with broader, rather than more specialized, undergraduate backgrounds. He also suggests that students might be admitted to graduate divisions rather than to specific graduate departments.

I would recommend that we must also appreciate that the granting system and administrative pressures increasingly reflect corporate business values in unwisely pressing for specialization, competitiveness, and grantsmanship. I hear bright young people saying that if they have to channel their energies narrowly and play politics endlessly in academics or in business alike, then they would just as soon be making money. We will have to work hard to make working conditions and academic freedom better for bright young scholars.

6. Ideological reductionists have often been criticized for attempting to reduce their critics all to "antiscientific" "mystics" and "vitalists." The "nothing-but" habit of mind makes it easy to ignore cogent criticism as well as complexity.

7. I only came upon Paul Feyerabend (1978) and Michael Polanyi (1962, and Grene 1969) after this manuscript was complete. They are both cogently critical of the intense faddishness among scientists and laypeople alike in the belief that scientific thinking must be an impersonal method. The reader will want to turn to them for further reflection on this subject. There are items where we agree and items where we disagree. But I have not expanded my manuscript to give details. I believe my position on this issue is consistent and clear enough in the text so that the concerned reader can make thoughtful comparison.

8. Spring (1972) details that American society adopted a corporate model of organization for dealing with the rapid industrialization and urbanization at the beginning of the twentieth century. He focuses on the reshaping of the school system to provide young skilled workers to meet the needs of the corporate state and business corporations—specialists who can fit into organizations and help in or conform to scientific planning.

This is not the same as earlier goals of bringing enlightenment to self-reliant yeomen and citizens. Or, Jefferson's ideal of education, as a bulwark against oppression so that the citizens "may be enabled to know ambition under all its shapes, and prompt to exert their natural powers to defeat its purposes" (*Bill for the More General Diffusion of Knowledge*, 1778).

Douglas (1986) discusses why institutions have difficulty in pursuing rational goals. Beetham (1987) reviews theories of bureaucracy.

9. There is no reason to question this as a legalistic statement. One cannot stand in the same stream twice, and so on. But the disclaimer would be more reassuring if one did not read statements such as the following from the editor of a leading "neoconservative" forum (Tyrrell 1982).

The rise of the *American Spectator* and the decline of Liberalism is no coincidence. It is God's will.

If our liberals think they suffered during McCarthyism I wince to think of how they are going to feel in the future. Yet they deserve it.

It is simple enough to find fault with each of the political philosophies. No one is or should be above criticism. Surely I have been quite critical both of leftist and rightist extremists and an uncritical middle and of their willingness to deindividualize (chapter 9). But it is not in the wisely founded traditions of an open society when a group decides that they have God and/or truth/history on their side and that the opposition should be hounded or punished. (Neither is it philosophically defensible. This is much the point of Karl Popper's extensive discussion, *The*

Open Society and Its Enemies, to which the reader may wish to turn for additional information.)

McCarthyism refers to a hysterical era in American history when reputations were ruined, jobs were lost, people were socially outcast, all without due process. Even though I was a child at the time, its unfairness was conspicuous, indeed fortunately, and eventually to most thoughtful Americans. People, particularly influential people, with liberal or sometimes even moderate ideas were branded as subversive or as enemies of society. There was a fear of voicing opinions that might be taken as suspect by militant conservatives and other activists, fear of associating with people who might one day be branded was created from the national to the community levels. For all of the injustice, hurt, and furor, very few true subversives were actually found and the nation eventually rejected the whole movement as a black day in our history.

It was a witch-hunt of a sort, during a time when the United States was angry and confused because of the rapid emergence of power among Communists abroad. How did the Soviets gain atomic weapons? Why did China fall to the Communists? Who could one blame? And so on. A mood was created in which rumors and suspicions, with no good rules of evidence or fair hearing, were sufficient reasons to harass and harm. As unfortunate as the unnecessary human suffering and damage to American traditions of justice and fairness, was that the Senate committee hearings diverted national attention from a sober analysis and debate of the world and national situation.

Passionate scapegoat hunting may make us feel that we are being clever and doing something helpful when a situation is bad, but it is an unlikely path to confronting and correcting the actual sources of trouble.

Amusingly, a fifteen hundred-page FBI report has been released of investigations of violinist and physicist Albert Einstein conducted primarily during that period of the 1950s (FBI filed reports of Einstein as a spy and kidnap plotter, *New York Times*, 9 Sept., 1983, sec. 4, 17). He was alleged to be behind a Communist plot to take over Hollywood (the film industry), to be the leader of a spy ring, to be a participant in the kidnapping of Charles Lindbergh's child, to have invented a robot capable of reading human minds, and to have invented a destructive ray. Einstein was cleared of each charge by the series of investigations. He was possibly the most widely known and respected man in the world and this astonishing scenario gives a vivid flavor of the extremes of serious concern that were possible in those regretable times.

The neoconservative magazine *The American Spectator* is not to be confused with Joseph Addison's *Spectator*—a famous newspaper that he used to poke fun at scientists in the early 1700s.

10. Galileo had in mind educated commoners, as distinct from church and throne. He had little hope for the "common people who are rude and unlearned" (Jacob 1988).

11. She also argues provocatively that conceptual science and scholarship have made their most valuable leaps at times when they have idealistically been tied to "emancipatory interests" that question the predominate ideologies.

CHAPTER 12

The Liberal Arts Agenda Reconsidered

The wayfarer,
Perceiving the pathway to truth,
Was struck with astonishment.
It was thickly grown with weeds.
"Ha," he said,
I see that no one has passed here
In a long time."
Later he saw that each weed
Was a singular knife.
"Well," he mumbled at last,
"Doubtless there are other roads."
        Stephen Crane, "The Wayfarer"

Can the liberal arts objective still be approached by individuals? *How* can critical-thinking skills be developed? Could the learning of science play a role in this?

This chapter portrays some important aspects of the current status of the disciplines of knowledge, of our institutions of higher learning and research, and of the world that individuals with strong "Liberal Arts Ideals," including students, professors, and others, must negotiate.

An Incredibly Resourceful Species

The sea is a deadly wilderness for people in small boats. Yet the Polynesian culture long ago spread from Fiji across the vast Pacific to Easter Island, and from Hawaii south to New Zealand. Dangerous travel between the farflung islands of the eastern Pacific was done without compass or sextant. This navigation skill has been one of the amazing accomplishments of humankind. The building of the pyramids seems trivial beside it.

That great skill at building and handling boats was necessary, goes without saying. Great patience, courage, and confidence in the navigator's judgment should also be obvious. The Pacific Ocean is an enormous place and an island is a small target. A slight error in course of only a few degrees could easily mean disaster. The sun is too large to pinpoint and too alone in an empty daytime sky to offer the precision required. The accurate navigation of Polynesians, Micronesians, and other Pacific peoples without a sextant, chronometer, or compass is simply astonishing, as anyone who thinks deeply about the problems of navigation can testify.

European explorers and colonialists, venturing out with their new scientific navigation devices, single-mindedly seeking riches and lands and souls to conquer, did not stop to probe carefully how the "naked savages" had found their way about over vast stretches of open ocean. This largely unstudied question became roughly a mystery for hundreds of years into our time. Their complex methods, about which they were often also secretive, have been more recently studied by Westerners and were described for the Puluwats in Micronesia, in *East Is a Big Bird*, by Thomas Gladwin (and references in Kyselka 1987).

The navigators learn a sort of mental map of their Pacific world and its islands, in relation to a mental map of the positions of various stars as they rise and set on the horizon and as this changes throughout the year. This is simply an enormous amount of information to master. Still, it is not nearly enough by itself to allow one to reach a destination safely.

One cannot set a direct course and be assured of not drifting or being blown off of it. One cannot depend on dead reckoning. If one tacks into the wind, weaving back and forth, navigation is further complicated. In navigation, one must *compensate* frequently. The people of the Pacific had many ways of doing this. They learned the currents by the different forms of life in this area of water or that. They learned to use the orientation of wave patterns to maintain a course. They actually learned to read, by their frequencies, regional swell patterns and to understand how subtle risings and fallings of their little boats related even to what we would understand today as the *reflections* of regional ocean wave patterns off of distant islands, or the *obstruction* of other waves by distant islands. It has been told that the male navigator could identify subtle patterns of ocean swells by learning to sense gentle changes in the weight of his testicles as he squatted on the deck of the boat. This last is difficult to verify, but it does make the point that very delicate discriminations must be made and unusual talents cultivated to sort out subtle wave patterns of mixed shapes, frequencies, directions, and amplitudes in the choppiness of the open ocean.

Then, when close to their targets, cloud formations might give them clues of land. Certain species of seabirds, such as terns, flying in a uniform direction to roost at dusk reveal the direction of land. Other species are useless.

While Europeans were living in mistrust of their senses and abilities, mostly humbled before kings and clergy and clinging to faith in authority, only a few daring to begin relearning to use individual resourcefulness and thought, naked Polynesians were trusting their lives to their senses and wits and moving systematically over an ocean world far more vast than all of Europe. While clearly not "noble savages" or utopians, they developed diverse social and economic systems and means of population regulation that often allowed them to live easily and to be extremely happy with each other and with the resources and beauty that nature offered.

We *can* be incredibly resourceful creatures.

But a great many Westerners have largely returned to self-identities as isolated, if prideful, fragments of great ideologies, religions, and/or states. The potential of patience and thoughtfulness, as shown by Gandhi, da Vinci, Franklin, Jefferson, or Wollstonecraft, can go out of focus. Perhaps Anglo-Americans have mostly made the shift in national character (predicted in 1950 by David Riesman in *The Lonely Crowd*), from the independent inner-directed personality, to the other-directed personality, finding our security in brands of conformity. In any event, one can forget how much was done by small communities like Athens, Florence, and Venice; by even smaller groups such as the Royal Society, the Pythagoreans, or tribes of Polynesians and Melanesians; or by individuals like Buddha, da Vinci, Confucius, Galileo, Gandhi, Clarence Darrow, Harriet Beecher Stowe, Henry David Thoreau, Alexander von Humboldt, and Helen Keller.

Indeed, many if not most of us are implicitly taught litanies of intellectual impotence. Beneath the commonplace that anyone can grow up to become president or get rich, the stronger social message may be that we are pawns so pushed about by history, by economic and philosophical forces, that we can only adapt and swim with the current and try to make some quick money. One is told that history never did require great thinkers and doers, and that anyway it is no longer possible to be effective because life is too complex.[1] Why the PTA buys such unproven ideas is beyond me. Perhaps great states and ideologies need to promote such beliefs to keep their followers following. Perhaps the dominant capitalist and socialist production and consumer systems need to keep people's energies focused on their economic productivity and buying habits. Perhaps the accomplishments of our technological age have put us in such awe of techniques that one can only think of mastering techniques for living that were devised by others, and we do not learn how to go about developing our own—a new Age of Faith of sorts. Certainly the emphasis in most schools is to teach techniques and information rather than to provide experiences thoughtfully fostering exploration, discovery, imagination, interpretation, self-actualization, and self-discipline. Perhaps we live in times when "people with courage and character always seem sinister to the rest," as Hermann Hesse put it. Who knows? But it is clear that behind the common highmotivation rhetoric there is a strong tide of conformity, timidity, and even defeatism.

Scholars similarly seem too often to suffer from an "epistemological hypochondria," to borrow Geertz's phrase. Though Gellner (1988) argues that much of this fashion of pessimism actually conceals dogmatism. "The argument tends to be: because all knowledge is dubious, being theory-saturated/ethnocentric/paradigm-dominated/interest-linked (please tick your preferred variant and cross out the others or add your own), etc., therefore the anguish-ridden author, battling with the dragons, can put forward whatever he pleases." Scratch a relativist and watch an absolutist bleed—perhaps sometimes true.

History has taught that our senses can deceive us. Our reason can deceive us. Our common sense can deceive us. Our languages can deceive us. Our philosophies can deceive us. Our systems of justice can deceive us. Our systems of government can deceive us. Our religions can deceive us. Science can improve certainty dramatically, but it is difficult to master, and science can deceive us, especially in its ideological manifestations. One should reserve some skepticism for everything that appears to offer certain knowledge.

Fearful, or too lazy to make their own judgments, people will join a herd, intuitively feeling that a dozen eyes or a million eyes are better than their own two. But history also teaches that the herd may and probably will make very stupid judgments at any given time. It is easily manipulated through its slavish trust in its symbols, stories, and images. Those the herd selects to be its thinkers may well have been seduced by one ideology or another that, if pressed seriously, they probably do not understand very well after all. One should certainly reserve skepticism for the judgments of the herd. Indeed, some argue that "group-think" is a sure path to bad decisions.

Yet, the intellectual accomplishments of our species have been simply remarkable. They were not won by fearfulness and timidity. It is one thing to be concerned with difficulties and it is another to feel that one cannot work out ways of improving one's judgment. It is one thing to be skeptical and appropriately cautious and it is another not to work toward a system of values and not to risk a qualified degree of trust for anything but money. Similarly, it is one thing to be open-minded and it is another to give equal weight generously to nonsense that comes one's way.

If an inexperienced party goes into the wilderness they should know of lightning and poisonous animals and plants, of the perils of getting lost, of polluted water and of earth slides. One should be able to judge realistically if one is in shape for a rugged thirty-mile hike. This is only being realistic about the environment and about one's limitations. But all of this knowledge should not make one shun the woods. We learn to respect such dangers in order to survive and enjoy the wilderness. When we prepare ourselves, the dangers become no more than part of the adventure. Without knowing well the dangers, and the resources, of the wilderness, the adventure could instead be misery, if not fatal. Humankind came from the actual wilderness. Indeed, it was our element and we mastered living in it. We needed to fear it only when we did not know it.

I see parallels with our stance vis à vis the intellectual landscape. To one hearing of its dangers it may seem that we are living in a cruel, unappealing, and hostile wilderness. But again, humankind has survived being born in an intellectual wilderness, and while we are still stumbling about in it, *never have we been in a better position to understand it and our place in it.* As a *society*, possibly we are at the brink of irrationality as some warn, but the intellectual resources and perspectives available to the *individual* are marvelous. We are in a wonderful position to grasp the outlines, at least, of how our brains, languages, common sense, and philosophies have conspired to tumble us so often into deep holes. The instructed and thoughtful person should be able to avoid many pitfalls and see many panoramas.

One may groan, what possible good is this individual potential when things seem to be getting so bad and if civilized consciousness, in its best sense, is indeed beyond salvaging? Perhaps we should just try to make money and go dancing and believe whatever feels right.

We do not know for certain that humanity is sliding into gloominess or even chaos, though, the signs are admittedly nontrivial. Much of this can be seen in the large Presidential Report, *Global 2*000 (Council on the Environment 1980).

The world's population is rising alarmingly fast. Some critical resources are becoming restricted in patterns that will even further exacerbate political and economic problems. We face growing ecological instability. We see the rise of multinational corporations and an international banking system with narrow interests, beyond the oversight or control of individual governments. People evermore depend on the mass media for their information about reality. Ownership of the mass media is becoming evermore concentrated. Television and other mass communications affect us in other ways that many feel distorts reality and selfawareness.

Literacy, certainly effective literacy, seems actually to be on the decline at a time when the world can least afford it. The potentially wise are channeled into specialties to supply "experts" who may have no overview except from their narrow vantage points of background and selfinterest. They are under warping pressures in often badly managed productivity units and career systems that can even sometimes bring out the worst in one. The complex socioeconomic systems maintained by modern technology are delicate and could break down, as Roberto Vacca describes in *The Coming Dark Age*.

It is hard to see how institutionalized economic and intellectual poverty can be eliminated. Organized crime insinuates itself ever more deeply into society. Techniques of psychological manipulation are becoming more sophisticated even than Aldous Huxley imagined when he wrote his essay, *Brave New World Revisited*, to update his novel *Brave New World*, or than when Vance Packard wrote the first edition of the *Hidden Persuaders* in 1957. Huge financial and technical resources and psychological manipulation techniques have become available to fanatical religious and political groups of the right, left, and center who can powerfully insert their beliefs into public debate on critical issues, and thus muddle any intelligent broader discussion. Lives overly dominated by television viewing, consumerism, sex, greed, gossip, or drugs can carry with them the illusions of participation in the world. There are argued to be serious sociological problems related to the character and philosophy of our managerial and professional classes (e.g., Douglas 1986, Lebedoff 1981, Maccoby 1976, Whyte 1956).

So if the general social discourse is indeed growing evermore incompetent, then it will be evermore important for thinking individuals to take personal initiatives to negotiate their own development.

It Is Not a Matter of Simple Technique to Become a Whole and Thoughtful Person

So how can persons who desire it develop their potentials? What set of steps or techniques could one lay out? This is a delicate point. There is good advice to be given, certainly. But wisdom, paradoxically, cannot necessarily come from following even wise advice. Following and discovering are not necessarily tied to one another. Make friends with paradox, the Chinese say. Take good advice, but do not slavishly follow it. Try to understand but do not feel that you have understood once and for all. Good advice and the best technique can become a trap and grow sour just as love or the best intentions can. And, moreover, the best advice at the wrong time or to the wrong person will be no more meaningful or profound than the sayings that we find in fortune cookies and chuckle over in Chinese restaurants.

We live in a technological age and many of us tend to stand in awe of techniques. Many of us would not begin seriously to play sports, write literature, make love, raise children, cook, or make music without first seeking some techniques. This is fine, but just knowing that techniques exist may also intimidate us. There is nothing at all wrong with using good techniques. But we have let atrophy the arts of observation, exploration, self-study, invention, and of teaching ourselves. In a sense, imagination and will, once esteemed, have become degraded into mere whimsy, desire, choice, and determination. And so it is necessary to emphasize that if one expects to become wise by following techniques doggedly, one will quite possibly end up only fooling oneself, concentrating mostly on the technique itself, becoming unjustifiably smug and self-satisfied when it is mastered.

One should strive for the delicate balance, to be able to discipline oneself without becoming trapped by discipline. In Zen, for example, any instruction will usually be balanced by paradoxically poking fun at the instruction, so as to emphasize the value of a middle course that is difficult to prescribe and that individuals must find themselves. There is the story of Baso, the disciple of the Chinese master Ejo, who is spending much time sitting in meditation.

> "Why do you meditate?" Ejo asks Baso. "To become a Buddha," Baso answers. Then Ejo picks a floor tile and begins to rub it very hard. "Why do you rub that brick?" Baso asks. "To make a mirror," Ejo answers. "But no amount of polishing will change a brick into a mirror," Baso protests. "Just so," Ejo replies, "and no amount of cross-legged sitting will make you into a Buddha."

East or West, delicate balances have always been hard to find, and even with good instruction one must in the end find them for oneself. Experience has always revealed the danger that any good rule or advice can be misunderstood; we come to such always with private and social preconceptions. The great spiritual leaders have used the attention-concentrating device of speaking in parables. One cannot hope to make sense of parables without careful reflection.

Philosophers would dearly love to abstract the presumed rules of mental discipline by which truth can be approached. We all would. It would seem ideal if fallible human judgment could be replaced by solid rules and techniques of thought and observation. Philosophers of science search for the *algorithms* that might give mathematical certainty to judgment. We would all like that. But, sadly, the status of knowledge is hopelessly far from it. There are *several* scientific methods, and the brilliant breakthroughs often come when conventions can be broken and judgments can be made on some new basis.

In one sense, great teachers have always set down prescriptions for living. Long ago, for example, Confucius said, "He who learns but does not think is lost; he who thinks but does not learn is in danger." This is obviously recommending a "technique" of pausing to reflect and of being open to changing one's thinking with experience, and one that is as valid today as ever. Such rules have been good enough to maintain the largest and oldest of all civilizations. Such traditional rules for thinking and experiencing can lead to spiritual peace and reassurance in day-to-day life, even though, as Confucius also said, "The common people may be made to follow the way, but may not be able to understand it."

In some ways our situation has changed little in over twenty five centuries. People need, want, and consciously or unconsciously use rules for using their minds. One problem, though, is that we may think of ourselves as having come far beyond the ancients and far beyond traditional cultures. We may see ourselves as being in a state of progress and advancement. There is a feeling that our customs are superior because they are supposedly based on powerful science. Even comparatively well-educated people can become intellectual consumers and choice makers, and in this sense passive, and may be surprisingly intimidated from exploring, experimenting, pursuing the wisps of original thought that come to their minds, inventing, constructing creative lives. In the education of children we may actually curtail their initiative, exploration, and invention and instead discipline them to learn a menu of facts and techniques at a theoretically correct pace instead of teaching them to find their own pace and discipline themselves in thoughtfulness, and yet to retain playfulness.

Simply by way of example, D. C. Beard's *The American Boy's Handy Book* from 1890 is a charming glimpse of a period when young people were eager to turn nature or old junk into playgrounds for the imagination. It is filled with tips on applying "good old American ingenuity" to having fun and getting along on one's wits. The once popular book fell into obscurity for half a century as national character changed, and it has only now been reprinted for its historical interest. Why has it lost its appeal? In the present context this reveals something of the changing national character. We seem to prize the possession of acquired things over the ingenuity and resourcefulness needed to build or invent our own. Having money to spend for recreation now seems more respectable than having imagination to apply. Conforming to the values and activities of one's peers, competing in fashions and in bought possessions, seems admired more today than striking out on one's own path. Using "the right" techniques in the short term seems a more esteemed goal than being able to think a problem out and solve it with some imagination in the long term. Activities in childhood today seem to prepare a person more to follow rules, compete in trends, and consume merchandise than to develop their imagination and wits.

Some might object that I here place too much emphasis on thinking. Yet even in the Eastern disciplines where there is an emphasis on the intuitive, this must come *through* thoughtfulness if one is to reach "pure consciousness" or "no-mind." This is a point where Westerners have great difficulty in fully appreciating Eastern attitudes. One should move beyond thoughtfulness to cultivate a high refinement of the intuitive powers, but it is just plain stupid to think that one can leap past thoughtfulness and have instant wisdom.

Despite some mistakes, such as the bloodiness of the French Revolution, and the popularity of Mandevillism, the Age of Reason was not merely cold calculation but was also a time of growing compassion. The abolition of slavery began, the higher education of women was urged, prison was substituted for torture and death as punishment for many crimes and insanities, and soldiers and workers were better treated. I think of Baron von Humboldt idealistically teaching geology to the ignorant but enthusiastic miners of Steben so that they might begin learning of their environment. The Enlightenment, of course, fathered our own society and system of government and we can see its emerging humanism reflected in the ideals of the Bill of Rights, or in the *Federalist Papers*.

When I was 20 I was fortunate enough to spend an afternoon with Aldous Huxley. Shamelessly, I asked him what was the main point that he was trying to make as a writer. Apparently he took my youth and sincerity into account, for he did not throw me out of his house and tell me not to come back until I had a more original question. Instead he kindly said, "That love without intelligence can be ridiculous. That intelligence without love can be terrifying."

Reading and study alone are no substitutes for experience. And all that I say about study and thoughtfulness are tempered by this belief. But, experience without thoughtfulness and compassion is no better.

The mysterious individualist, B. Traven, writes a story of a drowned Indian boy in Mexico, *The Bridge in the Jungle*. Traven's sharp eyes and mind weave a simple tragic incident into a rich tapestry, a window into the "soul" of a people. He comments in the text:

> Aside from the fact that philosophy actually pays if you know how to
> handle it right, experience has taught me that traveling educates only those
> who can be educated just as well by roaming around their own country. By
> walking thirty miles anywhere in one's home state the man who is open
> minded will see more and learn more than a thousand others will by
> running round the world. A trip to a Central American jungle to watch how
> Indians behave near a bridge won't make you see either the jungle or the

bridge or the Indians if you believe that the civilization you were born into is the only one that counts. Go around and look with the idea that everything you learned in school and college is wrong.

Negotiating the Knowledge Establishment

I imagine that Traven would agree with a chuckle if I next suggest that one can even learn at school and college if one keeps one's mind open as one circulates around a campus and probes the libraries. One would need to be skeptical both of cynical and idealistic preconceptions, rumors, and group-think. One can study the place carefully for what it is, and not for what one thought it was, or even for what it claims to be.

The sociology of institutions of knowledge is complex. The diversity and complexity can be interesting in their own right. Many people seem to believe, somewhere deep in their minds, that knowledge has been cut up into neat pieces, like slices of pie, that the edges fit, and that scholars have worked all this out and this is what the academic disciplines represent.

This idea is misleading. Disciplines evolve along crooked paths. They may easily forget their places and origins in the scheme of human concerns and forget the issues that gave birth to them. Some subjects get left behind and are little, if at all, studied. It is important to appreciate that often our disciplines do not study the most interesting or relevant subjects, but merely those that are easiest to study, or that are well funded because of pressing needs and interest of government, industry, or the public. (Then too there are fads that have had neither a solid socioeconomic *nor* intellectual base, as Kuper 1988, pp. 9–14 argues.) Intellectual issues perhaps are even *usually* today set by socioeconomic factors, rather than by objective evaluations about the status of knowledge and the needs of the discipline based on its most profound knowns and unknowns.

No one on my huge campus could be found who studies Eastern logic, for example. Many anatomy departments nowadays study little if any anatomy. The biochemistry of brain and muscle is more likely to be their emphasis. Even the average educated person is surprised to learn that what ecology departments study and teach has little if anything to do with what is commonly meant by ecology. Political science presently in the United States may deal primarily with rules, the structures of institutions, and perhaps theory of governments. It may focus little if at all on the critical dynamics of political bargaining and intrigue. It may ignore the politics and power relationships of family dynamics, sanity, race, sex, religion, knowledge — in the sense that the important but justly controversial Michel Foucault, for example, discusses some of these. (See, for example, Foucault 1980, but also Merquior

1985 for a sample of criticism.) Health sciences are often the studies of diseases and not sciences of health—but that is too easy. One can't tell a book by its cover, nor necessarily by its title.

The bottom line is that the individual who wants to make maximum use of the resources of schools, libraries, reading or discussion groups, or whatever, can't risk being passive and simply enroll in a series of courses or read a list of books with promising titles. One had best spend some time finding out what actual conditions exist and options are available and why. Since the average person is content simply to accept what the System offers, this may take a bit of lonely work. But understanding the System itself can be as interesting and as relevant as the knowledge generated by its components and can help put those components in perspective. So it is hardly time wasted.

We can get deeper into this subject of the nature of intellectual disciplines by moving on to some of the perspectives that would benefit a thoughtful person. Much of this has been hinted at earlier in this book. But now we will look at it in terms of the present issue.

Psychology

Socrates' advice was, "Know thyself." This general idea has been important in both Western and Eastern philosophies, such as forms of Buddhism. Some familiarity with formal *psychology*, in conjunction with some study of self, is essential to becoming an independently thoughtful person. Some schools have very good introductory courses in psychology and some books will give a very good introduction to the functioning of mind and to the tricks that we can play on ourselves and that others will deliberately or inadvertently play on us and on themselves.

But, there is no guarantee that any given course or text will be useful. There is often a tendency to promote the "cutting edge" of *any* discipline, even in introductory courses, and to deemphasize fundamentals as conceived of here. To the purely reductionist mind, the details of mechanism *are* the fundamentals—the basic particles, so to speak. *This is an extremely important point to grasp.* Thus, an introductory course or book on psychology could emphasize neurophysiology or conditioning and learning, the testing of personality stucture, and so on, and be of quite limited use in the context developed here.

The cutting edge of a discipline may be set by intellectual challenges, and certainly the cellular basis of brain function *is* intellectually interesting. The cutting edge of any discipline may also be set by external and internal economic incentives. But almost anything can become intellectually interesting, profitable, or career advancing. Many graduates in psychology aim to

work for industry, giving advice on how to sell cars and soap, or how to select, train, and manage personnel. They may aim to become clinicians who
will try to help people change particular undesirable habits and attitudes, to
offer anxiety relief. Thus, many psychologists are technique oriented and
the individual becomes a means to an end. Some of them may even view
people as constellations of personality and intellectual traits to be manipulated toward social "adjustment," consumerism, manageability. Enlightenment may take on a very particular, if important, meaning. Their view of humans may be astonishingly particularized, which is not to deny the
important power of their techniques, in the right hands, for circumscribed
goals.

For example, it was John B. Watson, president of the American Psychological Association in 1915, who coined the term *behaviorism*. He aimed
for social engineering and was "an advocate of a creative scientific reductionism that narrowed the subject matter of psychology while expanding
the possibilities of its application." He went on to become "an advertising
executive who devised enormously successful campaigns to sell cigarettes
and toothpaste through cynical manipulation of emotional insecurities"
(Reed 1989).

Thus, while a study of psychology is to be highly recommended, some
courses and texts in modern psychology will be of limited interest in the
present context. One might better spend the time reading Confucius,
Balzac, Cervantes, Tolstoy, Traven, Orwell, Proust, Buddha, Kierkegaard,
Mark Twain, or a dozen other outstanding thinkers, and couple readings
with astute observations of persons and events. Dig out thoughtful people to
talk with. Start a discussion or reading group of able, diverse, kind, and sincere people. Graciously change groups from time to time. Beware of cults.
Do not mix only with peers. Move across generations to engage both the
experience of the old and the freshness of the young.

The close relationships should be based on trust, respect, kindness, openness, sensitivity. Such relationships are important not only for personal support but can facilitate self-exploration and form a basis for better understanding of others as well.

One should be kind, caring, and open with oneself and come to view mistakes and crises as opportunities to learn about oneself and others. Otherwise one may encourage denial and/or narrowness in oneself. One should
grasp the difference between selfkindness and selfishness and defensiveness,
for the last two are forces for blindness.

Even a good study of psychology (which *is* completely possible) would
not, of course, tell you "who you are" in any final sense. It would not let you
"know thyself" in a complete way. Psychology can deal mostly with func-

tioning and personality. These are important aspects of each of us but they are surely not *all* of us.

For most humans, *values* and *inspiration* have come from tradition, the family and peers, and more recently from advertising and the mass media. For a minority who want something more, they come from perhaps the study of the humanities, religion, or philosophy (in the broad sense). For only a very few will they come mostly from inner resources of creativity and insight. Yet the job of merging individual insights and dispositions with traditional values or with immediate possibilities on a spectrum of cultural values is nevertheless a formidable task for most of us.

Each individual is a dynamic, changing, product of his or her biology, culture, individual experience, goals, determination, and talents. Knowing what we *are* involves understanding what we *might realistically hope and dream to become* as well as our *status quo*. It is persons who can grasp how little we understand ourselves who may be able to grow most. We may need to get in touch emotionally and intellectually with the individual uniqueness (from childhood pain to underdeveloped talents and feelings) that exists beneath what we have been taught to believe of ourselves. But then this discovery must be part of a *process* of enlightened creation and growth; a healthy reconstruction if necessary, and not on the one hand merely intellectualization, or on the other merely some simple "release" of a supposedly completely formed entity captive within us. One sees much foolishness and even pain grow out of the rambling attempts of people to "find themselves" as though there is some true "me" already hidden in the grass like an Easter egg needing to be discovered and liberated within some time limit, before it rots or gets stepped on. The negotiation of real personal growth is usually at the same time more involved and yet easier than that unlikely quest.

History

A *sense of history* and of *historical thinking* is also to be highly recommended. It may be largely true, as Aldous Huxley put it, "That men do not learn very much from the lessons of history is the most important of all the lessons that history has to teach." But that is a complex statement, worth pondering, and it does *not* suggest that we should ignore history. The individual who aspires to superior judgment and perspective can surely benefit from a sense of history even if governments and humankind in groups seem so modestly capable of foresight improved by hindsight. As G. K. Chesterton phrased it, "The disadvantage of men not knowing the past is that they do not know the present. History is a hill or high point of vantage from which alone men see the town in which they live or the age in which they are living." Not much of our life makes very much sense without history. We are pushed and pulled by circumstances that are profoundly historical in character. We are all part of political, economic, social, and philosophical visions that are due as much to past epic debates as to present contingencies. The future course of events will in large part be determined by the past and by *the way we view the past.* Even our patterns of sexual attitudes and sense of family, the expression of our desires for fair and just treatment for ourselves and others, our patterns of hatreds and prejudices are the results of historical happenings. Ideas have origins and histories as much as do political boundaries.

But again, while some history is very well taught and written, too often it focuses on small issues, dates of battles, and genealogies of rulers. Much of it nationalistically glorifies the string of political decisions, wars, and books that have led to the status quo. Much of it is myth and in need of revision. But, some of it is carelessly revisionary, attempting to reinterpret events and trends in terms of climatic change or some ideological perspective, and so on. Complexity and human dimensions, unresolved aspirations, will be filtered out as being irrelevant to so-called more scientific and simple schemes. Introductory texts commonly leave out embarrassing but critical political or religious details, as we have seen.

Historical method is difficult and has many pitfalls. But its study gives us a critical sense of the time scale and the dimensions of flux in which we exist. It gives us cause for optimism as well as concern. And properly discovered, it is just damned interesting.

Cultural Comparisons

A *comparative cultural perspective* can be extremely enlightening by helping us to see our own place in the rich spectrum of human potential. A perspective on the multiplicity of actual solutions that humans have applied to the problem of how to be human can be both awesome and inspiring. It can save us from the *hubris*, vanity, and chauvinism that can be bred if we know or can respect only our own culture. It can help us to appreciate the positive effort that may be necessary to maintain what is best in the systems in which we live. As, for example, Ortega y Gasset argued in *The Revolt of the Masses*, unless we appreciate the highly artificial character of Western civilization, the forces of idealism, organization, and dedication that have improved it, it is highly unlikely that its *desirable humanistic* aspects will survive entropy, unmindful transformation, and decay to slogans. A partially actualized dream can never be fully realized or even maintained by those who cannot grasp the vision or the elements of its implementation.

Character and individuality develop not simply from accepting an accident of birth, but after we confront difficult concepts and choices in life. A comparative cultural perspective can aid one to see just how much of life *is* *indeed* choice unseen and not mere inevitability in which one has little alternative but to imitate and conform. Here again, though, courses and books in cultural anthropology may or may not enrich our perspectives, depending on their focus.

Science

Scientific and mathematical thinking enter our values and behavior in many both conspicuous and subtle ways; they should be understood. Much has been said of them in previous chapters.

Ideally science should be well understood also so that its *misapplication* can be detected (as also mathematics and statistics, e.g., Huff 1954, Berger and Berry 1988). Science represents a monumental advance in human thinking and we should understand in what ways this is so and for what reasons. But we should also understand the limits and misapplications of science. We should be sensitive to the fact that science—not the facts so much, but the way it sets intellectual and social priorities and focuses our attention—can be a Trojan horse for ideology and metaphysics.

When a complex subject matter is reduced to numbers and structure, we have not necessarily made it scientific, though that is where the quest for scientific respectability has sent some disciplines. Critics call this "math envy" or "physics envy." But *a thirst for social status* is not necessarily the same as useful progress. Quantification and the proper use of mathematics and statistics can be extremely, *profoundly* important. Yet it can be that a discipline is actually heading off into a mediocre or even sterile direction by reducing its subject matter largely to items that can be quantified, schematized, or experimented with easily.

We should remember the story of the drunk crawling about on his hands and knees under the lamppost one night, looking for his keys. "Are you sure you lost them here?" asked a passerby. "No, I don't think so," the drunk replied. "Then, why are you looking here?" the passerby asked in amazement. "The light is better here," replied the drunk. Sometimes disciplines ignore more interesting questions in favor of those that can be easily studied by the more "respectable" techniques. It can be forgotten that the most well-funded, high-market-value, research programs are not necessarily the most interesting or important in the broader scheme of things.

What commonly passes for scientific thinking may make a discipline turn away from subjects that no one has found a way to study quantitatively.

Thus, the "primitive" mind, fascinating and critical to an understanding of the human mind in general, has been largely neglected because no one has found "scientifically respectable" ways to study it.

A related problem is with academics and laypeople who confuse concepts and formulas with reality. It is easy to confuse the symbol with the thing itself. In Platonism, much influenced by Pythagoreanism, there is even the formal philosophical opinion that reality is not in the material world significantly, but is in the world of eternal forms. Some philosophers and historians would use the word *essentialism* at this point. Generally, essentialism refers to positions that place greater or total emphasis on presumed essences rather than the experienced, material world. Most of us are very deeply affected by Platonic, Pythagorean, or other essentialist thinking, attitudes, and perceptions without having the slightest awareness of the fact, or of its context, history, and implications.

Many scientists see scientific truths, laws, or firm theories, not as valid and excellent ways of describing things, but as being *the* True realities in a quasi-Platonic sense. This problem, again, is not one endemic to science. It has to do with abstraction in general. For example, as Walter Lippmann noted, "It is so much easier to talk of poverty than to think of the poor, to argue the rights of capital than to see its results. Pretty soon we come to think of the theories and abstract ideas as things in themselves."

The Pythagoreans believed that there is a reality beyond the senses and that it is numbers. Physics similarly reduced reality to equations and numbers and this worked perhaps too well. Rudolf Carnap in *The Philosophical Foundations of Physics*, needed to belabor that an equation in physics is only an agreement to stop inquiry. In effect we say we know enough when we can describe a phenomenon mathematically. We do not really know what gravity *is* materially when we have formulated the laws of gravity. We know what it is only in the sense that we know enough about it to predict how it will behave under given conditions. But even my introductory physics professor, unconscious of his own Platonism, did not understand this. He was taught that the formula was the actual phenomenon, period, and that is what he insisted in class. But the best scientists appreciate that equations only describe. They are terribly valuable abstractions, but are not the things themselves.

In both the natural and social sciences the system of description may become mistaken for material entities and causes. This very common mistake is not good science, but is an uncritical reversion to the metaphysical origins of science in Pythagoreanism and Platonism. It is particularly pernicious when social scientists become structuralists who in effect see people as mere numbers on tally sheets, or as the messy, fleshy, material manifestations of essential Truths embodied in the boxes and arrows of flow charts.

Will and intellectual individuality drop out of the discussion and become unimportant forces at best, illusions at worst. The System, the organization, comes first. We may even subordinate ourselves and our children to some System enthusiastically. We may welcome it as the structuring reality in our lives. (In a related discussion in *The Organization Man*, Whyte observed, the organization man "is not only other-directed, to borrow David Riesman's concept, he is articulating a philosophy which tells him it is right to be that way.")

The feeling can develop that one is even being especially scientific to describe situations entirely in terms of *one* theory (the seductive "nothing-but-ness" of misused reductionism). One can too easily end up bending the facts to fit a monistic theory, and treat deviation from theory as mere noise that blurs the signal—exceptions are insignificant and will somehow be explained away. Surely there are class struggles, for example; surely there is greed; but some economic determinists may become so rigid in their preconceptions that they cannot see the realities of the variety of self-interests and conflict beyond simple stereotypes. Economic determinists and other structuralists may insist that they are being completely objective and are being realistic, when they are being indeed *partially* insightful and "Realistic" largely in a Platonic or Pythagorean sense, which is actually quite metaphysical. A proper understanding of science should help one to recognize what can and does commonly advance from quasi science to pseudoscience.

Biological determinism can also be a trap. Some advocates excuse compulsive competition, aggression, compassionlessness, and even dishonesty on the grounds that these are *the* "essential," basic, aspects of human nature, true to the genes. But one can just as easily argue, with plenty of developmental and cross-cultural evidence, that these are merely conditional personality traits and that *the* (supposed) basic drives would be for peace, companionship, curiosity satisfaction, and so on. The hypothesis of *any* "essential" human moral nature is a much abused expectation from politics, philosophy, and religion. This is not science. All that biologists can be quite certain of is that human *capacities allow* the broad range of human experience, for better or worse. Children will learn to do what they need to in their society to get care and as much love and attention as they can. This may involve suppressing or enhancing biologically existing capacities for selfishness or cooperation, aggression or gentleness, even learning complex forms of verbs.

We very much need abstractions to help us to see order and patterns in the complexity of life. But these useful abstractions are also filters that will exclude much imporant reality if we do not understand their nature.

Next it is important to realize that here are *several* scientific methods. There are *experimental methods* used in nearly all the sciences, and *com-*

parative methods used especially in astronomy, biology, geology, anthropology. The *thought experiments* of Galileo and Einstein have had their place in the advances of science. Moreover, much of scientific discovery has not taken place by simple application of method. There is much art and intuition involved in discovery. Formal method is only a *part* of verification, data gathering, and communication.

Unfortunately, a feeling for most of this is not easy to come by in ordinary sources of scientific knowledge. As a professional scientist, it pains me to have to write this. In the university, large introductory courses tend to be used to cram facts and formulas that can be easily tested into students who are being screened for admission to a department, medical school, etc. We even sometimes refer to "screening courses." Introductory books can reflect this tradition.

It might seem proper to turn to philosophy for wisdom of all sorts, perhaps rich insight into science; but it should be clear by now that this is far from the case. Philosophers, especially in the recent English-speaking tradition, *have* made some interesting findings, limited insights into science, but they have tended to be preoccupied more with the logic of formal method than with the process of discovery or with attitudes and procedures that have actually worked. The philosophical profession generally has needed to define science in a *limited* way—in part so that it can distinguish science from philosophy and discover and establish its own turf and special claim to knowledge. Its job is not to explain science in a complete and unbiased manner—as Piaget, for example, argues in *Insights and Illusions of Philosophy.* Indeed, our academic philosophy rarely includes even Eastern philosophy, certainly not Mesoamerican, but is enmeshed in its own very particular set of issues today.

The Crisis in Educational Vision

The sorry situation outlined here would not change unless readers—including students—could make clear to administrations and publishers a demand for different types of books and for institutional structures to study them. The consumers might make clear to administrators and publishers that they do not want only intellectual shopping malls with rows of specialty shops. In *On Higher Education,* David Riesman details that, instead, due to economics and other factors, the universities have become academic marketplaces and that education is being determined by the student-as-consumer, who typically does not know what he or she really needs. Apathetic or docile consumers are not likely to get the best educations, unless the trend changes. *Caveat emptor.* (See also chapter 11, note 8.)

Keep in mind too that the disciplines are odd places where professors can find security and prosper very well professionally by being specialists. People in disciplines tend to promote what has worked well for them. But the strategy that is optimal within disciplines may not be an optimal model for all of society.

Even well-meaning administrators these days would need powerful constituencies if they were to move against economic, sociological, and political forces. Rarely can one set one's own agenda. Job security and career ladders are at issue. I sit and have sat on important committees at my university and nationally and have done some good. But this is not really much power in the larger scheme. I would have to convince people, from my colleagues to state and federal legislators, that people "out there," such as students, *very* badly want the changes that many of us already agree are needed—for there are many persons and pressures strongly pushing in other directions.

In 1982 David Saxon, physicist and then president of the University of California, wrote in *Science* that "liberal education is and will continue to be a failed idea" if students cannot be helped to understand what scientific knowledge is and is not, and cannot use it to evaulate their own experience.

He does not propose teaching liberal arts students science as a skill or offering them watered down science courses. His proposal is more ambitious.

First, he says, students should be helped to grasp the nature of physical laws. They should be able to grasp their utility *and be sensitive to their misapplication.* They should be able to grasp how scientific laws have been arrived at and what is meant by their truth.

Second, they should be helped to understand the operations of probability and chance in our complex world.

Third, they should be helped to grasp that science is not a catalog of facts, but an integrated body of laws, observations, models, and methods that provide a basis for critical evaluation of life's potentialities, and of the facts and theories of nature.

Saxon argues that "the ability to distinguish sense from nonsense is an indispensable aspect of a liberal education." Science and the liberal arts, he says, have too long been isolated from one another. It is time to rejoin them.

Why did the president of the United States' largest university system not simply *order* such good changes at his own institution and write personal letters directly to other university presidents? Because, the modern multiversity serves many masters, and presidents have limited roles and little political power these days, as former University of California president Clark Kerr pointed out in his *Uses of the University.*[2]

Why publish such an article? To attempt to develop a constitutency for his ideas, it is safe to assume. To test the waters and see if he could motivate support from some of the major factions that have power in institutions of higher learning, and that believe that neatly packaged specialists are what we each should be and are what they and the world need — the departments, industry, business, the press, the military, professional societies, the medical and dental professions, the legal profession, agricultural interests.[3] Moreover, one must convince personnel offices to buy the new human product and must convince average students that they are better off or even more marketable if they take the new courses that might be offered.

Why was it necessary for former Columbia University Provost and Dean of Faculties Jacques Barzun to call for radical educational reforms (that did not happen) in a series of books such as *The House of Intellect?* Why was it even necessary for him to point out that "men have come to believe that they can link knowledge and action without a regular gradation of intellects to harbor and diffuse ideas, or a common concern about the welfare of Intellect as an institution"? Or, "All thinking educationists are troubled when they consider the present chaotic state of education in our society." One would need enormous support to institute major reforms.

It might take a long time to develop any sort of integrated institutional educational strategy, or it quite possibly will simply never happen at all.[4] It is sobering that university presidents and other leaders in postsecondary education have so often and so cogently written passionately of what we have let our institutions become, especially since World War II, while their faculties and trustees will instead busy themselves with other more "immediate" and "practical" concerns and will herald the upside as they promote their own career systems and the accomplishments of these systems. People will best be advised to be very careful and thoughtful about their studies in our institutions of higher learning. *There are simply enormous and wonderful resources available, but one must work out a strategy for using them effectively.* Perhaps our institutions are best viewed as large living libraries in which the student or other user must very carefully develop his or her own use strategies.

While some private colleges and universities are reputedly a bit better off than public colleges and universities, in fact they are victims to many of the same pressures. They too must deal with "departmentalism," and the "balkanization of knowledge." They must raise funds through outside grants, alumni, boards of trustees, and the tuitions of the students-as-consumers and herald loudly the up side to all of this. Money speaks loudly and relentlessly in the Ivory Tower.

Pioneering Self-Development

Study is in any event no substitute for thoughtful experience. The develop-

ment of the modern rational mind was of a mind that *combines thoughtfulness with experience*. Well-founded *theories* have proved to be critical
guides to thoughtfulness, but they are not substitutes for life itself or for
what is to be learned from open-minded but reflective living.

So one who wants to grow should resist anything that threatens to inhibit
or to demoralize one's basic level of curiosity and exploration. Curiosity
promotes experience and promotes analysis of experience. It is precious.
But not all of society treasures it. Many, indeed, are suspicious of it. On a
related subject, one might recall Mandeville's lengthy "pragmatic" attacks
on publically financed education (beyond the technical) for the salaried
masses (and chapter 6, note 6). This could create job dissatisfaction, his followers have feared. It could also allow subordinates to become too
independent—"and should a Horse know as much as a Man, I should not
desire to be his Rider."

Don Quixote (1605) by Cervantes is not merely the story of an impossi-
bly idealistic and romantic old man who attacked windmills and did other
such crazy things. Many claim that it remains the world's greatest novel.
There is much more to the story than the stereotype taken from the first few
chapters suggests.

For one thing it is perhaps the first fiction in which there is transformation of character and intellectual outlook. Quixote has the courage and determination to leave his comfortable if small life and ride out into the world
to have *experiences*. The *inner persons* of Quixote and Sancho are well de-
lineated and we follow them as *experience and rational discussion lead to
qualitative mental growth*. Philosophy is tied to experience, courage, and
reflectiveness, and a person can thusly become transformed and can expand. In this sense Quixote and Sancho are manifestations of the Renaissance struggle. Moreover, the thoughtful reader can grow with the characters. There is a compulsion of the author to show others the way. Quixote
reflected the birth of a type of human that Europe had not seen much of for
a thousand years while the emphasis was on growth in religious devotion.

The great neuroanatomist and Nobel Prize winner Santiago Ramon y Cajal
was especially remarkable in that he developed his mind, talents, and work
in *virtual academic isolation*, claiming that he owed everything to:

> ... a profound belief in the sovereign will; faith in work; the conviction that
> a persevering and deliberate effort is capable of molding and organizing
> everything, from the muscle to the brain, making up the deficiencies of
> nature and even overcoming the mischances of character—the most
> difficult thing in life. (*Recollections of My Life*)

Cajal mentions *Don Quixote* as an important part of his reading in youth.
He also focuses on Defoe's *Robinson Crusoe* (1719):

... what impressed me most of all was the noble pride of this man who, by his own unaided efforts, found that an uninhabited island, full of dangers and pitfalls, was capable of being transformed, by the miracles of determination and intelligent effort, into a delectable paradise. "What a supreme triumph it must be," I thought, "to explore a virgin territory, to gaze upon scenes untouched by the hand of man, adorned with their original flora and fauna, which seemed created expressly for the discoverer, as a reward for his outstanding heroism." (*Recollections of My Life*)

Despite the colonialist and racist aspects of *Crusoe*, Cajal developed into a very decent human being. He drew his inspiration for self-liberation from the general spirit of the Reniassance and the Enlightement—faith in experience, intelligent effort, determination, thoughtful organization. One of the world's great achievers began to flourish on the intellectually infertile landscape of nineteenth-century Spain, where there was little other scientific activity.

The accomplishment that was Cajal seems quite remarkable in our age and society, where life has instead been relatively supportive and attractive for scientists. But we should remember that the Renaissance and Enlightenment from which Cajal took the inspiration to develop nonconformistically were times of intellectual self-made individuals. The thinkers of Galileo's day were pioneers of a new character and intellect for Europeans, and they blazed thought trails in relative isolation even while the Inquisition was in progress. Cervantes wrote in jail. Some witch trials were held and alleged witches were killed in England as late as 1712 (until 1718 in France). Beccaria's *On Crimes and Punishments*, which led to great reforms in criminal justice systems, was published as late as 1764. The great movement for the legalization of free intellectual expression did not come until the late seventeenth and early eighteenth centuries. Science had led the way by showing the power of thoughtful experience.[5] And from this in turn came the great experiment in the intellectual permissiveness of modern democracy.

The modern university and information distribution systems demand progressively more responsibility of younger or older individuals to develop strategies to negotiate coherence in institutions that have been progressively bent, fragmented, and reshaped since World War II. The task of self-organization may seem formidable, but it is *much*, much easier than the task faced by the people of the Renaissance and the Enlightenment. We can certainly learn by reflecting on their efforts.

Mary Wollstonecraft, who wrote *Vindication of the Rights of Woman* (1792), was a self-educated person. Indeed she was rejected by the learned women of the bourgeois class of her day. Yet her extraordinary book remains often remarkably insightful and fresh and stands up quite well with today's books on the analysis of gender roles and class.

Another noteworthy person of this general era was Benjamin Franklin. He was famous for many accomplishments in business, government, science, philosophy, and diplomacy. Here we should consider his system for character perfection as outlined in his *Autobiography*. Some have seen it as a caricature, some as a paradigm, of self-study and self-discipline. It was an essential part of one's reading during the days of the self-made individual in the United States and elsewhere.

Franklin carefully listed and defined the virtues that he wanted to perfect—temperance, silence, order, resolution, frugality, industry, sincerity, justice, moderation, cleanliness, tranquility, chastity, humility. Then he actually made out score cards that he used for daily self-examination. Thus he adapted a much older Buddhist and Christian moral tradition and he *studied* his own behavior and attitudes, each week concentrating on a particular virtue, so that he could evaluate himself and change his habits. "I was surpriz'd to find myself so much fuller of Faults than I had imagined, but I had the Satisfaction of seeing them diminish."

Franklin also made out a schedule for himself each day, planning each hour for productive activity. At the head of each day's schedule was a question, "The Morning Question, What Good shall I do this Day?" At the end, "Evening Question, What Good have I done today?" He goes on to describe how difficult it is to conduct such an ambitious plan, but that the system *can* result in some valuable progress.

Franklin spent much of his life trying to bring the thinking of the Enlightenment to a semi-isolated, somewhat uncultured group of settlements of traders, trappers, farmers, and refugees from religious wars, who were building a new society in North America. Many of them were well aware that they were a people with opportunity, a people on the way up. And their survival as an economic, political, and ideological force was at issue as well. Those in our time who have focused on the overly puritanical and overly money-oriented values in the *Autobiography* have condemned it. However, those have praised it who have been more interested in it as an example of one who successfully took his destiny into his own hands in the spirit of the Renaissance and the Enlightenment. It was a creature of its time and place in these respects.

But thinking and being cannot be simply discipline. D. H. Lawrence campaigned against many of the suffocating aspects of Calvinism and Victorianism. He essentially urged a revival of the active, personal, Dionysian side of the European psyche that had been usually denied and suppressed since the fall of Rome: an exploration of the senses, intuition, spontaneity:

The soul of man is a dark vast forest, with wildlife in it. Think of Benjamin fencing it off! . . . I can't stand Benjamin. He tries to take away my wildness and my dark forest, my freedom. (*Studies in Classic American Literature*)

Franklin's system of self-discovery and self-discipline only applied to *already* stated and determined objectives and values. In this important sense it *lacked* the spirit of the Renaissance and cannot be compared to the notebooks of da Vinci or to *Don Quixote*, with their more open-ended exploration of nature, mankind, and oneself.

Mark Twain, with his characteristic flair for witty exaggeration commented:

> A malevolence which is without parallel in history, he would work all day and then sit up nights and let on to be studying algebra by the light of a smoldering fire so that all other boys might have to do that also or else have Benjamin Franklin thrown up to them. Not satisfied with these proceedings, he had a fashion of living wholly on bread and water, and studying astronomy at meal time—a thing which has brought affliction to millions of boys since, whose fathers had read Franklin's pernicious biography. ("The Late Benjamin Franklin")

Of course Twain is correct. Franklin was in fact never a saint himself (as he admitted), and too many generations of parents and teachers may have failed to portray his efforts in perspective. An excessive insistence on conventional puritanical or Calvinistic virtues, thrift, ambition, self-discipline can smother the spirit. Twain was a man who had "gone out West" and learned from experience, who had learned to laugh at his own mistakes and those of others, and who had learned to learn from laughing. He was a student of life. Twain had been adventuresome about learning from life itself, about *being thoughtful based on experience*. Indeed this may seem like a strange thing to say of a nineteenth-century journalist and humorist, but Mark Twain in his own way was himself as good an example of the Age of Reason as was Ben Franklin.

Despite such cogent criticisms of Franklin, he provides a good example of a self-made individual from the Age of Reason who was able to work out some means of self-discovery and self-discipline and move his life, works, and awareness beyond the norms of his people. But with some hindsight we can see that any good idea can go sour if it becomes merely a static model, a technique to accomplish specific goals, and the spirit behind it is not open to debate.

Conventional perceptions in one way or another usually stand in the way of the development of individual sensitivity and understanding. S. I. Hayakawa noted, "If you see in any given situation only what everyone else can see, you can be said to be so much a representative of your culture that you are a victim of it."

In what sense a victim? In the sense that, as mere representatives of popular or cliquish values and perceptions, we are denied glimpses of our potential individuality and objectivity and thus are denied the opportunity to explore and nurture them. We mistake the status quo of our present identity and narrowly informed thoughts and intuitions for something to be defended against change. Without the ability to empathize and to assume new points of view, we become trapped within our own conditioned minds and our accident of birth. We age and shrink rather than grow. Individuality becomes trivialized, will and control over one's own destiny become little more than comfortable illusions, platitudes, empty opportunities (as though it were made legal for fish to drive cars).

Actually, such are old ideas in the American tradition of individuality. Ralph Waldo Emerson, long ago, put it strongly in his important essay "Self-Reliance." "Society everywhere is in conspiracy against the manhood of every one of its members."

Stronger still are the words of Thoreau in *Walden* or in "On the Duty of Civil Disobedience." He argues that the worst form of servitude is when "you are the slave-driver of yourself." In the opening pages of *Walden*, Thoreau explains why he felt the need to remove himself from the routine of social life and go off to reflect for a time in the woods by the shore of Walden Pond.

> [I]n shops, and offices, and fields, the inhabitants have appeared to me to be doing penance in a thousand remarkable ways. What I have heard of Bramins sitting exposed to four fires and looking in the face of the sun; or hanging suspended, with their heads downwards over flames, . . . or dwelling chained for life, at the foot of a tree; or measuring with their bodies, like caterpillars, the breath of vast empires, . . . even these forms of conscious penance are hardly more incredible and astonishing than the scenes which I daily witness.
>
> I see young men, my townsmen, whose misfortune it is to have inherited farms, houses, barns, cattle, and farming tools. . . . Better if they had been born in the open pasture and suckled by a wolf, that they might have seen with clearer eyes what field they were called to labor in. . . . By a seeming fate, commonly called necessity, they are employed, as it says in an old book, laying up treasures which moth and rust will corrupt and thieves break through and steal. It is a fool's life, as they will find when they get to the end of it, if not before.

This discussion should not be mistaken to suggest that one cannot be ordinary and also happy. That would be a ridiculous position to maintain. But it can be argued that a free person in an open society should choose the conventional, if that is his or her conclusion, as the preferred option in the context of a variety of personal options that are honestly understood.[6] Otherwise freedom is largely an illusion for each individual. Life and our own outlooks and insights may be much the poorer if we cannot at least understand and sometimes see as adventure the rich diversity of expression around us, even if we finally prefer very conventional lives for ourselves.

Compelling arguments have been made that the "normalcy" that we are taught will make us happy is in any event often a myth, a mirage. It is obvious that unreflective striving to be "ordinary" in any event leads too often to frustration and neuroses, and to only superficial illusions of security and happiness.

Last, if we are not able to be sympathetic to people who do wish to explore their individuality we may in fact, if unknowingly, abandon our principles of freedom and individuality as a society. This issue of tolerance frames a major difficulty in building and maintaining an open society in which individuality can flourish when it desires.

Of course most of us become too busy with obligations and other plans for much thoughtful exploration. And thus there is indeed the risk of *unthoughtful* exploration that will become merely dissipation. The happiness that routine values offer too often may be a siren; yet this may appear restfully safer than alternatives. In "Ode to Duty," Wordsworth asks Duty to set him free from a past of vain temptations and strife, from trusting mostly himself in his love of freedom. "Thee I now would serve more strictly, if I may." And now "I supplicate for thy control;"

> Me this unchartered freedom tires;
> I feel the weight of chance-desires:
> My hopes no more must change their name.
> I long for a repose that ever is the same.

Writing and Self-Study

Writing can become a most valuable part of one's life. Emerson noted, "A poem, a sentence causes us to see ourselves. I be, and I see my being at the same time." Francis Bacon observed, "Reading maketh a full man, conference a ready man, and writing an exact man."

Self-study can take many forms. A young inexperienced hopeful of 23 years named Margaret Mead left family and friends to cross the Pacific by ship and spend years living among village people in Samoa and New Guinea. She went with the intention of finding out more about *her own* North American society. Were the troublesome behaviors of American teenagers the simple product of hormonal changes, as some argued, or were they produced by aspects of modern life? She took no methods with her, mostly the *spirit* of scientific inquiry. She made mistakes in those early years (Brady

1983) but she also gained valuable insight into this and other related questions, such as the development of male and female temperaments. She developed methods as she went along. And simply maintaining a sense of purpose and objectivity was a constant concern.

> [L]etters written in and received in the field have a very special
> significance. Immersing oneself in life in the field is good, but one must be
> careful not to drown. One must somehow maintain the delicate balance
> between empathic participation and self-awareness, on which the whole
> research depends. Letters can be a way of occasionally righting the balance
> as, for an hour or two, one relates oneself to people who are part of one's
> other world and tries to make a little more real for them this world which
> absorbs one, waking and sleeping. (*Letters from the Field 1925–1975*)

This is clearly a postmedieval rational mind writing—the self-conscious determination to be careful and try to maintain objectivity in order to learn and understand from experience, the very act of articulating the process for the benefit of oneself and others.

The keeping of private notebooks can be quite enlightening. It can help to clarify ideas the vagueness of which had been troubling. It can teach one how moods can affect one's thinking. It can show how easily good ideas can get forgotten if they are not written down and developed. It can teach that one does not understand a thing as well as one might have thought, or that one has a better grasp of another thing than one would otherwise have known. It can reveal in what ways the writer changes with age and experience, and in what ways not. It helps to identify subjects where one needs to read, discuss, or think more. Sometimes the unconscious more than the conscious mind seems to do the writing and one can learn that there is a lot more thinking going on in the subconscious, and in more interesting ways, than had before been appreciated. It can enormously help in understanding one's feelings. It can help one to laugh at oneself.

Once I found writing a chore, tedious and personally unnecessary. Now I see it as essential to the development of my thought and self-conduct. Now it is clear to me, as Alfred Kazin put it, that, "in a very real sense, the writer writes in order to teach himself, to understand himself, to satisfy himself; the publishing of his ideas, though it brings gratification, is a curious anticlimax."

Many people have found writing to be a valuable part of their lives—*not writing things to tell to or to please others*, but writing *to explore oneself*, using pencil and paper almost as one would use a mirror to study one's posture and movement, or as one could use a photographic album to study one's growth. Across time, from Leonardo's notebooks to Mead's letters and field notes, we see that writing has not merely been useful to help one remember, but it can be critical in the perception and organization of one's thinking.

One might experiment freely with all sorts of writing and no one style might be more right than the rest. It is a question of which style is convenient to the mood and the subject. A scientist may keep laboratory notes, or a careful journal on nearly a daily basis when working in the field, but otherwise simply jot down ideas, and even quotations from others, in notebooks. One may explore various sorts of ideas in crude poetic form, sometimes in more careful poetic form. Many have explored ideas by writing fiction, though the works sometimes reveal little to the reader of the idea that the writer was pondering. One who is musically competent might similarly use music to explore apparently disrelated conceptual problems. Somewhat as we explore with metaphors or in dreams?

The last items may be idiosyncratic. But this is not important now. The point is that there may be no one best way *always* to structure writing so as to explore ideas and one's own psyche. I can highly recommend writing as a vehicle for self-study and self-organization, but it may be useful to experiment with form. One should work to master the conventional essay form for maximum usefulness. But experimentation with more idiosyncratic forms might also be useful for some people in some circumstances.

A main point in this chapter has been that for those who want to improve their judgment, the path may seem very difficult. Indeed it seems to get more difficult as the times reorganize and we pass from the Age of Reason to the Age of Technique, as our universities transform into multiversities, as professional vocations become careers, as dialogue turns into debate, as ideas turn into ideologies, as information is trimmed, tailored, and packaged for commercialization, as people become less mindful of building a better world and become more preoccupied with security, survival, and escapism.

But these difficulties should not blind one to the fact that the history of our species shows that we are capable of being marvelously resourceful as individuals and in small groups. We can be very clever and inventive creatures. The people of the Pacific figured out for themselves how to navigate the vast open ocean. Out from the centuries dominated by faith, the Inquisition, the religious wars, and even the recent era of ideologies, there have climbed nonconformist individuals who figured out how to explore and discipline themselves and make better judgments than their families and neighbors were capable of. As science was able to manipulate nature in connection with a careful study of it, people of the Renaissance and Enlightenment similarly found that by understanding ourselves and those around us we can mold ourselves into something more than the pressures of society would seem to allow.

The rational mind at its most powerful can be used to elevate the *whole individual.* This is more possible today than ever, but one must first learn to recognize and appreciate the nature of the very serious obstacles that stand in one's way.

NOTES

1. A paradox of our society is that with all the pressures for conformity we are also taught that we are each inherently special (chapter 9). This creates an inner confusion that sometimes even surfaces.

My point here is that most of us can do a great deal to improve our thinking and to make better judgments, thusly to enjoy more secure, entertaining, interesting, and rewarding lives. I am *not* suggesting that we can all become rich, or that we can all become Napoleons, Tolstoys, Einsteins, or presidents.

2. Kerr, incidentally, felt that a limited solution might be to design a new campus and academic incentive system from the ground up. The Santa Cruz campus began with high hopes as such an experiment. Idealistic professors and students flocked to it. But the experiment lost its political support early when Kerr was removed as president by Governor Reagan in a show of force (Grant and Riesman 1978).

3. These groups do not necessarily realize the cumulative effect that they have had on higher education. It is not usually a matter of conscious manipulation, nor necessarily a conscious agenda. Indeed a first reaction of such groups to reading the above would be that I am exaggerating and being unfair to them. The interactions are in fact subtle, but powerful. Persons in the groups merely lobby for their own short-term interests without serious consideration of the overall impact. Most administrators are hired by regents, trustees, or faculty merely to *manage* programs already in place. They rarely have the vision or talents or political backing to correct the situation. They seldom have the political power base to stand up against the diverse forces with special interests in the institution. Usually, faculties are at odds and divided by parochial departmental and individual interests.

4. Some of the most *general* goals are easy to set out and for all to agree upon. It is easy to agree with Cornell University President Frank Rhodes, who in the May 22, 1985, *Chronicle of Higher Education*, wrote, in part:

A college graduate should:

Be able to read and listen with comprehension and to write and speak with clarity, precision and grace.

Have a sense of context—physical, biological, social, historical—within which we live our lives.

Have some insight into a time and culture other than our own.

Be able to reflect in an orderly way on the human condition and our beliefs, values and experience.

Be able to appreciate non-verbal symbols, including the creative and performing arts.

Be able to work with precision, rigor and understanding in a chosen discipline, so as to understand not only something of its content, but also its premises, its relationships, limitations and significance.

That's a tall order. And it represents the work of a lifetime, of course, not just of two or four years. But the attitudes and spirit that underlie these skills can be developed in the course of a college career. They can be developed but frequently they are not.

Some critics such as E. Z. Friedenberg, in *The Dignity of Youth and Other Atavisms*, warn that lofty rhetoric can be used, though, as a "trick" to lure students from different backgrounds into middle-class perceptions and ideologies rather than to help them to explore the meaning of their own lives. Of course it can be used to do just that.

Significantly, there has been a movement that calls for educational coherence but that defines this in terms of a certain set of Great Ideas that come from a certain set of Great Books that were long used in certain ways to construct middle-class perceptions and values.

A. Bloom (*The Closing of the American Mind*), for example, flies in the face of every authoritative analysis of the problems with our universities and attempts to place blame on issues centering on the radicalism of the 1960s. (Didn't we see it on television?) "Coherence" would then become the teaching of certain Great Books in certain Correct Ways, and the stamping out of experimental programs with liberal followings—women's studies, black studies, etc. (For one response see Simonson and Walker 1988; also see Hayakawa 1950, chapter 13.) This agenda quite fails to see the really difficult issues at stake for our civilization and the forces that have been at work since well before the 1960s.

Former Secretary of Education W. J. Bennett, part of an especially business-oriented administration, and seeming to overlook the vauable recent findings in the following academic areas, when chairman of the National Endowment for the Humanities, wrote that disciplines such as history, literature, and philosophy "have become frighteningly fragmented, even shattered." He urged scholars and teachers "individually and through their professional organizations," to "demand of educators, of public officials, of colleagues, of students, that the humanities be studied in a coherent and serious way." (*Chronicle of Higher Education*, December 1, 1982). Friedenberg and Bennett might define fragmented and coherent differently.

5. The story is, of course, complex and also involved the rise of a merchant and trading class and increased contact with the outside world. The priests, representing the old order, thought that they had good reasons not to trust their senses and so they refused to look through Galileo's telescope. What if they were to see only illusions? Every thoughtful and faithful person knew that the senses could not be trusted. The merchants, though, representing the new order, found that when they looked, they could get advance notice of ships' arrivals, and could make money. In such ways science found political support among the merchants. Science also found military applications even in Leonardo's day. Science helped to break the authority of the religious/aristocratic establishments and this was also to the advantage of the rising merchant class. Even increased trade with the East and the discovery of the New World contributed to revolutions of thought in Europe. But the independent force of scientific thinking, the *possibilities* it suggested to begin to improve justice, should not be underestimated.

6. There has been debate in opinion-making circles over this long-standing issue. "Individuality" is usually praised openly, but often condemned in smaller circles. Critics have argued, in the tradition of the greatly influential economic idealogue Mandeville, that economic ambition is "individuality," but that a philosophy of the exploration of human potential will evolve into an ideology that will raise the expectations of the masses of people too high—that there has already been *too much* individual exploration and that this threatens to destabilize the employee situation. We have heard Mandevillian complaints, especially during the 1960s and 1970s, that people were becoming "overeducated" and would not stay happily at their jobs. People should aspire even more to security and stability, and self-esteem compatible with the prevailing system of socioeconomic incentives and rewards.

> The Welfare and Felicity therefore of every State and Kingdom, require that the Knowledge of the Working Poor should be confin'd within the Verge of their Occupations and never extended ... beyond what relates to their Calling. ... A servant can have no unfeign'd Respect for his Master as soon as he has sense enough to find out that he serves a fool. (*The Fable of the Bees*)

Contrary to appearances, though, there is probably very little serious self-exploration. The powerful with Mandevillian leanings have probably been overreacting, even on their own terms. Much individual exploration has been superficial, unthoughtful, unenlightening, largely faddish, and actually quite conformist viewed broadly (for example, style and property competition within social subsets). Such merely feeds political, sociological, and religious cults and leads to a lot of consumerism as one exchanges the trappings of successive life-styles. Mandeville would be pleased.

Other individual exploration may be socially necessary given the opportunities opened up by new knowledge and technologies. One cannot ignore the fact that many groups—ethnic, women, men—have necessarily been forced to experiment with new life-styles by changes in immigration patterns, demographic patterns, job markets, transportation, and communication trends. This is innocent from a Mandevillian set of concerns to maintain a buyer's market of manageable employees.

The details and causation of an *apparent* surge in individuality exploration are quite complex, and whether it threatens to destabilize society is not at all clear, considering the array of other forces that perturb society. Adventuresome individuality can be highly visible and people who are "different" can easily irritate any given group. Pluralism among individuals should not become a scapegoat for social problems, though.

One can moreover turn the simplistic complaint around a bit to reveal more of the inherent complexity in this issue. In early rural and agricultural economies, people labored largely to feed and shelter themselves. But as nations have moved into luxury economies, sex, status, and even Christmas became important factors in creating a demand for goods, as Vance Packard (*The Hidden Persuaders*, 1980) and many others have detailed. Then the economy, employment, and opportunities for economic gain became based more on psychological needs than on physiological needs. Advertising and social values have interacted to heighten the attitude that we are *not* "all right" unless we keep up with friends and the Joneses and buy various soaps, clothing, records, new cars, soft drinks, cigarettes, and a better home; read certain books or films; and so on. The economic system may be energized by such needs, allowing more freeways, weapons, bureaucrats, hospitals; spiritual dissatisfaction and diffuse and unrealistic hopes and expectations may be the fall out from the System. Our strong but unexamined belief that we necessarily operate from "free-will" blinds us from seeing the connections, it can be argued.

Spiritual dissatisfaction may actually be *essential* to power economies (not merely capitalist) that fluorish in no small part by selling things that "make us feel better about ourselves," and that are powered by the sales of diversions and status symbols. ("When the going gets tough, the tough go shopping.") People who do not feel good about themselves but who are out of touch with their feelings can also be easier to manage. One can argue that much more anxiety, dissatisfaction, and experimentation with life-styles are due to commercialism and needs for national wealth than to philosophical ideals of individual development.

The last could be an economic determinist point of view. A right-wing, firm, economic determinist point of view might in Mandevillian fashion then support particular self-dissatisfactions that can benefit the marketplace and that do not threaten it. A radical left-wing, firm, economic determinist might see this as exploitation and would selectively support self-dissatisfactions that produce challenges to the System. One set of factions condones especially those brands of individuality that can be commercialized. Another set of factions condones those brands of individuality that can be radicalized. They can mean very different things when they praise the word, or when they condemn it.

In attacking a philosophy of self-exploration, its most aggressive critics have not targeted carefully and have ignored the diversity of behaviors that merely appear as philosophical self-exploration and the more extensive laying out of alternative hypotheses that a complete discussion would demand.

EPILOGUE

... a person's conscience ain't got no sense. ... If I had a yaller dog that didn't know no more than a person's conscience does, I would p[o]ison him. It takes up more room than all the rest of a person's insides, and yet ain't no good, nohow. Tom Sawyer he says the same.

Huck Finn

When I began to study the brain and behavior in the early 1960s there was a consuming optimism in the air among the scientific community. Tranquilizers had been developed, and with wonderful results, and in many ways this had revolutionized the outlook of the mental health sciences. There was completely serious talk that we might understand the brain one day well enough to go on to invent a pill that would block aggression and put an end to war, crime, and violence. These broad social problems had long been seen by many as the result of vestigial, ugly urges from the dawn of our evolution and bad genes. Primitive emotions were the enemy. They were seen as blocking some supposed innate, universal, good common sense.

It seems so inexcusably naive now. Of course, anyone who watches monkeys fight or mate in a zoo will see reflections of ourselves; it is obvious that we share some special kinship. As a behavioral biologist, I probably see more details of similarity between monkeys and humans than most do. Yet I also see that monkeys do not have great ideological and religious wars. They do not develop excuses to conquer that conceal economic exploitations. They do not drive themselves to ulcers and heart attacks, or even to murder, in order to acquire objects that are status symbols within the context of one system of beliefs or another. They do not invent arts to illustrate complex inner worlds to each other. They do not teach their children stories about the world and about themselves. They do not have elaborate rules of thought to try to decide if they are being fair with one another. And, in any event, primates in nature are not solitary and completely selfish but are social, and their fighting tends to be more organizational or squabbling than mean or genocidal.

Now it is more obvious that a lot of our social problems have as much to do with beliefs, economic interests, and lack of sophistication in teaching our children to deal with their feelings in healthy ways as with anything else. There will probably be no pill that will kill our primate capacity for anger and thus end war. One can take orders and push buttons and drop bombs even if one is not angry—even if one is on calming drugs, as we have seen. A pill that would end denial and increase genuine compassion might help, but even if it were scientifically possible to have one, those in power would persuade us not to take it if they suspected that the other side might somehow cheat, or that it might interfere with their options to pursue economic or political objectives militarily.

We are ready to fight other people usually because of what we believe about them and about ourselves, about authority, about our ability to coexist our life-styles, and our economic needs. These beliefs are products of our cultural traditions, the information we are exposed to, and the ways we have disciplined our minds (if at all) to deal with any of this.

I have made the point in this book that the particular biology of the brain *is a key element* in the ways we process information, and in the development of our cultural traditions. But the problems we face are not the simple result of uncontrollable, dark, vestigial reptilian urges from an ugly evolutionary history. Thought systems, economic systems, and social systems inform our emotions at levels of organization and complexity that have emergent properties not predictable from studies of brain chemistry and physiology alone. Emotions and selfishness can indeed cloud our reason, but we would still have many personal and societal problems even if our brains were huge and free from rage. And we can do much about our problems when we give thoughtful attention to our ways of living.

My students tell me that some schools and parents teach ethical development these days by asking students to ponder extreme situations and to define their self-interests in these, to weigh the costs and benefits in each choice. For example, "Your grandmother is dying and needs some medicine. You have no money and can only get the medicine by robbing a drugstore. What is the right thing to do?" This opens up cost/benefit discussions of, How close is grandma to dying anyway? And, What is the penalty for robbing drugstores? This is what some would call, "getting kids to think." It also

teaches a certain consciousness of self-interest, and of survival. But the subliminal message is a sort of social and individual hopelessness.

Such questions and discussions as they have been presented to me seem
to presuppose that we have little more control over our lives than to be
pawns in a system and to react to odd dilemmas that it presents to us. I am
not certain of what actually goes on in homes and in the schools, but this is
a sort of unspoken and *often spoken* attitude that seems real for a growing
number at the university level, and among persons in business and government. The world is seen as essentially one in which thinking and judgment
are merely questions of *decision making*, of how we react to immediate
"realities" by making simple choices, a sort of crisis management. This is
called "practical." Long-range practical concerns are brushed aside as being
"hypothetical" or "idealistic."

But this is only superficially "practical" in the sense of "useful"; it is
mostly "easy." When we discuss life in such easy terms we have in effect
bought certain cynical philosophies and perspectives of life. We have in effect agreed not to practice the full use of our imaginations to explore the
realistic possibility of, and practical steps to the construction of, a life in
which the necessity for extreme ethical choices would be kept to a bare
minimum. An interesting thing to teach and talk about with students would
be how to construct relatively peaceful lives that will allow us to avoid having difficult daily conflicts thrust upon us. How can we get from where we
are to a life or a world in which days can be decent and fulfilling?

This is not presumed to be simple and it would not be done quickly, but
it is not impossible. We would have to understand our own values, their origins, how they compare to other value systems, and figure out ways to improve them. We would have to learn how institutions work. We would have
to know ourselves well enough to see where we can fit in best without too
many compromises. We would have to figure out how to be more thoughtful. We would have to understand political, social, economic, and family systems well enough to see what improvement would be and how to do it.

Colin Turnbull described the Ik of East Africa as being thoroughly selfish,
unloving, nasty, mean, and cruel (*The Mountain People*, 1972). Famine and
dislocations made them that way. It became everyone for oneself. Children
would knock over old people and pry food out of their mouths. Mothers
would commonly laugh at their children's difficulties. Death went un-
mourned. Their civility was in pathetic ruin.

> On one occasion I saw two youths on a ridge high up on Kalimon
> masturbating each other. It showed some degree of conviviality, but not

much, for there was no affection in their mutuality; each was gazing in a
different direction, looking for signs of food. . . .

It is good common sense, some in our competitive society will say, that
people should become completely selfish when times get really tough. Dog
eat dog, social Darwinism, and such. I even see many antievolutionists to
whom this side of social Darwinism makes appealing and good common
sense. Yet not all peoples who are starving let their humanity slip so far. Icien
common behavior was even rare in the brutal Nazi concentration camps. We
must conclude that something in the Icien view of life, in their philosophical and perceptual substrate for social organization, in their definition of
practicality, was so inappropriate to the stresses that events carried them to
disaster as a people. The Ik do not illustrate the *foundations* of human nature, but they do illustrate a dark *potential* in all of us.

Not all peoples who have troubles let their reason slip so far. Circumstances even improved at one point, but that did not help the Ik. For various
reasons it did not make sense to the selfish Ik to cooperate in the growing of
crops when it began to rain and so they even let their fields rot. Thus, they
had a brand of common sense and had constructed value and perceptual
systems that became a mental labyrinth from which they could not escape.

> . . . it was obvious from the outset that nothing had really changed due to
> the sudden glut of food, except to cause interpersonal relationships to
> deteriorate still further if possible, and heighten Icien individualism beyond
> what I would have thought even Ik to be capable of. If they had been mean
> and greedy and selfish before with nothing to be mean and greedy and
> selfish over, now that they had something they really excelled themselves
> in what would be an insult to animals to call bestiality.

> The Ik had faced a conscious choice between being human and being
> parasites, and of course had chosen the latter.

We must consider how far the Ik had slipped. For most of human history,
well over 99 percent of it, we were *all* hunting-and-gathering people. Hunters and gatherers today show a lot of admirable humanitarian traits. As Turnbull put it:

> The result of a typical hunting-and-gathering social organization is a simple
> and effective system of human relationships, and this is what so strongly
> appeals to many of those who have worked with them. If we can learn
> about the nature of society from a study of small-scale societies, we can
> also learn about human relationships, and that seems fully as valuable and
> valid. The smaller the society, the less emphasis there is on the formal
> system, and the more there is on inter-personal and inter-group relations,
> to which the system is subordinated. Security is seen in terms of these
> relationships, and so is survival. The result, which appears so deceptively

simple, is that hunters frequently display those characteristics that we find so admirable in man: kindness, generosity, consideration, affection, honesty, hospitality, compassion, charity and others. This sounds like a formidable list of virtues, and so it would be if they *were* virtues but for the hunter they are not. For the hunter in his tiny, close-knit society, these are necessities for survival: without them society would collapse.

Westerners will not find every aspect of tribal life pleasant and agreeable, and I am not suggesting that we idealistically conjure up images of the pure noble savage. But tribal peoples on the whole, especially hunters and gatherers, are quite decent folks.

The Ik may have fallen very quickly, perhaps in only a matter of months. It is remarkable how quickly the consciousness and values of a people can get ugly. The Nazis needed only a few years to gain control of the values of the cultured German nation and mold its competitiveness and nationalistic pride into something terrible. It took only weeks to involve the pious people of Salem in an awful witch-hunt among friends and neighbors.

The Ik responded badly to serious problems and developed a brand of common sense and a notion of practicality that could not even deal with opportunities to liberate themselves. For their habits and consciousness, the opportunities did not exist. These were invisible.

I am concerned that many persons, whether in the university, among the general public, or in government, have accepted various metaphysical, or in some cases fatally oversimplified, notions such as: that simple competition alone organizes both nature and society; that conflict is both inevitable and essential to progress; that the basic emotional needs are simply for status, power, and sex; that self-interest is short-term self-interest; that personal objectivity is impossible; that intellectual individuality can only be a mirage or a danger. Such pseudoscientific basic assumptions, and denial, obscure the paths to deep thinking or to careful planning for peaceful and sane individual or collective futures. They can become self-fulfilling prophecies. The Ik are merely one of several worst-possible scenarios to ponder.

Still others see the world largely in terms of good and evil, and this does not raise the level of analysis and discussion of our times or elevate the status of knowledge. One of the important factors in the fall of Rome was that so many pagans and Christians alike saw problems of society merely in terms of sin and divine punishment. They could not see past religious implications of disasters to material causes and understanding that would allow them to correct the complex material and social problems that were weakening Roman society.

Science teaches us that we must understand the reality about how things function if we are to have much control over them. Myths and denial only give us the seeds for more misleading stories to make up when things go badly.

About one-third of the U.S. population is psychically damaged by the compulsive behaviors and denial systems that worsen when drug/alcohol abuse invade a relationship or family. Still more deal badly with their feelings through the compulsive uses of gambling, sex, consumerism, entertainment, food, religion, or work. These trends are paralleled in certain other "advanced" capitalist and socialist nations. Obviously conventional wisdom in those societies is inadequate to help many people to find healthy ways to meet their deeper emotional needs. Surely we could do enormously more to develop healthier patterns of human relationships and expression. If there can be an up side to these widespread social problems, it is that they grew so large during the economic difficulties of the 1980s that evermore people began to seek psychological treatment and entered self-help support programs. Perhaps a modest but growing number of Americans and others will become better able to recognize examples of denial and of compulsive behaviors and better able to pioneer alternative and more healthy ways of dealing with feelings and personal interactions.

It made Mark Twain's Huckleberry Finn sick to see a town of God-fearing and hospitable people turn into a mean mob and tar and feather a couple of humbugs, and have a good time at it. "Human beings can be awful cruel to one another." Huck was not much smarter than the rest and he was no saint, but there was some gleam of independence and decency in him. And he intuitively knew that it was precious for him to guard it. His story ends:

> I reckon I got to light out for the Territory ahead of the rest, because Aunt Sally she's going to adopt me and sivilize me, and I can't stand it. I been there before.

It is so innocent of youth to think that one can run off to some time or place where there will be harmony and adventure instead of petty denial, meanness, and unfairness. Any grown-up lawyer could have told him smugly that a boy can't best be civilized by staying uncivilized. The words won't allow a legal document.

Even a raft on the big river can be no true escape. One cannot remain forever peacefully suspended, quietly floating, in the detachment of innocent and carefree being, secure in summery rebellion, only touching ashore for occasional adventures among hypocrites and bunglers. One starts inevitably to become part of the traffic of the road. Even "the Territory" had developed its own problems by the time Twain wrote, as he well knew from his own early years there. The gold rush had drawn all sorts of greedy folks west, and the dream fever of getting rich had begun to sweep the soul of the entire Republic.

Yet we cannot help but admire the decency of Huck's dreams and determination. And there was very smart sense to his basic youthful intuition. We wear a mask until we come to fit it, as the saying goes. The best way to avoid getting corrupted would be to find ways around, to avoid growing into, situations that would force one to develop a compromised identity. Huck could not do that 100 percent, but he had the right idea, and he was willing to explore.

And, come to think of it, he had already done a pretty fair job of keeping clean, for a boy who'd been around some.

Additional Readings

ADDITIONAL READINGS

Obvious original sources that are simple to locate, the Bible, de Tocqueville, Newton, and so on, are not listed here. Obvious secondary sources such as the Durants' *History of Civilization* series were invaluable aids but likewise are not listed. The result is a partial list of more specialized references along with some suggestions for reading. In many cases I list a book merely because it seemed interesting to me in the present context, and so might also to some readers, and not because I necessarily endorse most or all of it.

Abell, George O., and Barry Singer. 1981. *Science and the Paranormal: Probing the Existence of the Supernatural.* Scribner's: New York.

Abercrombie, Nicholas, Stephen Hill, and Bryan S. Turner. 1986. *Sovereign Individuals of Capitalism.* Allen and Unwin, London.

Abir-Am, Pnina G. 1987. The biotheoretical gathering, trans-disciplinary authority and the incipient legitimation of molecular biology in the 1930s: New perspectives on the historical sociology of science. *History of Science* 25:1–70.

Abrahams, Gerald. 1975. *The Chess Mind.* Hodder & Stoughton: London.

Adam, Barry D. 1986. Age, Structure and Sexuality: Reflections on the anthropological evidence on homosexual relations. In *The Many Faces of Homosexuality: Anthropological Approaches to Homosexual Behavior*, Evelyn Blackwood (ed.), 19–33. Harrington Park Press, New York.

Adams, Frederick Upham. 1914. *Conquest of the Tropics: The Story of the Creative Enterprises Conducted by the United Fruit Company.* Doubleday Page & Co., New York.

Aristotle. 1941. *The Basic Works of Aristotle*, Richard McKeon (ed.). Random House, New York.

Armstrong, David M. 1978a. *Nominalism and Realism.* Cambridge University Press, Cambridge.

————. 1978b. *Universals and Scientific Realism.* Cambridge University Press, Cambridge and London.

Aronowitz, Stanley. 1988. *Science as Power: Discourse and Ideology in Modern Science.* University of Minnesota Press, Minneapolis.

Atkinson, Max. 1984. *Our Masters' Voices: The Language and Body Language of Politics.* Methuen, New York.

Axelrod, Robert. *1984. The Evolution of Cooperation.* Basic Books Inc., New York.

Ayala, Francisco José, and Theodosius Dobzhansky. 1974. *Studies in the Philosophy of Biology: Reductionism and Related Problems*, University of California Press, Berkeley.

Bailey, Helen Miller, and Abraham P. Nasatir. 1960. *Latin America: The Development of its Civilization*. Prentice-Hall, Englewood Cliffs, N.J.

Bannister, Robert C. 1988. *Social Darwinism: Science and Myth in Anglo-American Social Thought*. Temple University Press, Philadelphia.

Bateson, Gregory. 1958. *Naven*. Stanford University Press, Stanford.

Beard, D. C. 1983. *The American Boy's Handy Book*. Reprint of 1890 edition includes new foreword by Noel Perrin. Godine, Boston.

Beatty, John. 1989. Chance and life: Controversies in modern biology. In *The Empire of Chance*, G. Gigernzer, et al. (eds.). Cambridge University Press, Cambridge.

Bednarski, J. 1970. The Salem witch-scare viewed sociologically. In *Witchcraft and Sorcery*, M.Marwick (ed.). Penguin Books, Middlesex.

Beetham, David. 1987. *Bureaucracy*. University of Minnesota Press, Minneapolis.

Bellone, Enrico. 1980. *A World on Paper: Studies on the Second Scientific Revolution*. MIT, Cambridge.

Berger, James O., and Donald A. Berry. 1988. Statistical analysis and the illusion of objectivity. *American Scientist* 76:159–165.

Berger, Peter L., and Thomas Luckmann. 1967. *The Social Construction of Reality*. Doubleday, New York.

Berman, Ronald. 1981. *Advertising and Social Change*. Sage Publications, Beverly Hills and London.

Beveridge, William I. B. 1950. *The Art of Scientific Investigation*. Vintage Books, New York.

Bharati, Agehananda. 1975. *The Tantric Tradition*. Samuel Weiser, New York.

Birke, Lynda. 1986. *Women, Feminism and Biology: The Feminist Challenge*. Methuen, New York.

Boswell, John. 1980. *Christianity, Social Tolerance, and Homosexuality*. University of Chicago Press, Chicago.

Bowler, P. 1976. *Fossils and Progress: Paleontology and the Idea of Progressive Evolution in the Nineteenth Century*. Science History Publications, New York.

———. 1988. *The Non-Darwinian Revolution: Reinterpreting a Historical Myth*. The Johns Hopkins University Press, Baltimore.

Boyer, Carl B. 1985. *A History of Mathematics*. Princeton University Press, Princeton.

Bradbury, Malcolm. 1975. *The History Man*. Arrow, London.

Brady, Ivan. 1983. Special section: Speaking in the name of the real: Freeman and Mead on Samoa. *American Anthropologist* 85(4):908–947.

Broad, William, and Nicholas Wade. 1982. *Betrayers of the Truth: Fraud and Deceit in the Halls of Science*. Simon and Schuster, New York.

Brown, Harold I. 1979. *Perception, Theory and Commitment: The New Philosophy of Science*. University of Chicago Press, Chicago.

Brown, Phil. 1974. *Toward a Marxist Psychology*. Harper Colophon, New York.

Bullock, T., R. Orkland, and A. Grinnell. 1977. *Introduction to Nervous Systems*. W. H. Freeman, San Francisco.

Bunge, Mario. 1962. *Intuition and Science*. Prentice-Hall, Englewood Cliffs, N.J.

Burton-Bradley, B. G. 1975. *Stone Age Crisis: A Psychiatric Appraisal*. Vanderbilt University Press, Nashville.

Butterfield, Herbert. 1965. *The Origins of Modern Science 1300–1800*. The Free Press, New York.

Canfield, J. (ed.). 1966. *Purpose in Nature*. Prentice-Hall, Englewood Cliffs, N.J.

Carnap, Rudolf. 1966. *Philosophical Foundations of Physics*. Basic Books, New York.

Chadwick, Henry. 1967. *The Early Church*. Penguin Books, New York.

Chan, Wing-Tsit. 1963. *A Source Book in Chinese Philosophy*. Princeton University Press, Princeton.

Chargaff, Erwin. 1980. *Heraclitean Fire: Sketches of a Life Before Nature*. Warner, New York.

Chase, A. 1977. *The Legacy of Malthus: The Social Costs of the New Scientific Racism*. Alfred A. Knopf, New York.

Chase, Alston. 1980. *Group Memory: A Guide to College and Student Survival in the 1980s*. Atlantic Monthly Press, Boston.

Chattopadhyaya, Debiprasad. 1986. *History of Science and Technology in Ancient India: The Beginnings*. Firma KLM Private Ltd., Calcutta, India.

Chaudhuri, Nirad C. 1979. *Hinduism: A Religion to Live by*. Oxford University Press, New York.

Chippindale, Christopher. 1983. *Stonehenge Complete*. Cornell University Press, Ithaca, N.Y.

Church, Joseph. 1961. *Language and the Discovery of Reality*. Vintage Books, New York.

Churchland, Patricia Smith. 1988. *Neurophilosophy: Toward a Unified Science of the Mind/Brain*. MIT Press, Cambridge.

Cleary, Thomas. 1988. Translator's introduction to *The Art of War*, by Sun Tzu. Shambhala, Boston.

Cobban, Alfred. 1960. *In Search of Humanity: The Role of the Enlightenment in Modern History*. George Braziller, New York.

Cohen, Elie A. 1953. *Human Behavior in the Concentration Camp*. W. W. Norton, New York.

Collins, K. 1967. Marx and the English agricultural revolution: Theory and evidence. *History and Theory* 6(3):351–81.

Connery, Donald S. 1977. *Guilty Until Proven Innocent*. G. P. Putnam's Sons, New York.

Conway, Flo, and Jim Siegelman. 1979. *Snapping: America's Epidemic of Sudden Personality Change*. Delta Books, New York.

———. 1982. *Holy Terror: The Fundamentalist War on America's Freedoms in Religion, Politics and our Private Lives*. Doubleday, Garden City, N.Y.

Coren, S., and J. Stern Girgus. 1978. *Seeing is Deceiving: The Psychology of Visual Illusions*. Lawrence Erlbaum, Hillsdale, N.J.

Cosby, Michael R. 1984. *Sex in the Bible: An Introduction to What the Scriptures Teach Us About Sexuality*. Prentice-Hall, Englewood Cliffs, N.J.

Council on Environmental Quality 1980. *The Global 2000 Report to the President: Entering the Twenty-First Century*. Pergamon, New York.

de Bono, Edward. 1986. *De Bono's Thinking Course*. Facts on File Publications, New York.

Dewey, John. 1930. *Individualism Old and New*. Capricorn Books, New York.

Diamond, Stanley. 1974. *In Search of the Primitive: A Critique of Civilization*. Transaction Books, New Brunswick, N.J.

Dicks, D. R. 1970. *Early Greek Astronomy to Aristotle*. Cornell University Press, Ithaca, N.Y.

Dickson, David. 1988. *The New Politics of Science*. University of Chicago Press, Chicago.

Dixon, B. 1978. Unreasonable leaps. *Omni* 1(2):22.

Dobzhansky, Theodosius. 1960. *The Biological Basis of Human Freedom*. Columbia University Press, New York.

Dodds, Eric R. 1973. *The Ancient Concept of Progress: and Other Essays on Greek Literature and Belief*. Oxford University Press, New York.

Domhoff, G. William. 1985. *The Mystique of Dreams: A Search for Utopia Through Senoi Dream Theory*. University of California Press, Berkeley.

Douglas, Mary. 1986. *How Institutions Think*. Syracuse University Press, Syracuse, N.Y.

Drachmann, Anders B. 1922. *Atheism in Pagan Antiquity*. Gyldendal, London.

Drinnon, Richard. 1980. *Facing West: The Metaphysics of Indian-Hating and Empire-Building*. University of Minnesota Press, Minneapolis.

Eiduson, Bernice T. 1962. *Scientists: Their Psychological World*. Basic Books, New York.

Einstein, Albert, and Leopold Infeld. 1966. *The Evolution of Physics: From Early Concepts to Relativity and Quanta*. Simon and Schuster, New York.

Eliade, Mircea. 1972. *Shamanism: Archaic Techniques of Ecstasy*. Princeton University Press, Princeton.

_____. 1982. *A History of Religious Ideas*. University of Chicago Press, Chicago.

Ellul, Jacques. 1964. *The Technological Society*. Vintage Books, New York.

Emmet, E. R. 1960. *Thinking Clearly*. Longmans, London.

Fair, Charles. 1974. *The New Nonsense: The End of the Rational Consensus*. Simon and Schuster, New York.

Falco, Giorgio. 1964. *The Holy Roman Republic: A Historical Profile of the Middle Ages*. A. S. Barnes, New York.

Ferguson, John. 1978. *War and Peace in the World's Religions*. Oxford University Press, New York.

Feyerabend, Paul. 1978. *Science in a Free Society*. Verso, London.

Finley, M. I. 1980. *Ancient Slavery and Modern Ideology*. Penguin, New York.

Foster, Hal (ed.). 1985. *Postmodern Culture*. Pluto, London.

Foucault, Michel. 1980. *Power/Knowledge*. The Harvester Press, Brighton, Sussex.

Foulkes, David. 1982. *Children's Dreams: Longitudinal Studies*. John Wiley & Sons, New York.

Frankel, Hermann. 1973. *Early Greek Poetry and Philosophy*. Helen and Kurt Wolff, New York.

Frerichs, A. C. 1957. *Anutu Conquers in New Guinea*. Wartburg, Columbus, Ohio.

Friedenberg, Edgar Z. 1965. *The Dignity of Youth and Other Atavisms*. Beacon, Boston.

Fuerst, John A. 1982. The role of reductionism in the development of molecular biology: peripheral or central? *Social Studies of Science* 12: 241–78.

Furniss, Edgar S. 1957. *The Position of the Laborer in a System of Nationalism: A Study of the Labor Theories of the Later English Mercantilists*. Kelley & Millman, Inc., New York.

Futuyma, Douglas J. 1982. *Science on Trial: The Case for Evolution*. Pantheon, New York.

Gardner, Martin. 1981. *Science: Good, Bad, and Bogus*. Prometheus, Buffalo, N.Y.

Garfield, P. 1974. *Creative Dreaming*. Ballantine Books, New York.

Gatty, Harold. 1979. *Nature is Your Guide: How to Find Your Way on Land and Sea*. Penguin, New York.

Geertz, Clifford. 1984. Anti anti-relativism. *American Anthropologist*. 86:263–278.

Gellner, Ernest. 1988. The stakes in anthropology. *American Scholar*. 57(1):17–30.

Gierke, Otto. 1987. *Political Theories of the Middle Age*. Cambridge University Press, New York.

Gilbert, G. Nigel, and Michael Mulkay. 1984. *Opening Pandora's Box: A Sociological Analysis of Scientists' Discourse*. Cambridge University Press, New York.

Gilman, William. 1965. *Science U.S.A.* Viking, New York.

Gladwin, Thomas. 1970. *East is a Big Bird: Navigation and Logic on Puluwat Atoll*. Harvard University Press, Cambridge.

Goldstein, Thomas. 1980. *Dawn of Modern Science: From the Arabs to Leonardo Da Vinci*. Houghton Mifflin, Boston.

Goodall, Jane. 1986. *The Champanzees of Gombe: Patterns of Behavior*. Harvard University Press, Cambridge.

Graham, Loren, R. 1987. *Science, Philosophy and Human Behavior in the Soviet Union*. Columbia University Press, New York.

Grant, Gerald, and David Riesman. 1978. *The Perpetual Dream: Reform and Experiment in the American College*. University of Chicago Press, Chicago.

Grant, Michael. 1976. *The Fall of the Roman Empire: A Reappraisal*. Annenberg School Press (Thomas Nelson & Sons, London).

Greenberg, Daniel S. 1971. *The Politics of Pure Science*. New American Library, New York.

Gregory, R. 1970. *The Intelligent Eye*. McGraw-Hill, New York.

_____. 1974. *Concepts and Mechanisms of Perception*. Charles Scribner's Sons, New York.

_____. 1978. *Eye and Brain: the Psychology of Seeing*. McGraw-Hill, New York.

Gregory, R., and E. Gombrich. 1973. *Illusion in Nature and Art*. Charles Scribner's Sons, New York.

Grene, Marjorie. 1969. *Knowing and Being: Essays by Michael Polanyi*. University of Chicago Press, Chicago.

Griffin, Donald R. 1981. *The Question of Animal Awareness*. The Rockefeller University Press, New York.

Grunebaum, G. von, and R. Caillois. 1966. *The Dream and Human Societies*. University of California Press, Berkeley.

Gunn, J. A. W. 1975. Mandeville and Wither: Individualism and the workings of providence. In *Mandeville Studies*. Irwin Primer (ed.), 98–118. Martinus Nijhoff, The Hague.

Gupta, Avijit. 1988. *Ecology and Development in the Third World*. Routledge, London.

Guthrie, W. K. C. 1967. *A History of Greek Philosophy*. Cambridge, London.

Halifax, Joan. 1979. *Shamanic Voices: A Survey of Visionary Narratives*. E. P. Dutton, New York.

_____. 1982. *Shaman: The Wounded Healer*. Crossroad, New York.

Hanke, L. 1970. *Aristotle and the American Indians: A Study of Race Prejudice in the Modern World*. Indiana University Press, Bloomington.

Harding, Sandra. 1986. *The Science Question in Feminism*. Cornell University Press, Ithaca, N.Y.

Harner, Michael J. 1973. *Hallucinogens and Shamanism*. Oxford University Press, New York.

_____. 1980. *The Way of the Shaman: A Guide to Power and Healing*. Harper & Row, New York.

Harris, Marvin. 1979. *Cultural Materialism: The Struggle for a Science of Culture*. Random House, New York.

Hastings, James. 1961. *Encyclopaedia of Religion and Ethics*. Charles Scribner's Sons, New York.

Hayakawa, S. I. 1950. *Symbol, Status, and Personality*. Harcourt, Brace & World Inc., New York.

_____. 1978. *Language in Thought and Action*. Harcourt Brace Jovanovich, New York.

Heider, Karl. 1979. *Grand Valley Dani: Peaceful Warriors*. Holt, Rinehart and Winston, New York.

Heilbroner, Robert L. 1980. *The Worldly Philosophers: The Lives, Times, and Ideas of the Great Economic Thinkers*. Simon and Schuster, New York.

_____. 1988. *Behind the Veil of Economics*. W. W. Norton, New York.

Held, Robert. 1985. *Inquisition*. Qua d'Arno, Florence.

Herdt, Gilbert H. 1981. *Guardians of the Flutes: Idioms of Masculinity*. McGraw-Hill, New York.

_____. 1982. *Rituals of Manhood: Male Initiation in Papua New Guinea*. University of California Press, Berkeley.

_____. 1984. *Ritualized Homosexuality in Melanesia*. University of California Press, Berkeley.

Herrick, Jim. 1985. *Against the Faith: Deists, Skeptics and Atheists*. Prometheus Books, Buffalo, N.Y.

Herrigel, Eugen. 1971. *Zen in the Art of Archery*. Vintage Books, New York.

Hesse, Mary B. 1961. *Forces and Fields: The Concept of Action at a Distance in the History of Physics*. Thomas Nelson and Sons, New York.

Heyers, Conrad. 1984. *The Meaning of Creation: Genesis and Modern Science*. John Knox, Atlanta.

340   Additional Readings

Highwater, Jamake. 1981. *The Primal Mind: Vision and Reality in Indian America*. Harper & Row, New York.

Hofstadter, Richard. 1955a. *Academic Freedom in the Age of the College*. Columbia University Press, New York.

———. 1955b. *Social Darwinism in American Thought*. Beacon, Boston.

———. 1963. *Anti-intellectualism in American Life*. Random House, New York.

Holton, Gerald. 1986. *The Advancement of Science and its Burdens*. Cambridge University Press, New York.

Honour, H. 1975. *The New Golden Land: European Images of America from the Discoveries to the Present Time*. Pantheon/Random House, New York.

Horne, Thomas A. 1978. *The Social Thought of Bernard Mandeville: Virtue and Commerce in Early Eighteenth-Century England*. Columbia University Press, New York.

Houdini, H. 1924. *A Magician Among the Spirits*. Harper & Row, New York.

Hroch, Miroslav, and Anna Skybova. 1988. *Ecclesia Militans: The Inquisition*. Dorset, Leipzig.

Hsu, Francis L. K. 1952. *Religion, Science and Human Crisis: A Study of China in Transition and its Implications for the West*. Routledge & Kegan Paul: London.

———. 1961. *Psychological Anthropology: Approaches to Culture and Personality*. Dorsey, Homewood, Ill.

———. 1981. *Americans and Chinese: Passages to Differences*. University Press of Hawaii, Honolulu.

Huff, Darrell. 1954. *How to Lie with Statistics*. W. W. Norton, New York.

Hull, D. 1980. Sociobiology: Another new synthesis. In *Sociobiology: Beyond Nature/Nurture?* G. Barlow and J. Silverberg, (eds), #35, 77–96. Westview Press, Boulder, Colo.

Hurlbutt, R. 1965. *Hume, Newton, and the Design Argument*. University of Nebraska Press, Lincoln.

Hunt, E. K. 1986. *Property and Prophets: The Evolution of Economic Institutions and Ideologies*. Harper & Row, New York.

Huxley, Aldous. 1958. *Brave New World Revisited*. Harper & Row, New York.

Hyland, Drew A. 1973. *The Origins of Philosophy: Its Rise in Myth and the Pre-Socratics*. Capricorn Books, New York.

Ifrah, Georges. 1985. *From One to Zero: A Universal History of Numbers*. Viking Penguin: New York.

Inglis, Amirah. 1974. *Not a White Woman Safe: Sexual Anxiety and Politics in Port Moresby 1920–1934*. Australian National University Press, Canberra.

Jacob, Margaret C. 1988. *The Cultural Meaning of the Scientific Revolution*. Alfred A. Knopf, New York.

Jacoby, Russell. 1987. *The Last Intellectuals: American Culture in the Age of Academe*. Basic Books, New York.

Jahoda, Gustav. 1969. *The Psychology of Superstition*. Penguin Books, Middlesex, England.

Jennings, F. 1975. *The Invasion of America: Indians, Colonialism and the Cant of Conquest*. University of North Carolina Press, Chapel Hill.

Johns, R. J. 1978. A new approach to the construction of field keys for the identification of tropical trees. *Australian Journal of Ecology*. 3:403–409.

Johnson, B. C. 1981. *The Atheist Debater's Handbook*. Prometheus Books, Buffalo, N.Y.

Joravsky, David. 1983. Unholy Science. Review of *Betrayers of the Truth*, by Broad and Wade. *New York Review of Books* 30(15): 3–5.

Josephson, B. D. 1980. *Consciousness and the Physical World*. Pergamon, New York.

Josephson, Matthew. 1962. *The Robber Barons: The Great American Capitalists 1861–1901*. Harcourt, Brace & World, New York.

Kaegi, Walter Emil, Jr. 1968. *Byzantium and the Decline of Rome.* Princeton University Press, New York.

Kalweit, Holger. 1988. *Dreamtime and Inner Space: The World of the Shaman.* Shambhala, Boston.

Katz, Solomon. 1955. *The Decline of Rome and the Rise of Mediaeval Europe.* Cornell University Press, Ithaca, N.Y.

Kaufmann, Walter. 1979. *Tragedy and Philosophy.* Princeton University Press, Princeton.

Kay, Lily E. 1988. Laboratory technology and biological knowledge: The Tiselius electrophoresis apparatus, 1930–1945. *Hist. Phil. Life Sci.* 10:51–72.

Kaye, Howard L. 1986. *The Social Meaning of Modern Biology: From Social Darwinism to Sociobiology.* Yale University Press, New Haven.

Kaye, Marvin. 1975. *The Handbook of Mental Magic.* Stein and Day, New York.

Kevles, Daniel J. 1977. *The Physicists: The History of a Scientific Community in Modern America.* Alfred A. Knopf, New York.

Key, B. 1973. *Subliminal Seduction: Media's Manipulation of a Not So Innocent America.* Signet Books, New York.

Kim, Ha Poong. 1981. What do Zen masters do with words? *Philosophical Forum* 12(2):101–115.

Kitcher, Philip. 1982. *Abusing Science: The Case Against Creationism.* MIT Press, Cambridge.

———. 1985. *Vaulting Ambition: Sociobiology and the Quest for Human Nature.* MIT Press, Cambridge.

Klaw, Spencer. 1968. *The New Brahmins: Scientific Life in America.* William Morrow & Co., New York.

Kline, Morris. 1985. *Mathematics and the Search for Knowledge.* Oxford University Press, New York.

———. 1953. *Mathematics in Western Culture.* Oxford, London.

Koestler, Arthur. 1973. *The Call Girls.* Random House, New York.

Koestler, Arthur, and J. R. Smythies (eds.). 1969. *Beyond Reductionism. New Perspectives in the Life Sciences.* Beacon Press, Boston.

Kohn, Alexander. 1986. *False Prophets: Fraud and Error in Science and Medicine.* Basil Blackwell, Oxford.

Krebs, Hans A., and Shelley, Julian H. 1975. *The Creative Process in Science and Medicine.* American Elsevier, New York.

Kroll, Jerome, and Bernard Bachrach. 1982. Medieval visions and contemporary hallucinations. *Psychological Medicine* 12:709–721.

Krupp, E. C. 1983. *Echoes of the Ancient Skies: The Astronomy of Lost Civilizations.* Harper & Row, New York.

Kuhn, Thomas S. 1970. *The Structure of Scientific Revolutions.* University of Chicago Press, Chicago.

Kuper, Adam. 1988. *The Invention of Primitive Society: Transformation of an Illusion.* Routledge, New York.

Kuznick, Peter J. 1987. *Beyond the Laboratory: Scientists as Political Activists in 1930 America.* University of Chicago Press, Chicago.

Kyselka, Will. 1987. *An Ocean in Mind.* University of Hawaii Press, Honolulu.

Lander, E. 1981. In through the out door. *Omni* 3(5):44–107.

Larsen, Stephen. 1976. *The Shaman's Doorway: Opening the Mythic Imagination to Contemporary Consciousness.* Harper Colophon, New York.

Lapham, Lewis H. 1985. *High Technology and Human Freedom.* Smithsonian Institution Press, Washington, D.C.

Latour, Bruno. 1987. *Science in Action.* Harvard University Press, Cambridge.

Latour, Bruno, and Steve Woolgar. 1979. *Laboratory Life: The Social Construction of Scientific Facts.* Sage, London.

Lawlor, Robert. 1982. *Sacred Geometry: Philosophy and Practice.* Crossroad, New York.

Lebedoff, David. 1981. *The New Elite: The Death of Democracy.* Watts, New York.

Lenoir, Timothy. 1982. *The Strategy of Life: Teleology and Mechanics in Nineteenth Century German Biology.* D. Reidel, Dordrecht.

Levin, Margarita. 1988. Feminism and Science. *American Scholar* 57(1):100–106.

Levy-Bruhl, Lucien. 1978. *The Notebooks on Primitive Mentality.* Harper Torchbooks, New York.

Lieberson, Jonathan. 1982a. The "Truth" of Karl Popper. *New York Review of Books* 29(18):67.

———. 1982b. The Romantic Rationalist. *New York Review of Books* 29(19):51.

Lindsay, Robert B., and Henry Margenau. 1936. *Foundations of Physics.* Wiley & Sons, New York.

Lloyd, G. E. R. 1970. *Early Greek Science: Thales to Aristotle.* Chatto & Windus, London.

Loftus, Elizabeth. 1979. The malleability of human memory. *American Scientist* 67:312–320.

———. 1984. *Eyewitness Testimony.* Cambridge University Press, New York.

Lovejoy, Arthur O. 1936. *The Great Chain of Being.* Harvard University Press, Cambridge.

———. 1961. *Reflections on Human Nature.* Johns Hopkins University Press, Baltimore.

Lowe, Donald M. 1982. *History of Bourgeois Perception.* University of Chicago Press, Chicago.

Luck, Georg. 1985. *Arcana Mundi: Magic and the Occult in the Greek and Roman World.* Johns Hopkins University Press, Baltimore.

Lukes, Steven. 1973. *Individualism.* Basil Blackwell, Oxford.

Maccoby, Michael, 1976. *The Gamesman.* Simon and Schuster, New York.

Mackay, C. [1841] 1932. *Extraordinary Popular Delusions and the Madness of Crowds.* L. C. Page, New York.

MacMullen, Ramsay. 1984. *Christianizing the Roman Empire A.D. 100–400.* Yale University Press, New Haven.

Macpherson, Crawford B. 1962. *The Political Theory of Possessive Individualism.* Oxford University Press, New York.

Maeroff, G. 1982. *Don't Blame the Kids: The Trouble with America's Public Schools.* McGraw-Hill, New York.

Mandeville, Bernard. 1924. *The Fable of the Bees: or Private Vices, Public Benefits.* (A construction from several editions by F. B. Kaye with commentary.) Oxford University Press, Oxford.

Mangan, J. A., and James Walvin. 1987. *Manliness and Morality: Middle-Class Masculinity in Britain and America, 1800–1940.* St. Martin's Press, New York.

Mannheim, Karl. 1936. *Ideology and Utopia: An Introduction to the Sociology of Knowledge.* Harcourt Brace Jovanovich, New York.

Marks, John. 1980. *The Search for the "Manchurian Candidate": The CIA and Mind Control.* McGraw-Hill, New York.

Marsden, George M. 1980. *Fundamentalism and American Culture: The Shaping of Twentieth-Century Evangelicalism 1870–1925.* Oxford University Press, New York.

Martin, Mike W. 1986. *Self-deception and Morality.* University Press of Kansas, Lawrence.

Mathews, Jerold. 1985. A neolithic oral tradition for the van der Waerden/Seidenberg origin of mathematics. *Archive for History of Exact Sciences.* 34:193–220.

Maturana, Humberto R., and Francisco J. Varela. 1987. *The Tree of Knowledge: The Biological Roots of Human Understanding.* New Science Library, Boston.

Mayr, Ernst. 1982. *The Growth of Biological Thought.* Harvard University Press, Cambridge.

Mazzarino, Santo. 1966. *The End of the Ancient World.* Faber and Faber, London.

McLellan, David. 1979. *Marxism After Marx.* Houghton Mifflin, Boston.

Merquior, J. G. 1985. *Foucault.* University of California Press, Berkeley.

Metzger, Walter P. 1955. *Academic Freedom in the Age of the University.* Columbia University Press, New York.

Miller, Alice. 1984. *For Your Own Good: Hidden Cruelty in Child-Rearing and the Roots of Violence.* Farrar, Straus, Giroux, New York.

———. 1986. *Thou Shalt Not Be Aware: Society's Betrayal of the Child.* New American Library, New York.

Mitchell, Stephen. 1976. *Dropping Ashes on the Buddha.* Grove Press, New York.

Montagu, Ashley. 1978. *Learning Non-Aggression: The Experience of Non-literate Societies.* Oxford University Press, Oxford.

Morris, H. 1963. *The Twilight of Evolution.* Baker Book House, Grand Rapids, Mich.

Murray, Gilbert. 1955. *Five Stages of Greek Religion.* Doubleday Anchor Books, Garden City, N.Y.

Needham, Joseph. 1981. *Science in Traditional China.* Harvard, Cambridge.

Nelkin, Dorothy. 1987. *Selling Science: How the Press Covers Science and Technology.* W. H. Freeman and Co., New York.

Neugebauer, O. 1969. *The Exact Sciences in Antiquity.* Dover, New York.

Oliver, Chad. 1981. *The Discovery of Humanity.* Harper & Row, New York.

Oppenheimer, F. 1974. The study of perception as a part of teaching physics. *American Journal of Physics* 42:531–537.

Orienti, Sandra, and René de Solier. 1976. *Hieronymus Bosch.* Crescent Books, New York.

Packard, Vance. 1981. *The Hidden Persuaders.* Washington Square Press, New York.

Payne, Peter. 1981. *Martial Arts: The Spiritual Dimension.* Crossroad, New York.

Pelikan, Jaroslav J. 1984. *Scholarship and Its Survival: Questions on the Idea of Graduate Education.* Princeton University Press, Princeton.

Pelletier, J. A. 1951. *The Sun Danced at Fatima.* Caron Press, Worcester, Mass.

Pellicani, Luciano. 1988. On the genesis of capitalism. *Telos* 74:43–61.

Perry, William G., Jr. 1970. *Forms of Intellectual and Ethical Development in the College Years: A Scheme.* Holt, Rinehart and Winston, New York.

Pirenne, Henri. 1958. *A History of Europe.* Doubleday, Garden City, N.Y.

Piaget, Jean. 1971. *Insights and Illusions of Philosophy.* Meridian, New York.

Plato. 1961. *The Collected Dialogues of Plato Including the Letters.* Edith Hamilton and Huntington Cairns (eds.). Princeton University Press, Princeton.

Polanyi, Michael. 1962. *Personal Knowledge: Towards a Post-Critical Philosophy.* University of Chicago Press, Chicago.

Pollard, Sidney. 1968. *The Idea of Progress: History and Society.* C.A. Watts, London.

Popovsky, Mark. 1979. *Manipulated Science: The Crisis of Science and Scientists in the Soviet Union Today.* Doubleday, Garden City, N.Y.

Popper, Karl R. 1950. *The Open Society and Its Enemies.* Princeton University Press, Princeton.

———. 1960. *The Poverty of Historicism.* Routledge & Kegan Paul, London.

———. 1982. *The Open Universe: An Argument for Indeterminism.* Rowman and Littlefield, Tolowa, N.J.

Pospisil, Leonard. 1958. *Kapauku Papuans and Their Law. Yale University Publications in Anthropology* 54:1–294.

———. 1963. *Kapauku Papuan Economy. Yale University Publications in Anthropology* 67:1–502.

Powers, Jonathan. *Philosophy and the New Physics.* Methuen, New York.

Price, Derek J. de Solla. [1963.] 1986. *Little Science, Big Science . . . and Beyond.* Columbia University Press, New York.

Radhakrishnan, Sarvepalli, and Charles A. Moore. 1957. *A Sourcebook in Indian Philosophy.* Princeton University Press, Princeton.

Radin, Paul. 1927. *Primitive Man as Philosopher.* D. Appleton and Co., New York.

_____. 1972. *The Trickster: A Study in American Indian Mythology.* Schocken, New York.

Rajneesh, Bhagwan Shree. 1976. *Meditation: The Art of Ecstasy.* Harper Colophon, New York.

Raju, P. T. 1953. *Idealistic Thought of India.* Harvard University Press, Cambridge.

_____. 1985. *Structural Depths of Indian Thought.* State University of New York Press, Albany.

Ramón y Cajal, Santiago. 1989. *Recollections of My Life.* MIT Press, Cambridge.

Randi, James. 1980a. *The Magic of Uri Geller.* Ballantine Books, New York.

_____. 1980b. *Flim Flam: The Truth About Unicorns, Parapsychology and Other Delusions.* Lippincott & Crowell, New York.

Ravetz, Jerome R. 1971. *Scientific Knowledge and its Social Problems.* Clarendon Press, Oxford.

Reale, Giovanni. 1987. *A History of Ancient Philosophy. I. From the Origins to Socrates.* State University of New York Press, Albany.

Reed, James. 1989. A psychologist of the '20s. *Science* 244:1386–1387.

Regal, Philip J. 1985. The ecology of evolution: Implications of the Individualistic Paradigm. In *Engineered Organisms in the Environment: Scientific Issues,* H. O. Halvorson, D. Pramer, M. Rogul (eds.), 11–19. American Society for Microbiology, Washington, D.C.

Reit, Seymour. 1980. *Masquerade: The Amazing Camouflage Deceptions of World War II.* New American Library, New York.

Riesman, David. 1980. *On Higher Education: The Academic Enterprise in an Era of Rising Student Consumerism.* Jossey-Bass, San Francisco.

Ripley, Julien A. 1964. *The Elements and Structure of the Physical Sciences.* Wiley & Sons, New York.

Robertson, John M. 1969. *A History of Freethought.* Dawsons of Pall Mall, London.

Robinson, Daniel N. 1982. *Toward a Science of Human Nature: Essays on the Psychologies of Mill, Hegel, Wundt, and James.* Columbia University Press, New York.

Ronan, Colin, A. 1982. *Science: Its History and Development Among the World's Cultures.* Facts on File Publications, New York.

Rose, Hilary, and Steven Rose. 1980. *Ideology of/in the Natural Sciences.* Schenkman, Cambridge, Mass.

Ross, Helen E. 1974. *Behaviour and Perception in Strange Environments.* George Allen & Unwin Ltd., London.

Rothmyer, Karen. 1981. Citizen Scaife. *Columbia Journalism Review* 20(2): 41–50.

Roy, Rustum. 1981. *Experimenting With Truth: The Fusion of Religion With Technology, Needed for Humanity's Survival.* Pergamon, New York.

Rudhyar, Dane. 1982. *The Magic of Tone and the Art of Music.* Shambala, Boulder, Colo.

Ruse, Michael. 1982. *Darwinism Defended: A Guide to the Evolution Controversies.* Addison-Wesley, Reading, Mass.

_____. 1986. *Taking Darwin Seriously.* Basil Blackwell, New York.

Sacks, Oliver. 1985. *The Man Who Mistook His Wife for a Hat: and Other Clinical Tales.* Harper & Row, New York.

Sargent, Lydia. 1981. *Women and Revolution: A Discussion of the Unhappy Marriage of Marxism and Feminism.* South End Press, Boston.

Sarton, George. 1970. *A History of Science.* The Norton Library, New York.

Schadewald, Robert. 1981. Scientific Creationism, Geocentricity and the Flat Earth. *Skeptical Inquirer,* Winter 1981–82

Schiffman, H. 1976. *Sensation and Perception: An Integrated Approach.* John Wiley & Sons, New York.

Schwartz, Barry. 1986. *The Battle for Human Nature: Science, Morality and Modern Life*. W. W. Norton, New York.

Segre, Michael. 1980. The role of experiment in Galileo's physics. *Archive for History of Exact Sciences* 23:227–52.

Seidenberg, A. 1962a. The ritual origin of geometry. *Archive for History of Exact Sciences* 1:488–527.

————. 1962b. The ritual origin of counting. *Archive for History of Exact Sciences* 2:1–40.

————. 1978. The origin of mathematics. *Archives for History of Exact Sciences* 18:301–42.

————. 1983. The geometry of the Vedic rituals. In *Agni: The Vedic Ritual of the Fire Altar*, vol. 2, Frits Staal (ed.), 95–126. Asian Humanities Press, Berkeley.

Sekida, Katsuki. 1975. *Zen Training: Methods and Philosophy*. Weatherhill, New York.

Shapere, Dudley. 1974. *Galileo: a Philosophical Study*. University of Chicago Press, Chicago.

Sharma, Chandradhar. 1952. *Dialectic in Buddhism and Vedanta*. Hafner, New York (Nad Kishore and Bros., Banaras, India).

Sheehan, Bernard. 1980. *Savagism and Civility: Indians and Englishmen in Colonial Virginia*. Cambridge University Press, Cambridge.

Siegel, R. 1981. Accounting for "afterlife" experiences. *Psychology Today* 15 (1):65–75.

Simberloff, D. 1982. A succession of paradigms in ecology: essentialism to materialism and probabilism. In E. Saarinen (ed.), Conceptual Issues in Ecology, 63–99. D. Reidel Publishing, Boston.

Simonson, Rick, and Scott Walker. 1988. *Multicultural Literacy*. Graywolf, St. Paul, Minn.

Simpson, G. 1947. *This View of Life: The World of an Evolutionist*. Harcourt, Brace and World, New York.

————. 1967. The crisis in biology. *American Scholar* 36:363–377.

Sirois, F. 1974. Epidemic Hysteria. *Acta Psychiatrica Scand. Suppl.* 252:1–46.

Smith, Alan G. R. 1972. *Science and Society in the Sixteenth and Seventeenth Centuries*. Harcourt, Brace Jovanovich.

Solomon, P. (ed.). 1961. *Sensory Deprivation: A Symposium*. Harvard University Press, Cambridge.

Sophocles. 1947. *The Theban Plays*. E. F. Watling (trans.). Penguin Books, New York.

Spring, Joel H. 1972. *Education and the Rise of the Corporate State*. Beacon, Boston.

Starkey, M. 1959. *The Devil in Massachussetts: A Modern Inquiry into the Salem Witch Trials*. Alfred A. Knopf, New York.

Stcherbatsky, F. Th. 1962. *Buddhist Logic*. Dover, New York.

Stone, I. F. 1988. *The Trial of Socrates*. Little, Brown & Company, Boston.

Suzuki, Daisetz Teitaro. 1965. *The Training of the Zen Buddhist Monk*. University Books, New York.

Tabori, Paul. 1959. *The Natural Science of Stupidity*. Chilton Co., New York.

Tawney, R. H. 1962. *Religion and the Rise of Capitalism*. Peter Smith, Gloucester, Mass.

Taylor, F. Sherwood. 1949. *A Short History of Science and Scientific Thought*. Norton Library, New York.

Taylor, R. 1966. Comments on a mechanistic conception of purposefulness. In *Purpose in Nature*, J. Canfield (ed.). Prentice Hall, Englewood Cliffs, N.J.

Tedlock, Barbara. 1987. *Dreaming: Anthropological and Psychological Interpretations*. Cambridge University Press, New York.

Teitelbaum, Michael S. 1988. Demographic change through the lenses of science and politics. *Proceedings of the American Philosophical Society* 132:173–184.

Thilly, Frank, and Ledger Wood. 1957. *A History of Philosophy*. Henry Holt, New York.

Thoburn, James M. 1906. *The Christian Conquest of India*. American Baptist Missionary Union, Boston.

Thouless, Robert H. 1939. *How to Think Straight: The Technique of Applying Logic Instead of Emotion.* Simon and Schuster, New York.

_____. 1942. *Straight Thinking in Wartime.* Hodder and Stoughton, London.

_____. 1971. *An Introduction to the Psychology of Religion.* Cambridge University Press, London.

Thrower, James. 1971. *A Short History of Western Atheism.* Pemberton Books, London.

Toulmin, Stephen, and June Goodfield. 1961. *The Fabric of the Heavens: The Development of Astronomy and Dynamics.* Harper & Row, New York.

Tuan, Yi-Fu. 1968. *The Hydrologic Cycle and the Wisdom of God.* University of Toronto Press, Toronto.

_____. 1979. *Landscapes of Fear.* University of Minnesota Press, Minneapolis.

Tucker, D. F. B. 1980. *Marxism and Individualism.* St. Martin's, New York.

Turnbull, Colin M. 1972. *The Mountain People.* Simon and Schuster, New York.

Turton, David, and Clive Ruggles. 1978. Agreeing to disagree: the measurement of duration in a Southwestern Ethiopian Community. *Current Anthropology* 19:585–600.

Tyrrell, R. Emmett, Jr. 1982. The voice grows louder. *American Spectator* 15 (12): 4–5.

Tytell, John. 1976. *Naked Angels: The Lives & Literature of the Beat Generation.* McGraw-Hill, New York.

Urton, Gary. 1981. *At the Crossroads of the Earth and the Sky: An Andean Cosmology.* University of Texas Press, Austin.

da Vinci, Leonardo. 1970. *The Notebooks of Leonardo da Vinci.* Jean Paul Richtes (ed.). Dover, New York.

Vogt, Evon Z., and Ray Hyman. 1959. *Water Witching U.S.A.* University of Chicago Press, Chicago.

de Waal, Frans. 1982. *Chimpanzee Politics: Power and Sex Among Apes.* Unwin Paperbacks, London.

Wade, Ira O. 1938. *The Clandestine Organization and Diffusion of Philosophic Ideas in France From 1700 to 1750.* Princeton University Press, Princeton.

_____. 1969. *The Intellectual Development of Voltaire.* Princeton University Press, Princeton.

Waerden, B. L., van der. 1983. *Geometry and Algebra in Ancient Civilizations.* Springer-Verlag, New York.

Watson, Peter. 1980. *War on the Mind: Military Uses and Abuses of Psychology.* Penguin Books, New York.

Watts, Alan A. 1957. *The Way of Zen.* Pantheon Books, New York.

Watzlawick, Paul. 1977. *How Real is Real?* Random House, New York.

_____. 1984. *The Invented Reality.* W.W. Norton & Company, N.Y.

Weber, Max. 1958. *The Protestant Ethic and the Spirit of Capitalism.* Charles Scribner's Sons, New York.

_____. Science as a vocation. In *From Max Weber*, H. H. Gerth and C. Wright Mills (eds.), 129–56, Oxford University Press, New York.

Westfall, Richard S. 1958. *Science and Religion in Seventeenth-Century England.* Yale University Press, New Haven.

_____. 1971. *The Construction of Modern Science: Mechanisms and Mechanics.* Wiley & Sons, New York.

White, Andrew Dickson. [1896] (1960). *A History of the Warfare of Science with Theology in Christendom.* Reprint. Dover Publications, New York.

White, Edward Lucas. 1927. *Why Rome Fell.* Harper & Brothers, New York.

White, Gerald T. 1968. *Scientists in Conflict: The Beginnings of the Oil Industry in California.* The Huntington Library, San Marino, Calif.

White, Lynn Jr. 1962. *Medieval Technology and Social Change.* Oxford University Press, Oxford.

_____. 1966. *The Transformation of the Roman World: Gibbon's Problem After Two Centuries.* University of California Press, Berkeley.

Whyte, William H. Jr. 1956. *The Organization Man.* Simon and Schuster, New York.

Wiener, Norbert. 1954. *The Human Use of Human Beings: Cybernetics and Society.* Doubleday Anchor Books, Garden City, N.Y.

Williams, Roger J. [1956] (1972). *Biochemical Individuality: The Basis for the Genetotrophic Concept.* University of Texas Press, Austin.

Wilken, Robert L. 1984. *The Christians as the Romans Saw Them.* Yale University Press, New Haven.

Wilson, Bryan R. 1973. *Magic and the Millennium: A Sociological Study of Religious Movements of Protest Among Tribal and Third-World Peoples.* Harper & Row, New York.

Wilson, Colin. 1956. *The Outsider.* Houghton Mifflin, Boston.

_____. 1967. *The Mind Parasites.* Oneiric, Berkeley.

Witcutt, W. P. 1958. *The Rise and Fall of the Individual.* Macmillan, New York.

Wittreich, W. 1959. Visual perception and personality. *Scientific American* 200(4):56–60.

Wollstonecraft, Mary. [1792] (1982). *Vindication of the Rights of Woman.* Penguin Books, New York.

Wood, Ellen Meiksins. 1972. *Mind and Politics: An Approach to the Meaning of Liberal and Socialist Individualism.* University of California Press, Berkeley.

Woodfield, A. 1976. *Teleology.* Cambridge University Press, New York.

Wright, L. 1976. *Teleological Explanations: An Etiological Analysis of Goals and Functions.* University of California Press, Berkeley.

Yoxen, Edward, 1982. Giving life a new meaning: The rise of the molecular biology establishment. In *Scientific Establishments and Hierarchies*, N. Elias, H. Marins, and R. Whitley (eds.), 123–43. D. Reidel, Dordrecht.

Index

Index

Abell, G., and B. Singer, 185
Abercrombie, N., S. Hill, and B. Turner, 206
Abir-Am, Pnina, 198
aborigines, Australian: 162–63; dreams and reality, 181
absolutism, 49, 73, 126, 156, 161, 176, 196, 207, 208, 219, 220, 297. *See also* monism; Truth
academic freedom, 126, 292
acculturation: partial, 100-101
Adam, Barry, 108
Adams, Upton, 151–53
adaptive advantages: of behaviors/thought modes, *xi*, 32, 36, 42, 43–44, 52, 72, 160, 191, 236; of common sense, 88–110; of homosexual customs, 94, 108; of language alteration, 144–49; of Reductionism, ideological, 85; of shamanism, 182–83; of skepticism (*see* critical thinking). *See also* individual advantage
Addison, Joseph, 293
adversarial systems, 133. *See also* polarization
alchemy, 57, 58, 186, 261
alcohol and drugs, 50, 82, 86, 95, 289, 308, 330. *See also* compulsive behaviors; denial

Alexander the Great, 18, 119, 229, 260
Alexandria: destruction of antiquity, 175, 262. *See also* Christians, impact on pagan empiricism
alienation, 69–72, 124, 208, 217–20, 243–44
altruism, 241
American Boy's Handy Book (Beard), 301
Amerindians, 80, 81, 164, 264
anarchy: rational, of Godwin, 124
Anasazi, 162
Anaximander, 262
animal awareness, 69–71
Anthony, Susan, 53
anthropology, 52–53, 66–68, 83, 91, 92, 181, 182, 201, 207, 213, 289, 307–8, 319–20
anthropomorphic causes in nature, 116
Aristarchus, 259
Aristotle, 17, 18, 20, 24, 29–30, 113, 115–20, 127–30, 134, 135, 176, 188, 193, 195, 222, 224, 247, 257–62, 272, 275, 276
arms race, 131
Armstrong, David, 167
Aronowitz, Stanley, 288
Arrowsmith (Lewis), 231–32

73, 292; of women, 195. *See also* critical thought; liberal arts; universities
egalitarianism, 53
ego: constructions of, 45, 124, 168, 206–31, 191, 210–20, 306; costs and benefits, 182, 191; escape from, 182, 190, 203, 251; provincializes, 171, 189, 234, 235. *See also* self-consciousness
egoism, 123, 206–26, 219. *See also* middle class; selfishness
Egyptian science, 258, 267, 268
Eiduson, B. T., 203, 283
Einstein, Albert, 33, 106, 148, 311, 322; and L. Infeld against complete reductionism, 82, 156; and FBI investigation, 293; on common sense, 88
El Topo (Jodorowski), 166
Ellul, Jacques, 285
emergentism. *See* holism
emergent properties, 83–85
Emerson, Ralph Waldo, 318, 319
emic, 53
empathy, 56, 70–71, 182, 183; and art, 56–58, 74
Empedocles, 261
empiricism, 17, 118, 172–204, 250–54, 259, 262
enlightenment, 61, 166, 188–92, 203, 209, 292, 305. *See also* wisdom
Enlightenment, the, 195–96, 208, 218, 220, 225, 287, 288, 302, 315–17, 321
environmental crisis, 167, 195, 198–99, 287, 298
Epicurus, 24
"epistemological hypochondria," 297
epistemological modes: philosophical, 186–92; corporate scientific, 197–200, 278–92; modern scientific, 193–98, 269–78, 282–85, 289–90; "scientific" divination, 183–86; shamanistic, 180–83, 201
epistemology, 84, 188–90, 197
Equus (Shaffer), 54–55
Eratosthenes, 259
escapism, 159, 181, 220, 251, 291, 299, 321, 324, 330
Essay on Man, An (Pope), 111–12
essentialism/essences, 82, 108, 113, 114, 116–20, 129, 130, 135, 138, 156, 162, 196, 221, 232, 258, 309, 310

ethics in modern academia, 192
etic, 53
Euclidean geometry, 157, 260
eugenics, 139–40
Euripides, 211
European history and empiricism, 172–80
Eusebius, 175
evidence, 8, 67, 249, 255
evolution, 36, 194, 241, 257, 270, 274, 326; of brain, 44, 214; causation in, 119; and Creationists, 101–3, 135, 257; of language, 143–49; local genetic populations in, 81; misuse of, 114 (*see also* design; naturalistic ethics; social Darwinism; sociobiology; teleology)
Exodus, 143
experience (and personal growth), 254, 290, 294–324
experiment, 289, 310; as buzz word, 148; critical functions of, 75–77; among primitive peoples, 263. *See* critical thinking; scientific methods
exploitation, 73, 124, 251, 327; and language, 151–53; by denial of education, 80, 314, 323. *See also* subjugation; thought control
extinction, 134, 135, 241
Eye in the Sky (Dick), 71–72
Eyewitness Testimony, 63

Fable of the Bees, The (Mandeville), 137–39, 159
faith, 50, 178. *See also* literalism; reductionism, ideological; reality, as a subjective construction
Falco, Giorgio, 175
false consciousness, 22, 137, 145, 189, 205–31, 243–44. *See also* perception; reality
Fatima, Portugal: miracle, 46, 271
Faust (Goethe), 130
Feeling for the Organism, A (Keller), 253
feminism, 113, 114, 145, 191, 195, 289–90, 302, 315, 323. *See also* women
feminist ideological struggles, 167–68
Ferguson, John, 143
Feyerabend, P., 237, 292
Final Causes (Aristotle's), 119, 120, 123
Finley, M. I., 175

stable worldviews, 97, 100–107
Stalin, Joseph, 74, 224, 240
Stansfield, Ronald, 46
Starkey, M., 13
statistics: and essentialism, 130;
    misapplications, 308
status. *See* respectability
Stcherbatsky, F. Th., 188
Stein, Gertrude, 205
stereotypes, 71–72, 125, 127–30, 155, 156,
    159–69, 208, 244
Stevenson, Robert Lewis: and dreams, 183
Stewart, Kilton, 201
Stone, I. F., 17, 260
structuralism, 239, 309; as a reductionism,
    85
style: mistaken for content, 63, 149–51
subconscious, 61
subjective reality, 31–53, 54–86, 92, 189,
    210–20, 328. *See also* reality
subjugation, 114–15, 127–29, 138, 145,
    152–53, 160, 176, 195, 208, 219,
    221–22, 236, 239, 260–63, 286–90,
    292–93, 299, 314, 317, 323. *See also*
    inferiority; mercantilism; poverty;
    thought control
Sumner, W. G., 135
survival of the fit (natural selection), 121.
    See also evolution
survival of the fittest. *See* social Darwinism
superorganism: organizations compared to,
    158–59; paradigm for ecological
    communities, old, 125
Suzuki, D. T., 168
synchronization of minds, 58–65, 211
synthesis, 83–85, 252. *See also* lateral
    thinking
Szent-Gyorgyi, Albert, 279

Tabori, Paul, 36
Tagore, Rabindranath, 50
Taoism, 120, 191, 203, 260
Tantra, 191
tastemaking, 147
Taylor, F. S., 265
Taylor, Gordon R., 46
technique-oriented mentality, 296, 299–302
technocracy/technocrats, 78, 286, 288
technological dystopia, 130–31, 168, 272

technology, 79, 195–99, 272, 273, 285, 299;
    runaway, 285–86
Tedlock, Barbara, 183
Teitelbaum, M., 139
teleology, 114–40, 158, 220, 222, 240. *See
    also* design
Ten Commandments, 142–44
Thales, 258, 261–62
Thatcher, Margaret, 138, 231
theism, 120, 121, 123, 130, 269–78, 290
theology, 103, 183. *See also* philosophy
theory (defined), 256–57
Thilly, F., and L. Wood, 136, 224
think tanks, 225
Thoburn, James, 243
Thomas Aquinas, Saint, 178, 272
Thoreau, Henry David, 296, 318
thought control, *xi*, 14, 73, 100–106, 131,
    208, 251, 297, 299, 314, 323–24; and
    language, 144–69; limits on, 211–14;
    and media, 146–47; and science,
    288–89. *See also* brainwashing; social
    confusion
thought experiments, 229–31, 247, 252,
    311. *See also* scientific methods
thought systems, 161, 171–204, 255, 326.
    See also common sense; cybernetics;
    thought-system organisms
Thrower, James, 277
time, 33, 43
de Tocqueville, A., 228
torture, 8, 57, 73, 126, 213, 287, 302
totalitarianism, 81, 260, 286. *See*
    authoritarianism; cultural imperialism;
    monism; Nazis; Stalin
Touch of Evil, A (Wells), 8
tragedy, 16, 20
Traven, B., 163, 302–3
Truth: beyond the senses, 171–204, 256–70;
    competing claims to, 187–88, 260–63;
    in Eastern disciplines, 182–92; exclusive
    (Platonic), 24, 26, 49, 50, 52, 78, 85,
    122, 156–58, 166, 173, 178, 187, 191,
    193, 195, 196, 230–31, 259, 261, 265,
    309. *See also* reality
Tuan, Yi-Fu, 178
Tucker, D. F. B., 208, 239
Turnbull, Colin, 242, 327–29
Turton, D., 265; and C. R. Ruggles, 266
Twain, Mark. *See* Clemens, Samuel

Philip Regal is professor of ecology, evolution, and behavior and a former member of the board of directors for the center of humanistic studies at the University of Minnesota. He studies the mechanisms and patterns of adaptation of organisms to their environment. During his pre- and postdoctoral work at UCLA, he was the coordinator's appointee to the Brain Research Institute's Mental Health Training Program. Regal was a Distinguished Scholar and Professor at the Universities of Miami and Western Australia and the Australian National University. He has consulted with the National Science Foundation and other agencies on national science policy issues, and has served on several national and international committees, including the UNESCO Man in the Biosphere Program, the AAAS Project 2,061, "Science for Every American," and the A. P. Sloan Foundation's Committee for Molecular Evolution.